CHIEF ENGINEER
OF THE UNIVERSE

Albert
Einstein

**EINSTEIN'S LIFE
AND WORK IN CONTEXT**

History of Knowledge is a new series from Wiley-VCH.
Internationally acclaimed experts bring new perspectives to the history of knowledge and introduce readers to hitherto unknown worlds of research and its conflictual history. The series is published in cooperation with the Max Planck Institute for the History of Science.

The three volumes *Albert Einstein – Chief Engineer of the Universe:*
Einstein's Life and Work in Context
One Hundred Authors for Einstein
Documents of a Life's Pathway
have been published to accompany the exhibition of the same title: *Albert Einstein – Chief Engineer of the Universe*,
which was conceived by the Max Planck Institute for the History of Science on the occasion of the Einstein Year 2005.

Editor	Jürgen Renn
Authors	Katja Bödeker, Elena Bougleux, Reiner Braun, Paolo Brenni, Jochen Büttner, Peter Carl, Giuseppe Castagnetti, Peter Damerow, Wolfram Dolz, Lucio Fregonese, Rainer Friedrich, Carmen Hammer, Dieter Hoffmann, Stefan Iglhaut, Horst Kant, Christoph Lehner, Dierck-E. Liebscher, Peter McLaughlin, Wolf-Dieter Mechler, Philipp Muras, Markus Pössel, Jürgen Renn, Simone Rieger, Violetta Sánchez, Matthias Schemmel, Michael Schüring, Stefan Siemer, Urs Schoepflin, Giorgio Strano, Kurt Sundermeyer, Matteo Valleriani, Milena Wazeck, Jörg Zaun
Assistant Editor	Giuseppe Castagnetti
Bibliography	Sabine Bertram, Lindy Divarci
Translators	Isabel Cole, Gloria Custance, Anthony B. Heric, Anita Mage, Susan Richter, Ann Robertson, Pamela Selwyn
Illustrations	Laurent Taudin, Paris
Image Editors	Hartmut Amon, Edith Hirte, Tanja Starkowski
Design / Production	Regelindis Westphal Grafik-Design, Berlin Antonia Becht, Berno Buff, Anja Gersmann, Norbert Lauterbach
Image Editing	Satzinform, Berlin
Print / Binding	NEUNPLUS1 – Verlag + Service GmbH, Berlin

This publication was made possible due to the kind support of the
Klaus Tschira Stiftung gGmbH, Heidelberg

KLAUS TSCHIRA STIFTUNG
GEMEINNÜTZIGE GMBH

Booktrade edition
ISBN-10: 3-527-40571-2
ISBN-13: 978-3-527-40571-8

The catalogue accompanying the exhibition *Albert Einstein - Chief Engineer of the Universe* is published in German under the title *Albert Einstein – Ingenieur des Universums. Einsteins Leben und Werk im Kontext.* Book trade edition ISBN- 3-527-40573-9

Further titles:
Jürgen Renn (Ed.): *Albert Einstein – Ingenieur des Universums. Hundert Autoren für Einstein*, Berlin: WILEY-VCH, 2005.
Book trade edition ISBN-3-527-40579-8
Jürgen Renn (Ed.): *Albert Einstein – Chief Engineer of the Universe. One Hundred Authors for Einstein*, Berlin: WILEY-VCH, 2005.
Book trade edition ISBN-3-527-40574-7
Jürgen Renn (Ed.): *Albert Einstein – Ingenieur des Universums. Einsteins Leben und Werk im Kontext* together with
Dokumente eines Lebensweges, two-volume-set. Berlin: WILEY-VCH, 2005. Book trade edition ISBN-3-527-40569-0
Jürgen Renn (Ed.): *Albert Einstein - Chief Engineer of the Universe. Einstein's Life and Work in Context* together with
Documents of a Life's Pathway, two-volume-set. Berlin: WILEY-VCH, 2005. Book trade edition ISBN-3-527-40571-2

Jürgen Renn (Ed.)

CHIEF ENGINEER
OF THE UNIVERSE

Albert
Einstein

EINSTEIN'S LIFE
AND WORK IN CONTEXT

WILEY-
VCH

WILEY-VCH Verlag GmbH & Co. KGaA

The exhibition is under the patronage of
the Federal Chancellor Gerhard Schröder.

Albert Einstein
Chief Engineer of the Universe
Exhibition in the Kronprinzenpalais, Berlin
from 16 May to 30 September 2005
www.einsteinausstellung.de

Organizers

MAX-PLANCK-GESELLSCHAFT

Max-Planck-Society
for the Advancement of Science

**MAX PLANCK INSTITUTE
FOR THE HISTORY OF SCIENCE**

within the framework of the Einstein Year 2005

A joint initiative of the Federal Government,
science, industry and culture

Design and Implementation

IGLHAUT PARTNER
+

www.iglhaut-partner.de

Sponsors

KULTURSTIFTUNG
DES
BUNDES

 **Federal Ministry
of Education
and Research**

**Stiftung
Deutsche Klassenlotterie Berlin**

An Exhibition without Walls –
interactive and online –
with the kind support of the
Heinz Nixdorf Foundation

BASF

**SIEMENS**

Fritz Thyssen Stiftung
FÜR WISSENSCHAFTSFÖRDERUNG

 KTF
THE KLAUS TSCHIRA
FOUNDATION gGMBH

ROBERT BOSCH STIFTUNG

as well as the Wilhelm and Else Heraeus
Foundation, and the Central European
University, Budapest

Exhibition Associates

Deutsches Museum

 HEBREW UNIVERSITY JERUSALEM

UNIVERSITÀ DEGLI STUDI
DI PAVIA

Media Associates

DW-TV
DEUTSCHE WELLE

rbb ①
RUNDFUNK BERLIN-BRANDENBURG

3sat

CONTENTS

Preface

Gerhard Schröder
Chancellor of the Federal
Republic of Germany

The exhibition *Albert Einstein – Chief Engineer of the Universe* is a centerpiece of the Einstein Year. Upon its opening I would like to send my warmest greetings and my congratulations on the success of this ambitious project.

With the Einstein Year 2005, the federal government, along with the spheres of science, trade and industry, and culture, would like to arouse interest and enthusiasm in research and technology, and strengthen the awareness of the mutual responsibility borne by science and society. This exhibition is an important contribution toward this goal. It is a reflection of the breadth and diversity of Einstein's scientific achievement and of his remarkable personality. In an impressive way it reminds us how a single person can leave his enduring mark on our world through talent, commitment, and responsible activity directed toward the good of society as a whole. The exhibition also shows that the history of science is indispensable for our understanding of the way knowledge fits together. In coming to terms with processes of innovation we understand how new ideas and technology emerge – and can use this knowledge for the future.

To everyone who participated in the realization of this project, I would like to express my appreciation for their hard work. To all its visitors I wish inspirational hours in the exhibition.

Greetings

Edelgard Bulmahn
Federal Minister of
Education and Research

"Einstein Year 2005" continues the tradition of the "Years of Science" which Germany has been holding for five years now. Organized jointly by the "Science in Dialogue" initiative and the major scientific organizations, the "Years of Science" present a broad program of events, focusing on a different field each year. It is our intention to make the public curious – curious about the new ideas which science opens up for us time and again. With the Einstein Year, which is part of the "World Year of Physics," we are celebrating an outstanding scientist and a great personality. And we are also contemplating the injustice which Einstein experienced in National Socialist Germany and which forced him to emigrate.

The exhibition *Albert Einstein – Chief Engineer of the Universe* has been developed by the Max Planck Institute for the History of Science. It is a centrepiece of the Einstein Year, a portrayal not only of Einstein's revolution, but also of its prerequisites and its consequences. The exhibition addresses an audience which is interested in science, as well as cultural dilettantes of every provenance, families, tourists, and school-children, and transcends the usual boundaries between the categories of science, art, and cultural history. The exhibition is based on questions that scientists and natural philosophers have argued about for centuries, showing Einstein's world in the context of his scientific and cultural environment. It thus encourages the visitor to see beyond specialized fields and to imagine science as an open system, in which even the simplest questions can have great consequences.

The development of the theory of relativity was the achievement of a young scientist who turned the theories prevailing at the time upside down. In our scientific system too, young scientists, especially, need the freedom to think in new directions and to rethink accepted knowledge. But curiosity and the joy of thinking are more than just the motivation for scientists – they are of existential importance for our country's future. The spirit of research of many great thinkers of earlier generations has made progress possible and has shaped our world today. Einstein in particular teaches us that philosophy and the history of knowledge can be preconditions for innovative research.

This year Albert Einstein, as a world citizen of Jewish faith, represents the countless people who were driven out of Germany by National Socialism. The expulsion of these people not only represented capitulation before any semblance of humanity; at the same time it led to an exodus of knowledge, of intellect and culture from Germany. A loss that can still be felt in our country today. Einstein knew that there is no purely German or French science. Science and research do not know international borders, they are international. The scientific community is a global community, in which Germany plays an active role. I am glad that many young scientists are once again coming to Germany today – to perform research, to teach, and to live.

Research secures our future. The Einstein Year should make us eager for this future. It is a year for asking questions, for thinking ahead, and, above all, for getting involved. Many events are aimed especially at young people, Einstein's "heirs." These are the very people whose interest we would like to arouse – not only in research and scientific topics, but also in the exciting adventure of the history of knowledge.

Christina Weiss
State Minister to the
Federal Chancellor,
Federal Government Deputy
for Culture and Media

Albert Einstein stands for an epoch-making transformation in the way we look at the world. Not only did he revolutionize the foundations of physics, in doing so he also changed our intellectual life as a whole. Einstein Year 2005 is thus not only an anniversary of science: the ingenious physicist's discoveries gave us a different, more complex view of reality. At nearly the same time, the historical movements of the avant garde in the arts expressed this new complexity. One hundred years after these events, the parallelism of the paradigm shifts in arts and sciences is impressive: around 1905 a radical renunciation of conceptions held valid up to that time takes place not only in the natural sciences; a break with tradition occurs at this time in the arts as well. Consider, for instance, the transition to abstraction in the plastic and graphic arts, and the emergence of atonality in music. Against this background it is only consistent to re-examine and honor Einstein's work in a broader context in the year 2005 as well.

These interactions between arts and science are interesting not only for historical reasons. Especially in recent years, a new mutual attention can be observed between these fields, an approach that transcends categories and disciplines. The Federal Cultural Foundation is one of the major sponsors of projects with a transdisciplinary orientation. Thus I am very pleased that it was possible to provide substantial support for the great exhibition *Albert Einstein – Chief Engineer of the Universe* with funds from the foundation. I am certain that the exhibition will find a large audience: in cultural and scientific circles and far beyond. For both the exhibition and its accompanying catalog, I wish many attentive visitors and viewers.

Hortensia Völckers
Artistic Director of the
Federal Cultural Foundation

The work of Albert Einstein is always honored predominantly as an outstanding achievement in the area of science. Yet Einstein's importance for society in the twentieth century goes far beyond the field of science: Einstein's life path and aura are above all a cultural phenomenon of the first order.

Einstein's contribution is known for "culture" in the metaphoric sense, both for the culture of debate and for political culture. Hardly any other German in the twentieth century drew so much attention to himself through courageous independence from any authority as Einstein. His independence – in taking a stand against World War I back in 1914 and later warning against the atomic bomb – makes him a model political citizen even today. Probably less well known is how complex the roots of Einstein's influence and personality are, and in how many ways they reach back into the culture of his time.

For this reason the Federal Cultural Foundation especially embraces the fact that the exhibition *Albert Einstein – Chief Engineer of the Universe* extends far beyond Einstein's scientific achievement, allowing the viewers to grasp these complex connections. In so doing, the exhibition also shows the extent to which the system of natural sciences as a whole must also be understood as a cultural phenomenon. Yet which factors of cultural surroundings favor the further development of scientific theories as a whole, and facilitate outstanding innovation by individuals?

It is an important objective of the Federal Cultural Foundation to explore this creative border area between culture and innovation. This is why the foundation promotes direct, strictly defined initiatives to create space for innovative cultural developments, space in which coordinate systems are regrouped, old signposts can be turned to point in a new direction, and journeys can be launched without a rigidly fixed itinerary.

Part of this goal is to learn more about the conditions required to open such space for innovation in culture. The Einstein exhibition makes an important contribution to this, and the achievements and life of Albert Einstein provide an excellent illustration. Therefore I am very pleased that the Federal Cultural Foundation was able to contribute to its realization.

Menachem Magidor
President of the Hebrew
University of Jerusalem

We at the Hebrew University of Jerusalem are proud to be the partners of the Max Planck Society in the exhibition *Albert Einstein – Chief Engineer of the Universe*. The exhibition tells the multifaceted story of Albert Einstein, the great intellectual pillar of Western civilization, who was responsible for one of the two most profound revolutions in our understanding of the world in which we live, and was the major initiator of the second revolution. But more than the revolution in physics associated with Einstein, he has become an icon and a symbol of the twentieth century. It is no accident that he is commonly titled "Man of the Century."

There is no major development in the first half of the century that Einstein is not connected with, either as a player, as in the development of the League of Nations, nuclear weapons, Zionism, and the establishment of the state of Israel, or as an important commentator expressing a unique voice. The list of subjects Einstein wrote about reads like the history of the twentieth century in a nutshell: World War I and its aftermath, Russia and communism, anti-Semitism and Zionism, Hitler's rise to power and the Holocaust, nuclear weapons and world peace, McCarthy in the United States and the development of Germany after World War II.

Even more meaningful than the subjects themselves is their spirit. Einstein, like the century he represents, was full of paradoxes and contradictions. He was one of the great scientific revolutionaries, yet he refused to accept the revolution of quantum mechanics; he was a pacifist who played a major role in the development of nuclear weapons; a citizen of the world who abhorred any national feeling and yet was a major supporter of the Jewish national movement. I believe that it is these contradictions and the multidimensional nature of Albert Einstein's personality which have contributed to turning him into an icon. Einstein is one of the founding fathers of the Hebrew University of Jerusalem. He was very involved in charting its academic ethos and mission. In the article *The Mission of Our University* he wrote: "Unfortunately, the universities of Europe today are for the most part the nurseries of chauvinism and of a blind intolerance... On this occasion of the birth of our University, I should like to express the hope that our University will always be free from this evil, that teachers and students will always preserve the consciousness that they serve their people best when they maintain its union with humanity and with the highest human values." His deep attachment to the University was expressed by means of his aid and support, but also by the emotional disputes he had at times with the leadership of the University.

To Einstein, the Hebrew University represented a combination of his commitment to his Jewish identity and his belief in the universal values of the pursuit of truth and the respect for every human being. Therefore, it is only natural that he chose the Hebrew University as the guardian of his archives, which are deposited in Jerusalem. It gives us great pleasure to make some of this material available to the public in Berlin. Understanding the story of Albert Einstein, his science, his social commitment, and the historical context of his life, is to understand our civilization and our world.

Roberto Schmid
Rector of the
University of Pavia

It is with great enthusiasm that the University of Pavia accepted the invitation to be a partner in the exhibition *Albert Einstein – Chief Engineer of the Universe*, a main event among the 2005 celebrations of the World Year of Physics.

Historical ties of a very special kind link this year's celebrations of Albert Einstein with our University's history. Pavia played no minor role in the life of the young Albert: from 1894 to 1902 Einstein's parents lived in Lombardy. In Pavia Albert's father and uncle founded an electrotechnical firm and built a factory which can still be seen along the Naviglio Ticinese canal that links Milan with Pavia and the Ticino river.

The young Albert spent most of the year 1895 in Pavia, privately preparing for his entrance examination to the Zurich Polytechnic. In Pavia he wrote his first scientific essay, helped to solve some technical problems at the factory and spent his free time with a number of friends and his sister Maya. He would correspond in Italian with one of these friends fifty years later, remembering those happy years. Three letters by Albert Einstein, kept in the University Museum, are on display at the exhibition.

What scientific environment did the Einstein firm and the young Albert encounter in Pavia? There is evidence of a contract between the Einstein & Garrone firm and the university for the electrification of a university building, signed in January 1896. The rector at the time was Camillo Golgi, who in 1906 became the first Italian to receive the Nobel Prize (in medicine, for his neurological discoveries). Although not a registered student, the young Albert, who lived only a few hundred meters from the university's main buildings, might easily have taken advantage of the huge, up-to-date collection of physics books in the university library, many of which were in German; he might have visited the wonderful collection of physics instruments in Volta's Cabinet, and he might have attended the lectures of the physics professor Adolfo Bartoli, a pioneer in physical chemistry.

The University of Pavia fully recognizes the importance of the history of science and its role in the diffusion of a real scientific culture. We are strongly committed, even in these times of limited resources, to the improvement of our museum system which hosts a number of collections with more than 600,000 items. When the Einstein exhibition comes to Italy, it will be housed in the new Museum for the History of Electrical Technology, a suitable environment for the "Engineer of the Universe." Additional partners in the Italian cooperation include the Institute and Museum of the History of Science in Florence and the Universities of Bologna and Bari.

I would like to thank the Max Planck Institute for the History of Science, the other partners, and all of the German institutions and sponsors that have made this event possible. On the Italian side I would like to thank the Ministry of Education and Research for its generous support.

Wolfgang M. Heckl
General Director of the
Deutsches Museum, Munich

Newton and Kant believed in outer space as a containing space for the stars, with a universal time that flowed homogeneously everywhere within it. Accordingly, space "in itself" was a prerequisite for everything else. With the theory of relativity, Einstein opened up a new – and revolutionary – understanding of space and time: events that are measured at the same time on the Earth must not necessarily be observable at the same time from a spaceship moving at high speed, and vice versa. The front of a flash of light emitted from a rocket moving at nearly the speed of light moves in the same direction as the rocket, but its velocity is not nearly twice the speed of light, but rather exactly the speed of light. Nature is strange that way. With Einstein's theories and mathematics, we are able to predict such events correctly. Yet we cannot "comprehend" them – not even the physicists among us. Our senses and our imagination are not amenable to such ideas, having developed on our planet at "velocities" around 1 billion times slower than light.

To date not a single experiment has "contradicted" the theory of relativity. One of its consequences is that space, time, and matter are inextricably connected with each other. Space is even "curved," that is, the theorems of plane geometry are not valid.

A further prediction of the theory of relativity is the formula known all over the world, $E = mc^2$, the theoretical understanding of atomic and hydrogen bombs.

Einstein's now famous paper about the light quantum hypothesis, too – light transmits energy not like a wave, but in packets of exactly determinable size – opened the gate to quantum mechanics. As a consequence it became apparent that the concept of causality in our everyday world cannot be applied to the micro-level without adjustments – with difficulties of "comprehension" similar to those for the theory of relativity.

But Einstein lived not only in the world of physics. The tragic events of the twentieth century: both world wars, increasing anti-Semitism, the atrocities of Nazi terror in Germany and Europe, the Cold War between West and East, all of this dug deep tracks in his life path. He fought actively for justice, against National Socialism, for the atomic bomb as a means of checking Nazi power, for pacifism – when the superpowers had achieved atomic "overkill" capability in the Cold War.

Einstein wandered between the worlds, becoming a citizen of the world whom the times of his life begrudged an enduring home. He also became a symbol of the power of natural science knowledge. A fate in which many of the greatest aberrations of the twentieth century are reflected.

The Deutsches Museum has a direct relationship with Albert Einstein: from 1920 to 1934 he was a member of the museum's directorial board. For this reason, too, the Deutsches Museum was proud to contribute to this exhibition and make available many valuable historical objects from its collections. May Einstein and his time become a little more "comprehensible" as a result.

Peter Gruss
President of the Max Planck
Society for the Advancement
of the Sciences, Munich

At the opening of the Einstein Year, Federal Chancellor Gerhard Schröder called for a "new culture of science" and expressed the wish that the broad public should grant more attention to the topics of science. The Max Planck Society can endorse this entreaty without reservation. The exhibition *Albert Einstein – Chief Engineer of the Universe* is an important contribution to this goal. Historical objects, experimental setups, animations, and media installations demonstrate clearly how Albert Einstein's discoveries influence our view of the world today. The exhibition further shows how new discoveries change our worldview, thus also emphasizing the key role that theoretical research plays in modern society.

The Max Planck Society feels especially indebted to Albert Einstein. Not only because a number of our institutes continue Einstein's research, but also because Albert Einstein was a member of the Kaiser-Wilhelm-Gesellschaft, our predecessor organization. However, in 1933 he was forced to emigrate – a dark chapter in the history of our society, and one we may not suppress in the Einstein Year. It is all the more important to remember all aspects of Einstein, not just the outstanding scientist, but also the committed democrat and politically responsible person. Einstein took a stand not only on scientific issues, but also on social, ethical, and religious matters. His ability to convey science to a broad public makes him a role mode especially in our time. With the exhibition conceptualized by the Max Planck Institute for the History of Science I link the hope that some of Einstein's enthusiasm for science will spread to its visitors.

Basic research forges one key for the welfare of our society. Although most of its results cannot be applied directly at first, in the long term they may bear fruit in the form of new technologies and products. Einstein's discoveries are a particularly good example of the enduring effect of fundamental research: without his theories we would have neither the laser nor commercially applicable satellite navigation systems. Science thus not only fulfills the human drive to investigate. It is essential for the future of our society, for its results are the point of departure for Germany's ability to hold its own in global competition. This background of scientific work is examined in the exhibition as well.

My thanks are due to the numerous institutions in Germany and abroad that participated in the preparation of the exhibition. Without the Hebrew University in Jerusalem, the Deutsches Museum in Munich and the University of Pavia it would never have been possible. The European Lab for Network Collision CERN, the German Electron Synchrotron DESY, and the Center of Applied Space Technology of the University of Bremen also made important contributions. Through the committed participation of research institutions from without and within the Max Planck Society, the exhibition also offers insights into modern science and identifies open questions researchers are working on today.

In Einstein's tradition, the exhibition attempts to make science accessible to all. Einstein also represents the dedication to an open and democratic society, which relies on independent research and the free availability of knowledge. Therefore the exhibition also opens up to discussion issues of historical and current science policy. The Max Planck Institute for the History of Science's approach of regarding science from the historical and cultural contexts marks the spirit of this exhibition. With this it becomes clear: science is a human enterprise concerning all of us.

Editor's Introduction and Acknowledgments

Young Einstein was a passionate reader of books that gave him an overview of science, among them Humboldt's *Cosmos* and popular science literature with interdisciplinary orientation and philosophical commitment. Here he found knowledge without the restriction of disciplinary borders, but with reference to the historical development of knowledge, and to the questions still open and the challenges they held. This book aims to make a small contribution to an overview adequate to the current state of science and its exciting history. Einstein's life and work constitute a juncture in the history of knowledge, at which many lines of development are bundled. Numerous threads proceed from his activity, on which decisive issues of the present depend, regarding both the worldview of physics and the responsibility and social role of science. Einstein's role in the history of science, along with his life's path – which led him through very different kinds of worlds, from the optimism of the nineteenth century all the way to the world torn by the Cold War in the twentieth – thus serve as an ideal point of departure to pursue such issues. *Albert Einstein – Chief Engineer of the Universe*? This characterization, taken from an envelope addressed to him, at first glance corresponds to the myth that endeavors to remove Einstein from the sphere of normal people. However, this myth veils that very peculiarity of Einstein's fate and character – including his scientific achievements, his moral courage, but also his human weaknesses. In this way his greatness seems less objectionable; it becomes somehow excusable, and at the same time ceases to be a standard for our own behavior. After all, a genius, especially a weird genius, cannot be a role model. *Albert Einstein – Chief Engineer of the Universe*. Primarily this is meant as an indication of the fact that our worldview is a product of human construction, as is the world it aims to reflect. Insofar as they are made by people, they can also be changed by people. Einstein's life and work are a symbol for this possibility of changing the world, through science and social involvement in the Enlightenment tradition. Perhaps this volume can not only contribute to the understanding of the prerequisites, the history, and the consequences of Einstein's activity, but also encourage us to pick up on the traditions beyond the myth.

The book is divided into three parts, each of which approaches Einstein's role for our understanding of the world today in a different way, corresponding to the three main parts of the exhibition it was written to accompany. The first part is dedicated to the relationship between *Worldview and Knowledge Acquisition* and concerns the great, revolutionary changes in the history of knowledge since the beginning of science. What invisible forces rule our world? How is our world constructed on a large scale? How do we know anything about the nature of our world, and what discoveries have changed our worldview? How does scientific knowledge fit in with our everyday conceptions and with our technological possibilities? Such questions are pursued on the basis of topics that have their point of departure in Einstein's biography, but at the same time offer an overview of important aspects of the history of knowledge since antiquity.

The second part is dedicated to *Einstein's Life Path* and deals with the scientific, political, and personal challenges he faced. What kind of world did he live in at the end of the nineteenth century, in which classical physics appeared to be nearly complete? Which open questions constituted the point of departure for Einstein's scientific revolution? In what environment did he grow up, and who helped him ask the right questions? How does a scientific revolution actually work, and what consequences does it have for science and beyond? Why was Einstein's political behavior so different from that of most of the other contemporary scientists? What did Judaism mean for Einstein? These topics are at the core of the second part, which proceeds with the introduction of the history of the emergence of the two major theories of modern physics, the theory of relativity and quantum theory.

The third part is dedicated to *Einstein's World Today* and concerns the scientific, technical, and political heritage he left us. Does the theory of relativity hold up to the new findings of physics and astronomy?

What importance does it have for our everyday life? Was Einstein right to be skeptical of quantum theory? What is the relationship between theoretical research and application? What worldview is science working on today? What political challenges is science currently confronted with? These and similar questions build a bridge between our world and Einstein's. They make clear that it is worth taking a look back occasionally, even in the face of the urgent problems of the present, to avoid succumbing to blind optimism about progress or being misguided by an irrational antagonism to science.

The individual parts of this volume contain short essays dedicated to the main topics of the exhibition and are arranged to follow their plotline. They are complemented by explanations of basic scientific principles, historical contexts, and individual personalities. The spectrum ranges from topics of physics like four-dimensional spacetime, via philosophical topics like post-modernity, all the way to technical issues like satellite navigation. In this manner the volume becomes a compendium, from which an overview can be gleaned about the origins, contexts, and consequences of the Einsteinian revolution. The essays and explanations are followed by illustrations of objects and documents described briefly. Although only a small selection of the exhibited objects could be considered here, these pages do give a vivid impression of the diverse currents in the historical context and the material culture on the basis of which the Einsteinian revolution and its consequences were possible. Worth a special mention are the perceptive drawings by Laurent Taudin. Thanks to their ironic verve, even the most abstract issues are bestowed with concrete meaning.

This volume is complemented by two others, structured in very much the same way. The volume *Albert Einstein – Documents of a Life's Pathway* offers a comprehensive collection of resources on Albert Einstein, selected on the basis of the emphases of the exhibition. The volume *One Hundred Authors for Einstein* contains easy-to-read texts by prominent authors, who elaborate on the topics of history and physics that could only be touched upon in this book. Taken together, the three volumes constitute an introduction to the history of knowledge, accessible to readers without academic training, a history in which Einstein played the key role.

This volume would not have been possible without the involvement of the entire team of authors, whose concise contributions created a work of reference for the history of science focusing on Albert Einstein. Editorial supervision was in the hands of Giuseppe Castagnetti, Carmen Hammer, and Simone Rieger. Editing of the graphics and arrangement of copyrights for both the publications and the exhibition were taken care of by Hartmut Amon, Edith Hirte, Tanja Starkowski, and Stephanie Giese. Lindy Divarci and Sabine Bertram put together the bibliography. The catalogue team worked under intense time pressure and with great commitment to process the complex material, supported and advised at every turn by the director of the library of the Max Planck Institute for the History of Science, Urs Schoepflin. Without the days and nights of tireless teamwork, many a hurdle would not have been cleared.

For the graphic design I thank Regelindis Westphal and her staff, who adjusted to the tight time framework with no less commitment and flexibility. Their graphic signature marks the exhibition and publications, advertising and outdoor installations.

Alexander Grossmann and the staff of the Wiley-VCH publishing house, above all André Danelius, I thank for the uncomplicated cooperation and the perspective on a new, joint series dedicated to the history of knowledge. The project emerged from a personal conversation with Alexander Grossmann and has been sustained since by mutual trust.

The publications and the exhibition *Albert Einstein – Chief Engineer of the Universe* emerged from an intensive cooperation by scientists and exhibition makers, headquartered at the Max Planck Institute for the History of Science, but involving many other institutions, above all our three exhibition partners, the Hebrew University of Jerusalem, the Deutsches Museum in Munich, and the University of Pavia. From the Hebrew University, Hanoch Gutfreund, Charlotte Goldfarb, Barbara Wolff, and Ira Rabin were especially involved in its realization. For important suggestions we are also indebted to the historian of science Mara Beller; her early death was a tragic loss, not only for our project. At the Deutsches Museum it was primarily Alto Brachner, Gerhard Hartl, Christian Sichau, and Stefan Siemer who contributed tirelessly to the success of the joint project. The University of Pavia contributed a strong team of historians of science, comprising Fabio Bevilacqua, Lucio Fregonese, Lidia Falomo, Carla Garbarino, Lea Cardinali, Enrico Giannetto, and Stefano Bordone, without whom the project would have lacked an important dimension. The project owes to these three institutions and their staff a great pool of items on loan, valuable historical sources, and a discourse on the subject that strengthened and focused the project. For their cooperation and support I would like to convey my special thanks.

At this juncture I also want to thank the *Collected Papers of Albert Einstein* and especially the director of the project, Diana Kormos Buchwald, for the faithful and ever benevolent cooperation. This project was so generous as to allow us to use transcriptions and archive materials and also supported us in many other ways.

For over two years numerous participants have been involved in this project – scientists, curators, planners and designers, implementers and sponsors – to whom I would like to express my gratitude and thanks. May those I neglect to mention here forgive me. All participants know that such a complex undertaking can only emerge from collaboration among many, and that cooperation on every level of the project was decisive for its success. The appendix includes as complete as possible a list of the persons and institutions involved in the structure of the project; unfortunately, I cannot mention all of their names here.

For its constructive attention and effective direction, as well as the support for public relations, I would like to offer my warmest thanks to the administrative headquarters of the Max Planck Society in Munich, especially Hardo Braun, Wolf Seufert, Bernd Wirsing, and Manfred Zimpel. We profited in many ways from their personal commitment, their competence, and their considerable experience in project organization. The trust within the Max Planck Society and the personal involvement of not only those mentioned here, on top of their usual tasks, was very motivating for us and indispensable to the success of the project. Last but not least, the financial backing from Munich granted our undertaking the necessary stability. The project administration staff of Claudia Paaß and Sabine Steffan from the Max Planck Institute for the History of Science provided especially valuable services in interfacing with the general administration in Munich.

The team of staff of the Max Planck Institute for the History of Science involved in the project constituted the backbone of the project's content: all demonstrated not only scientific creativity and competence, but also outstanding commitment and loyalty. I am especially thankful to my companion and friend of many years, Peter Damerow. The first conception of the exhibition and accompanying publications was a brainchild of ours, which underwent constant development in our mutual dialogue. To his wise advise and critical view the project owes decisive impulses and orientation in all of its dimensions. He made no compromises in demanding accuracy and understandability. The idea of the online stations was his, realized in collaboration with Malcolm Hyman, Dirk Wintergrün, Robert Casties, Michael Behr, and Julia Damerow. This realization received active support from the library's digitalization group.

My warmest and most personal thanks for the many ideas, the great dedication, and the time invested are due to the scientific team at the institute. Members of the project group were: Katja Bödeker, Elena Bougleux, Reiner Braun, Jochen Büttner, Peter Carl, Giuseppe Castagnetti, Lindy Divarci, Carmen Hammer, Dieter Hoffmann, Horst Kant, Christoph Lehner, Andreas Loos, Wolf-Dieter Mechler, Markus Pössel, Simone Rieger, Matthias Schemmel, Sandra Schmidt, Urs Schoepflin, Michael Schüring, Ekkehard Sieker, Kurt Sundermeyer, Matteo Valleriani, Jörg Zaun, and Milena Wazeck as project coordinator. The scientific responsibility for *Worldview and Knowledge Acquisition* was borne by Peter Damerow and Matteo Valleriani; *Einstein – His Life's Path* was designed by Giuseppe Castagnetti, Dieter Hoffmann, Christoph Lehner, Wolf-Dieter Mechler, Milena Wazeck, and Jörg Zaun; Elena Bougleux, Peter Carl, Markus Pössel, Michael Schüring, and Kurt Sundermeyer took care of the science behind *Einstein's World Today*. The considerable tasks of communication, organization, and correspondence were mastered confidently by Ursula Müller, Shadiye Leather-Barrow, Petra Schröter, and Carola Grossmann – far beyond the measure of quantity and quality to be expected from a secretarial office. It is a privilege to work as part of such a team.

The Max Planck Institute for the History of Science was supplemented by numerous scientific partners who offered additional resources and helpful cooperation, above all the members of the Scientific Advisory Board, to whom I am obliged for their many helpful suggestions. Several institutes of the Max Planck Society participated in the preparation of the exhibition and the accompanying volumes, above all the Institute for Astronomy in Heidelberg, especially Jakob Staude; the Institute for Astrophysics in Garching, especially Volker Springel; the Institute for Dynamics and Self-Organization in Göttingen, especially Dirk Brockmann; the Institute of Colloids and Interfaces in Potsdam, especially Markus Antonietti; the Institute for Gravitational Physics (Albert Einstein Institute) in Potsdam and Hannover, especially Bernhard F. Schutz, Karsten Danzmann, Elke Müller, Markus Pössel, Peter Aufmuth, and Werner Benger; the Institute for Nuclear Physics in Heidelberg, especially Dirk Schwalm; the Institute for Extraterrestrial Physics in Garching, especially Günther Hasinger, Elmar Pfeffermann, and Reiner Hofmann; the Institute for Quantum Optics in Garching, especially Herbert Walther, Gerhard Rempe, Thomas Becker, and Stephan Dürr; the Institute for Plasma Physics in Garching, especially Isabella Milch; and the Institute for Radio Astronomy in Bonn, especially Anton Zensus, Rolf Schwartz, and Axel Jessner. Thanks are also due to the Center for Astronomy at the University of Heidelberg, especially Joachim Wambsganss; the Astrophysical Institute in Potsdam, especially Dierck-Ekkehard Liebscher; the Comenius Garten in Berlin Neukölln, especially Henning Vierck; the Albert-Einstein-Gymnasium in Berlin Neukölln and its director Klaus Lehnert; the Albert-Einstein-Grund- und Realschule in Caputh; the German Electron Synchroton DESY in Hamburg and Zeuthen, especially Thomas Naumann and Ulrike Behrens; the Institute for Theoretical Astrophysics at the Eberhard Karls University in Tübingen, especially Hanns Ruder and Marc Borchers; the European Organization for Nuclear Research CERN in Geneva, especially Maximilian Metzger, Hans Falk Hoffmann, Rolf Landua, and Michael Hauschild; the Institute for Experimental Physics at the University of Vienna, especially Anton Zeilinger and Markus Aspelmeyer; the Instituto de Astrofísica de Canarias in Tenerife, especially John Beckman and Robert Watson; the Istituto e Museo di Storia della Scienza in Florence, especially Paolo Galluzzi and Giorgio Strano; the Chinese Academy of Science, Institute for the History of Natural Sciences, especially Liu Dun and Zaiqing Fang; the Mathematisch-Physikalischer Salon of the Dresden State Art Collections, especially Wolfram Dolz and Peter Plaßmeyer; the Museum of the History of Science in Oxford, especially James Bennett; the Nobel Foundation in Stockholm, especially Svante Lindqvist; the Centrum Judaicum in Berlin, especially Hermann Simon; the association of companies and institutes OpTec Berlin-Brandenburg, especially Bernd Weidner and Karl-Heinz Schönborn (Clyxon Laser

GmbH), Hans Schieber (Crystalix GmbH), and Mohammadali Zoheidi (FiberTech GmbH); the SensoMotoric Instruments GmbH, especially Niklas Mascher; the Hahn-Meitner Institute in Berlin, especially Gabriele Lampert and Axel Neisser; the department of physics at the Philipps University in Marburg, especially Hans-Jürgen Stöckmann and Ulrich Kuhl; the Physikalisch-Technische Bundesanstalt in Braunschweig and Berlin, especially Fritz Riehle, Andreas Bauch, and Frank Melchert; Siemens AG Corporate Technology, especially Dietmar Theis, along with Michael Schumann and Henning Hanebuth; the SiemensForum, especially Karl-Heinz Stritzke; Siemens Information and Communication Networks in Berlin, especially Michael Tochtermann, Detlef Rinder, and Rolf-Dieter Kraft; Siemens VDO Automotive AG in Regensburg and Schwalbach, especially Werner Seier and Urs Weidermann; the Center of Applied Space Technology and Microgravity ZARM at the University of Bremen, especially Claus Lämmerzahl and Hansjörg Dittus; the Rechen- und Medienzentrum of the University of Lüneburg, especially Martin Warnke, and the Konrad-Zuse Institute in Berlin. I thank them all and am pleased by the clear confirmation that scientific work and its conveyance can succeed thanks to a core of cooperation and networking.

Key roles in the original initiative for the exhibition project were played by Dieter Simon of the Berlin-Brandenburgische Akademie der Wissenschaften, Hans Ottomeyer of the Deutsches Historisches Museum in Berlin, and Susan Neiman of the Einstein Forum in Potsdam. Susan Neiman and Yvonne Leonard held the partner events to the exhibition at the Einstein Forum. For the fruitful cooperation on the occasion of the exhibition *A Tower for Einstein* in Potsdam I thank Gert Streidt of the Haus der Brandenburgisch-Preußischen Geschichte in Potsdam and the curator Hans Wilderotter.

The exhibition received important political support, above all from Frank Walter Steinmeier, head of the Federal Chancellor's Office, and from State Secretary Wolf-Michael Catenhusen of the Federal Ministry for Education and Research. Wolf-Michael Catenhusen took charge of the Einstein Year 2005 and thus our exhibition as well, and was key in launching the entire project. He chaired a committee to coordinate the Einstein Year, which also included Hartmut F. Grübel, Susan Neimann, Dieter Simon, and Gert Streidt, along with Joachim Treusch, the chairman of the steering committee of the initiative "Science in Dialogue," and Knut Urban, president of the Deutsche Physikalische Gesellschaft. I thank all of them for the exchange and advice. Gerd Weiberg and his Bureau Einstein-Year 2005, with the committed support of Franka Ostertag, contributed political savvy; he was constantly available for advice and a helpful mediator. We owe a lot to his diplomatic wisdom and warm communication. I direct very personal thanks to my old school friend Jürgen Neffe, at the time director of the Berlin office of the Max Planck Society and today the author of an important Einstein biography. He set up important contacts and advised the project in its early phase. Stephan Steinlein, Ralph Tarraf, and Eva Büttner of the Federal Chancellor's Office always accompanied the project with good will and wise advice and were also willing to help with minor problems and questions. Here we were encouraged to pursue the project ambitiously and many a hurdle was removed from our path.

The project found generous sponsors, supporters, and allies: without our financial partners' trust and willingness to take risks, an undertaking like *Albert Einstein – Chief Engineer of the Universe* would be unconceivable. In addition to the Max Planck Society, I must mention here the Federal Cultural Foundation, in the persons of Hortensia Völckers, Alexander Farenholtz, Fokke Peters, and Friederike Tappe-Hornbostel; Hartmut F. Grübel, Andrea Noske, and Sabine Baun of the Federal Ministry for Education and Research; Hans-Georg Wieck, Hans-Jürgen Reißinger, Falko von Falkenhayn, and Ralf Graßmann of the Stiftung Deutsche Klassenlotterie in Berlin; Horst Nasko of the Heinz Nixdorf Foundation; Claus Weyrich and Dietmar Theis of the Siemens Corporation, Klaus-Philipp Seif of the BASF Corporation; Jürgen C. Regge of the Fritz Thyssen Foundation; Klaus Tschira and Beate Spiegel of the Klaus Tschira Foundation; Ingrid

Hamm, Ingrid Wünning, and Rafael Benz of the Robert Bosch Foundation; Joachim Treusch of the Wilhelm and Else Heraeus Foundation; and Yehuda Elkana of the Central European University in Budapest.

Thanks is also due to the following companies: Wall AG supported us generously by providing advertising space, Zumtobel Staff GmbH helped us to provide sophisticated lighting for the exhibition, Fujitsu Siemens Computers GmbH supported us with computer equipment.

The television broadcasting companies rbb, especially Johannes Unger and Werner Voigt; 3sat, especially Gert Scobel and Daniel Fiedler, and DW-TV, especially Elke Opielka and Manuela Kasper-Claridge supported us as media partners, not only helping us draw public attention to the project, but even making a joint contribution to a public culture of science.

Our own team for press and public relations, with Ursula Schmidt, Reiner Braun, Ekkehard Sieker, and Nicole Schuchardt, spared no effort, idea or time in providing the most comprehensive information and communication possible – not only to the public, press and media, but with particular attention to schools and universities as well.

I thank all donors who entrusted us with their pictures, objects, and documents on loan – without them such an exhibition could not have been achieved. I thank the composer Georg Graewe and the video artist Karin Leuenberger for their installation; I thank Vladimir Ivachkovets for his film compositions and Luca Lombardi for his *Einstein duo*; I thank the musician Zeitblom for the sound landscape in *Einstein's World Today*. For the loan of his collection of original graphics of Einstein portraits, I thank Jan Murken, and C.V. Vishveshwara for his original caricatures. Neither do I wish to leave unmentioned here the impressive involvement of the artist Alexander Polzin. Finally, I thank Eryk Infeld and other contemporary witnesses for their willingness to share their memories with us.

Scientific ideas may be the point of departure for such a project, but the appearance of the exhibition and publications lie in the hands of planners and designers. Their achievement in rendering these ideas is responsible for generating an interface between the public and science that is still unusual for the latter, with popular representations in all media, from the scenography of the exhibition to its graphic design, from films and animations to the utilization of the exterior space. Stefan Iglhaut put together an outstanding team for this purpose, in addition to Serge von Arx for the scenography, neo.studio with Tobias Neumann, Moritz Schneider, and Gesa Glück for the exhibition architecture, Michael Dörfler for the films and animations.

The expert planners, implementers, and producers were led by Bernd Tibes and his office DGI Bauwerk, especially Ingo Bunk, as well as the exhibition producer Rainer Kaufmann. Their organizational and technical overview made the exhibition in the Kronprinzenpalais on Berlin's Unter den Linden possible, in which the majority of the required infrastructure had yet to be erected. Michael Flegel was in charge of the light planning, Thomas Buck for the media infrastructure, and Gerlinde Fasse-Müller for the ventilation and air conditioning technology, to name just a few of the outstanding specialists. With their help, a temporary museum was created of exceptional quality.

The FührungsNetz of the Museumspädagogischer Dienst Berlin, especially Christiane Schrübbers and Almut Weiler, as well as Katja Bödeker from our institute and Wendy Coones from Iglhaut+Partner, I thank for the execution of the educational program of the exhibition; for consulting I thank Christine Keitel-Kreidt. The educational program, with its guided tours, workshops and the implementation of "explainers," is directed especially toward a young audience, in accordance with contemporary models. Thanks to the commitment of the consultants and "explainers," many of them students of my Einstein seminar at Humboldt University, and Katja Bödeker's outstanding coordination and organization of their activities, the exhibition has achieved a very personal feel. The important task of organizing of the exhibition's daily operation was mastered by Julia Wendt and Corinna Conrad. For the both sophisticated and entertaining selection at the exhibition shop we thank the Walther König bookshop; for the culinary refreshment after the exhibition round, the Einstein Coffeeshops of Berlin.

I would deliberately like to emphasize my thanks to the curator and exhibition maker Stefan Iglhaut. The form of the entire project matured in intensive conversation with him and Peter Damerow, in mutual inspiration and uncertainty, in shared visions and controversies. He played a key role in networking the participants and convinced experienced individuals to take on important positions. His presence at all levels of the project and his equalizing moderation between scientists, project partners, designers, and producers made a significant contribution to the success of the undertaking *Albert Einstein – Chief Engineer of the Universe*. For this he is due my special gratitude. The entire team of experienced and dedicated staff at his exhibition office worked on the design and coordination of the project: Serge von Arx, Wendy Coones, Edith Hirte, Elke Kupschinsky, Anne Maase, Wolf-Dieter Mechler, Vanessa Offen, Christoph Schwarz, and Jörg Zaun. The absolutely intimate connection between the Institute for the History of Science and the exhibition offices was characteristic of the project.

For valuable initial conversations about exhibition planning I thank Martin Roth of Dresden and Martin Heller of Zurich. Through their specialized knowledge of the history of science, Peter Galison, Gerd Graßhoff, Michel Janssen, Andreas Kleinert, Robert Schulmann, John Stachel, Alfredo Tolmasquim, Helmuth Trischler, and Scott Walter provided helpful suggestions; over and again, Ranga Yogeshwar demanded generally understandable portrayals in the sense of a modern public relations for science; to Horst Bredekamp we are indebted for a discussion of the topic "Einstein and the Art of the Avant-Garde." Yehuda Elkana was my mentor and consultant in one. Countless suggestions and tips from such conversations flowed into the project.

Many friends and colleagues, including my co-directors at the Max Planck Institute for the History of Science, Lorraine Daston and Hans-Jörg Rheinberger, but also the families of all participants, I thank for their understanding that the project consumed all their energy during this time. It became a project that is dedicated not only to the great personality Albert Einstein, but hopefully makes history of science an issue for a long time to come.

WORLDVIEW AND KNOWLEDGE ACQUISITION

"Science as something extant and complete is the most objective, impersonal thing we humans know of. Science as something nascent, as a goal, however, is something just as subjective and psychologically conditioned as all other human aspirations."

What is the nature of the world and what can humanity understand of the world? These fundamental questions have occupied humanity for thousands of years. Natural scientists, theologians, philosophers, and laymen have approached these mysteries from various angles and developed competing pictures of the world. But how do worldviews emerge and fade away, and how are they connected with the acquisition of new insights and with developments in society?

In the tradition of scientific inquiry, epochal changes in living conditions and culture are brought about by science and experience. This tradition was continued by Einstein's work, which radically transformed our understanding of basic terms like space, time, matter, and radiation.

His theories provide answers to questions about the structure of the world, most of which concern invisible dimensions that may seem abstract and far removed from everyday experience. In truth, invisible forces have always played a decisive role in the history of human culture. But how are such forces connected with the nature of the world, both on the micro level and on the grand scale, and how do we really know anything about what constitutes these forces?

On the Shoulders of Giants and Dwarves

Einstein: Aristotle, the Greek philosopher, I presume? A true naturalist!

Newton: A charlatan, that's what he is. The armchair scholars believed his fairy tales instead of checking them in experiments. He even had the cheek to declare that heavy bodies fall faster than light ones.

Aristotle: Of course they do, any child knows that. Greater causes have greater effects. And who are you, you dwarf, who dares deny the obvious?

Newton: I was the first to formulate the law of universal gravity; the force that everything in the universe is subject to – from the orbits of the planets to the apple that falls from the tree.

Einstein: The esteemed Sir Isaac Newton! So is it true then, the story about the apple?

Newton: Of course it's true, just like everything else people praise me for. Mr.?

Einstein: Einstein, Albert Einstein. I had the honor of standing on your shoulders and thus saw a little farther.

Newton: Hey, just a minute – that's my line!

Aristotle: Gentlemen, you digress. In the universe everything has its order and its natural place. Heavy bodies fall straight down because they strive to reach their natural place, the center of the Earth. Therefore, when the apple falls from the tree, it is only trying to move toward its natural place.

Newton: Rubbish. When I saw that famous apple fall, I realized that its movement followed the same law as the movement of the Moon. The Earth attracts both, like a magnet attracts iron. Gravity is an invisible force that rules the entire universe.

Aristotle: And this force hurls the Moon round the Earth in circles, correct?

Newton: No. The Moon falls towards the Earth, just like the apple.

Dialogue on
Mount Olympus

Aristotle: First he insults me, and then he fobs me off with such nonsense!

Newton: It's like this: actually, the Moon would like to move straight on, because that is its natural movement – always straight on and always constant for all eternity, or at least until some external force stops it from moving in this way. This is what I have called the principle of inertia. Then, gravity comes along and pulls the Moon down a little. At the same time, however, the Moon strives to go straight on, and then falls down a little bit more towards the Earth because of its gravity. The result of this zigzagging, which happens so fast that you don't see it, is the Moon's near-circular orbit.

Aristotle: Why keep it simple when you can have it complicated! It is as though the Moon were a slave and gravity a sort of shackle.

Einstein: It never really quite made sense to me either. After all, in reality one hardly notices the effects of these two invisible forces. According to my theory of general relativity, on the other hand, the Moon and apple move along their trajectories because our world is constituted in such a way that these are the simplest and most direct for them, even though they are not straight lines. Thus, gravity is not an invisible force but simply the expression of how our world is, with all its irregularities, to which the distribution of solid bodies in our universe also belongs.

Newton: Was the heresy of Aristotle victorious again in your time, which states that there are places in space of different quality? And I had put physics on an irrefutable footing.

Einstein: Seekers of scientific truth must be prepared for change. And should not stop asking questions!

○ PERSONALIA

Aristotle

In Aristotle (384–322 BCE) we encounter one of the most influential philosophers of antiquity. Continuing the work of his teacher Plato, Aristotle seeks to synthesize all knowledge of his era. He is reported to have written around 150 treatises, of which only thirty survive.

Justus van Gent: Aristotle, around 1470

These texts cover an enormous range of knowledge, and also treat themes of contemporary science, psychology, ethics, politics, and rhetoric.

In his work entitled *Physics*, Aristotle engages primarily with the processes, changes, and motions of the material world that surrounds us and, in doing so, introduces fundamental terms and ideas that shape scientific thought up to the beginning of the early modern era. With regard to spatial motion, Aristotle distinguishes between natural and violent motion and, in contrast to the modern view, assumes that all motion presupposes a cause of motion, a mover.

Isaac Newton

Isaac Newton (1643–1727) is one of the decisive actors in the radical upheaval in knowledge that takes place in the sixteenth and seventeenth century, which is known as the scientific revolution. During its course, natural philosophy is transformed into an empirically based science, formulated predominantly in the language of mathematics. In his work *Philosophiae naturalis Principia Mathematica*, Newton attempts to synthesize findings in the field of mechanics reached by vari-

ous scientists before him, such as Copernicus, Kepler, Galilei, and Huygens, into a system that builds upon few assumptions. Unlike many of his predecessors, in doing so Newton does not differentiate between celestial and terrestrial mechanics. The three fundamental laws of his science of motion, "Newton's laws," and the universal law of gravitation, which he formulates in the *Principia*, are of enormous importance for the development of astronomy and physics.

Albert Einstein

In 1905, Einstein (1879–1955) publishes not only his dissertation, but also four articles in quick succes-

Albert Einstein, 1947

sion, which change physics. In particular his paper *On the Electrodynamics of Moving Bodies* ushers in a far-reaching change in the categories which physics, but also our everyday thinking, uses in trying to understand the world.

In this paper Einstein develops his special theory of relativity from the premise that the physical laws in all systems, which move uniformly relative each other, should be the same. In this way, Einstein can offer solutions to questions that have arisen in border areas of various branches of physics.

In 1915, his attempt to extend the theory of relativity to the problem of gravity leads him to formulate the general theory of relativity. In this theory, space and time fuse to form a unified, four-dimensional construct called spacetime, which is structured by the distribution of matter. Spacetime, in turn, defines the motion of matter on the basis of its structure.

Center: Godfrey Kneller Isaac Newton, 1702

Invisible Forces

Forces are the means with which we shape our environment. For thousands of years mechanical forces predominate: human physical strength, the strength of working animals, the force of flowing water, the power of the wind that turns the sails of a windmill.

Forces are invisible but, in the case of mechanical forces, we can easily see for ourselves the way they work. We can feel the strength of our body through the resistance of objects, when we try to move them. We can see an ox straining when it pulls the plough. We can feel the force of the wind on our skin.

There are, however, also other forces, which present puzzles that remain unsolved for centuries. Which invisible force pulls or pushes a stone toward the ground when it falls? Why does a stone continue to fly after it has left the hand that throws it? Why does a flywheel carry on driving the machine after the motor has been switched off? How does fire produce light and heat? Who moves the stars in the sky? What force turns ore into metal during smelting? What turns barley into beer or fruit into wine? How does a magnet attract metal without coming into contact with it? Why does amber start to attract objects when it is rubbed?

Energy-saving use of a lever to move a colossal statue, Assyrian relief, 7th century BCE

People have been asking these questions repeatedly since ancient times. However, it is the lot of the nineteenth century to deliver satisfactory answers. Only then is it possible to generate electricity and to store, transport, and transform it into mechanical forces. With the help of electricity magnets are constructed that can be switched on and off. Substances are decomposed into their chemical components which are then reassembled into the same or a different substance. Classic natural science develops, and gives its answers to the centuries-old questions, answers that we know from our schooldays. Success in controlling the invisible forces of nature changes the world. The growth of knowledge goes hand in hand with its application by rapidly developing industries. This success also nurtures the growing conviction that the goals of the Age of Enlightenment have finally been achieved. A flood of popular science publications promotes a scientific worldview, which will supposedly determine the future life of humankind.

Yet the discovery and control of more and more new forces and effects – from electromagnetic induction, the effects of cathode and x-rays, to the radioactive decay of elements – comes at a price. The edifice of science has only just been built, yet the cracks are already showing. Science is expected to establish a new and secure worldview; but irritated by increasingly unfulfillable expectations, science retreats behind the walls of specialized subdisciplines.

This world – caught between an optimistic, future-oriented scientific worldview and science that is disintegrating into high-powered specialized disciplines, which too often ignore the problems of integrating the vast stockpile of accumulated knowledge – is the stage upon which fate places Einstein.

○ HISTORICAL CONTEXT

◉ FOUNDATIONS

Science and Society

In the ancient world, science originates as a byproduct of the division of intellectual and manual labor. It offers interpretations of the world

Education and Industry

The Age of Enlightenment brings forth an ideal of education. Its essential components are access to knowledge for all and the ability to

Force and Energy

To begin with, the term "force" describes the experience of our own muscle power. The notion of the force of wind or fire originates from

Left:
Machine hall in the Leipzig building of the F. A. Brockhaus publishing house, around 1870

Right:
Steam engine at the Philadelphia World Exhibition, 1876

and intellectual training; however, to begin with, science has little impact on technology or the economy. Transmission of scientific learning depends on written texts and schools, whose continued existence is perpetually under threat. The legacy of classical antiquity is taken up by the Arab world and handed down through medieval monastic traditions. In the Renaissance, the science of the ancient world enjoys a revival. Engineers of the early modern era return to this knowledge to master the technical challenges of large-scale undertakings. Science receives support in society through the foundation of academies. Scientific disciplines form. During the industrial revolution, science becomes an indispensable part of the social system forming the basis for new technologies such as steam engines and electric motors.

make informed judgments. Improved technical education is fundamental to the military successes of the French Revolution. During the industrial revolution, however, schools have the primary task of producing disciplined workers. At the same time, demand grows for increased mathematical and technical knowledge and this leads to the foundation of new types of schools. After the revolution of 1848 education in the natural sciences attains emancipatory significance. Workers' educational associations and popular scientific literature are committed to participation for all in a better future that can be achieved through scientific and technological progress. For Einstein, the encounter with this tradition is a formative educational experience.

the transferral of this experience to nature. Muscle power, moreover, can be replaced by natural forces. Tools such as the lever make it possible to achieve the same effect with less effort. In the early modern era, the insight that machines cannot trick nature becomes increasingly recognized. The dream of perpetual motion dies. The idea of the world as a machine is linked to the idea of conservation of force. But is force not consumed during labor? Against the background of experience with the transformation of forces, the term "energy" as stored work is introduced in the nineteenth century. Thus, the consumption of force can be interpreted as the transformation of energy. Einstein recognizes that mass, too, is only a form of energy – the starting point for the utilization of a new natural force.

Thomas F. Dalibard proves that thunderclouds contain electricity, 1752

Apparatus for Producing Invisible Forces

Any use of natural forces requires that they are available in a controlled form. The use of electricity begins with the construction of electrostatic generators, while the use of atomic energy begins with the splitting of the atomic nucleus of uranium

1 (Plate Electrical Machine
Second half of the 18th century
The production of electricity with electrical machines uses the effect known since antiquity, that when certain materials are rubbed together they become electrically charged. Today we know that during this process one of the materials tears off negatively charged electrons from the other material so that one becomes negatively and the other positively charged.

The plate electrical machine replaces older, less efficient constructions which usually rely on friction produced on globes or cylinders made of glass. Until the first half of the nineteenth century it is the standard type of machine used for electricity generation.

In the English type of the plate electrical machine, built around 1750, the rotating glass plate rubs against four leather pads to generate electricity on the surface of the glass.

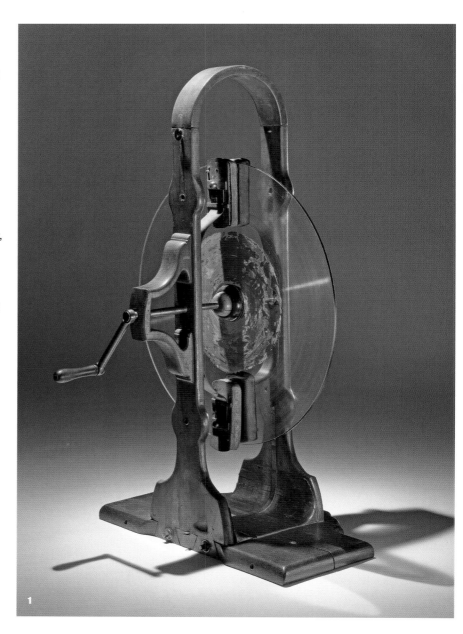
1

2

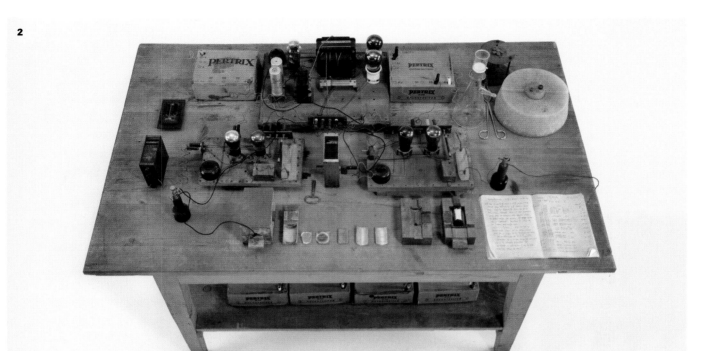

2 (Table with Otto Hahn's Experimental Appliances

Reconstruction

In 1938 Otto Hahn and Fritz Strassmann of the Kaiser Wilhelm Institute for Chemistry in Berlin experiment with the irradiation of uranium by neutrons and observe the production of an element that behaves like barium. Their discovery contradicts all expectations, and at first they notify only one person about it: Lise Meitner, who had fled from Nazi persecution to Sweden. Together with her nephew, Otto Robert Frisch, she finds a convincing explanation which simultaneously makes the experiment one of the first empirical confirmations of Einstein's formula $E = mc^2$: when the uranium nucleus is split into two parts, they fly apart, due to their equal charge, with precisely the amount of energy that corresponds to the mass difference (one fifth of the mass of a proton) between the uranium nucleus and its two split particles.

Looking back on his work, Otto Hahn writes in a text for the *Deutsches Museum* to accompany his laboratory bench exhibit: "And then something completely unexpected happened: [...] they [Hahn and Strassmann] were finally forced to assume that barium had been created during the bombardment of the uranium with neutrons. A process that physicists and chemists had previously considered impossible had been discovered: the bursting of the uranium in to two elements of average weight, the element barium and – as was soon to be seen – the inert gas krypton. [...] This process formed the basis for the exploitation of atomic energies. The additional neutrons emanating from the fission of the atom, as was later shown by others, can react again with uranium resulting in a chain reaction."

The subsequent age is characterized by efforts to control this chain reaction and the emited energy for civil and military purposes. In addition to electrical energy the public now becomes aware of a second "invisible force": "atomic energy" – and its role in the generation of electrical energy.

Gravity and Inertia

Aristotle: All bodies tend to move in natural motion to their natural places. Heavy bodies tend to move downwards toward the center of the Earth in a straight line, light bodies to move upwards in the opposite direction. Heavenly bodies are neither light nor heavy. Their natural motion is circular.

Newton: All bodies are inert. They maintain their state of rest or uniform, straight motion unless acted upon by an external force. All bodies are heavy with respect to each other. They mutually attract each other. The force that makes the apple fall from the tree is the same force that deflects the heavenly bodies from their straight path so that they orbit each other. Everywhere, motion takes place in a uniform and immovable, absolute space. Whether a body is in motion relative to this space can be concluded from the causes and effects of the movement. If a bucket filled with water is made to rotate, the water pushes outwards because of its inertia and rises up the sides of the bucket the more it is itself set into rotation by the rotating bucket. The rising of the water reveals that the bucket is in motion relative to the absolute space.

Mach: The law of inertia has often been discussed, and nearly always the questionable, hollow idea of absolute space has interfered to obscure matters. Absolute space is merely a figment of the mind, which cannot be evidenced by experience.

By referring to the experiment with the rotating water vessel, Newton believes that he is able to establish an absolute rotation. All that Newton's experiment with the rotating water vessel demonstrates is that the relative rotation against the sides of the vessel does not evoke any noticeable centrifugal forces, but that these are evoked by the relative rotation against the mass of the Earth and the other celestial bodies. No one can say what the qualitative and quantitative outcome

Aristotle, Einstein, Mach, Planck, and Newton measuring gravity

of the experiment would be if the walls of the vessel were to become progressively thicker and more massive until, finally, they are several miles thick.

Planck: Where Mach tries to proceed independently on the basis of his epistemology, he often errs. This includes Mach's persistently advocated, but physically useless idea that the relativity of all translational motions corresponds to a relativity of all rotational motions so that, for example, in principle it cannot be excluded that the starry sky is rotating around a stationary Earth.

Einstein: It is certainly not merely idle play, if we practice analyzing long-familiar concepts and practice demonstrating upon which conditions their justification and usefulness depends. I myself have derived considerable profit from Mach. Mach has clearly recognized the weaknesses of classical mechanics, and came very close to demanding a general theory of relativity. His considerations regarding Newton's bucket experiment show just how near his thoughts came to postulating relativity in the general sense (relativity of acceleration). Unfortunately, he lacked the awareness to see that we cannot decide through experiment whether the fall of a body relative to a coordinate system has to be attributed to the existence of a gravitational field or to an acceleration of the coordinate system.

Ernst Mach

In addition to a large number of important contributions to physics, of which his ground-breaking explorations on the velocity of projectiles that move faster than sound are the most famous, the work of

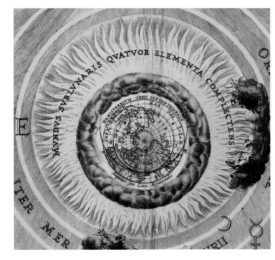

Planisphaerium Ptolemaicum, in: Andreas Cellarius, *Harmonia macrocosmica*, 1708

the Austrian phycisist Ernst Mach (1838-1916) also includes a series of significant studies on the physiology and psychology of perception. Above all, his work stands out because of his unusual and rigorous critique of the historical and epistemological foundations of physics.

In his book *The Science of Mechanics: A Critical and Historical Account of Its Development*, Mach presents an astute and unbiased analysis of the basis of mechanics. He interrogates the basic concepts of Newtonian mechanics, such as absolute space, which are generally believed to be irrefutable at this time. Mach's critique has a profound effect and for Einstein it represents an important starting point in the development of his general theory of relativity.

The Physics of Aristotle

The works of the Greek philosopher Aristotle, especially his doctrine on spatial motion, influence physics into early modern times. According to Aristotle, there are two simple kinds of motion, straight and circular. All other motions are a mixture of these two. Aristotle distinguishes further between natural and violent motion.

Each body has a unique, natural motion. Heavy bodies like stones tend to move downwards, light ones such as smoke tend to move upwards, and celestial bodies move in circles. Natural motions in the terrestrial sphere are determined by the four elements earth, water, air, and fire, whereas in the celestial sphere they are determined by a fifth element, also known as aether. Violent motions are caused by force. Twice the force moves a twice-as-heavy body equally fast, or an equally heavy body twice as fast.

Newton's Laws of Motion

Towards the end of the seventeenth century, many types of motion are described by mathematical laws. However, there does not appear to be any connection between the various motions. Yet is there a common foundation, some assumptions, from which motion can be deduced?

By examining the question how the motions of objects result from the causes of motion, so-called forces, Newton succeeds in establishing such a foundation. He reduces the connection between force and motion to three basic axioms, Newton's laws of motion. According to these laws it is not the motion, but rather its change, thus acceleration, which is caused by a force. The change is proportional to the force and proceeds in the same direction in which the force acts. Every action of one object on another corresponds to a reaction of equal size but in the reverse direction.

Newton's laws of motion not only make it possible to explain motion from known forces, but also to infer from observed motions which forces are operating. From Kepler's laws, which describe the motions of planets in elliptic orbits around the Sun, Newton deduces the attractive force acting between the Sun and the planets. Through generalizing this dependence, Newton arrives at the law of universal gravitation, according to which any heavy bodies attract each other with a force that is proportional to their mass and which decreases with the square of their mutual distance.

The Rotating Bucket – A Thought Experiment

When a bucket filled with water is set in rotation, the water rises up its sides and forms a concave surface. If the universe were completely empty and nothing existed apart from the bucket, could one still speak of rotation? Would the water form a concave surface in an empty universe as well, and would this be evidence of its rotation? What would happen if one held the bucket fast and, instead, set the entire universe in rotation? Would the water rise?

In the course of time, scientists repeat these thought experiments over and over, and come to very different conclusions. Newton sees in the rising of the water an effect of motion against absolute space. According to his opinion, the water, set in rotation in an empty universe, forms a concave surface exactly like in the real experiment and thus proves the existence of absolute space. However, if the universe rotates around the water, the surface stays flat.

For Mach, absolute space does not have any physical reality. In an empty universe the surface remains flat, one cannot establish whether the water rotates. The rising of the water is attributable precisely to a motion in relation to other masses, which is why it makes no difference to Mach whether the bucket or the universe rotates.

Einstein shares Mach's position; however, the meaning of his general theory of relativity for answering this question remains controversial.

Measuring Gravity and Inertia

1 (Double Balance for Measuring the
Dependance of Gravity on the Elevation
above the Earth's Surface
Philipp von Jolly, around 1880

During the 1870s Jolly tries directly by
weighing to confirm the changes in weight
dependent on height. He suspends a very
accurate balance in the tower of Munich's
Technical University. To the scale pans he
attaches twenty-one-meter-long wires with
another two pans at the end. An experimental object then weighs more in one of
the lower pans because it is closer to the
Earth's center than in an upper pan. Jolly
is very careful to note any possible factors
that can interfere with his measurements.
In particular he tries to minimize any expansion of the balance components due to
temperature change. Based on the results
of 500 measurements Jolly calculates a
mean weight difference of 31.686 mg
which within the measurement accuracy
corresponds with the theoretical value of
33.059 mg calculated according to Newton's law of gravitation.

2 (Device for Demonstrating the Inertia
of a Rotating Body
Eberhard Gieseler, 1905
From 1874 to 1914 Gieseler is professor of
physics and agricultural machinery science
in Bonn-Poppelsdorf. He constructs a demonstration device for schools which has
a flywheel with two strings wound round
its axel. When the wheel is released it de-
scends until the strings are fully extended.
Due to its inertia the movement continues
and the string winds back up again. The flywheel returns almost to its original height,
because of the principle of conservation
of energy.
He got the basic idea in Cambridge where
he was shown a popular children's toy that
was reputedly "Maxwell's favorite apparatus" for demonstrating the conservation of
energy. He says he then named his device
"Maxwell's apparatus" to promote sales.

Mach and Einstein

3 (Die Mechanik in ihrer Entwicklung
(The Science of Mechanics) by Ernst Mach
*First edition in the series Internationale
Wissenschaftliche Bibliothek, Leipzig, 1883.*
Mach writes his history of mechanics with
a theoretical objective. It contains an em-
piricist criticism of the dogmatism prevailing in physics at the time. A chapter is de-
voted to an in-depth criticism of Newton's
principle of inertia. The tendency of a body
to continue its movement in a straight line
is not a property of the body, Mach argues,
but rather an effect exerted on the body by
all other bodies in the universe.

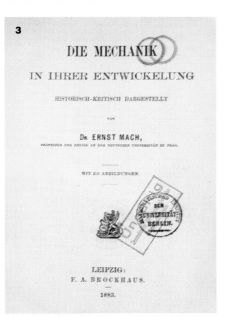

3

4

PHYSIKALISCHE ZEITSCHRIFT

No. 7.   1. April 1916.   17. Jahrgang.
Redaktionsschluß für No. 8 am 8. April 1916.

Ernst Mach.
Von A. Einstein.

4 (Albert Einstein's Obituary for
Ernst Mach
*Physikalische Zeitschrift,
vol. 17 (1916), no.7*
In his obituary Einstein emphasizes the
significance of Mach's critique for the
development of the general theory of
relativity. He maintains that Mach was
not far from calling for a general theory
of relativity.

Chemical Changes

Common sense tells us that movements of material objects in space cannot cause all the changes we see around us; some must be caused by qualitative changes in matter itself. In the course of the development of human culture, such observations become the basis of techniques to change material things so that they better suit the purposes of humankind. Conservation and preparation of food, brewing beer and making wine, extraction of copper, tin, lead, silver, gold, and iron for the manufacture of jewelry and powerful tools of bronze and steel are examples of complex chemical processes. Control of these processes is achieved by knowledge and experience that has been acquired over thousands of years.

Ancient philosophers already attempt to find general explanations for changes in matter. Aristotle's ideas are particularly influential. According to his doctrine, all materials are composed of an indestructible and formless "prime matter" which acquires specific qualities from different mixtures of the four elements, earth, water, air, and fire. These elements or "principles," which are not identical to the real materials, can not only be mixed in a multitude of ways but, under certain conditions, can even be transformed into one another.

In late classical antiquity and in the Arabic and Latin Middle Ages, Aristotle's doctrine became the most important starting point for the development of alchemy. Since according to him even the elements can be transformed into one another, his doctrine nourishes the hope of turning base metals into precious metals through "transmutation," or even of isolating the prime matter itself and to obtain from it the Philosopher's Stone, a universal elixir of transmutation.

In early modern science, Greek atomism – which was repressed by Aristotelism – increasingly gains significance and turns into a basic source of a radical critique of alchemy. In the course of the seventeenth and eighteenth centuries, the assumptions of formless prime matter and the transmutation of elements are abandoned in favor of the assumptions suggested by atomism that all substances are composed of a limited number of transmutable elements and that all chemical changes are processes of decomposing and recomposing compounds of these elements.

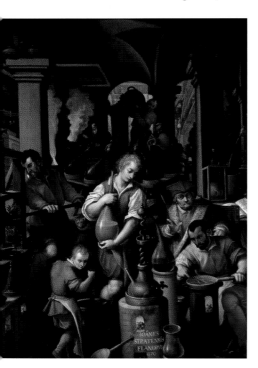

Above:
Jan Stradanus:
Alchemy, 1570

Right:
Analysis of air by
Antoine Lavoisier,
19th-century
engraving

The rise of modern chemistry is the result of meticulous detective work guided by these assumptions. In the chemical laboratories of the eighteenth and nineteenth centuries, countless experiments are performed to determine which substances react with each other in which proportions. After many futile efforts, researchers finally comprehend changes of substances qualitatively and quantitatively on the basis of assumptions about unchangeable "atoms" of the elements and the bonding of these atoms to "molecules" of the remaining substances.

In Einstein's time, new discoveries challenge this chemical view of the world. Electrochemistry, radioactivity, and nuclear fission compel researchers to admit that atoms themselves consist of even smaller particles. By changing their composition, the supposedly unchangeable elements can be changed into others, and new elements can even be created.

○ HISTORICAL CONTEXT

Chemical Industry

In the latter half of the nineteenth century, the chemical industry as well as the steel and electrical industries develop rapidly, especially in Germany and the USA. Particularly the advances in organic chemistry, necessary for the synthesis of artificial dyes, help as of 1869

to establish the worldwide preeminence of Germany's chemical industry.

A reaction to the requirements of industrial production is the stepping-up of academic and private research. The result is that soon the chemical industry represents an integral part of all areas of everyday life.

The significance of the chemical industry also influences the development of scientific theories, such as thermodynamics, and provides a decisive impetus for the revolution in physics that takes place at the beginning of the twentieth century. In 1911, the Kaiser Wilhelm Society is founded with considerable financial support from the chemical industry.

○ PERSONALIA

Theophrastus Paracelsus

Paracelsus (1492/3–1541) studies medicine in Ferrara and travels around Europe as an army surgeon. In 1527, he receives an appointment at the University of Basle. But his rejection of accepted authoritative opinions provokes the medical faculty and soon causes a scandal.

Left: Chemistry laboratory, early 19th century

Right: Marie and Pierre Curie in the laboratory, around 1900

His doctrine of "principles" combines Aristotelian and neo-Platonic thinking with the tradition of alchemy. According to Paracelsus, a natural body forms from an element, in the Aristotelian sense, and three principles: sulfur, salt, and mercury. These function as a kind of soul of the body and they determine its properties. The principles inherent in a body can be recognized through burning. In contrast to modern ideas, though, the body is not divided up into separate component parts. Rather, according to the doctrine of principles, a body is a composite unity of matter and principles.

Jöns Jakob Berzelius

Present-day knowledge that all materials are composed of a finite number of chemical elements has been furthered to a large extent by the Swedish chemist Jöns Jakob Berzelius (1779-1848).

Berzelius establishes the law of constant proportions, which states that all chemical compounds always contain the same proportion of the chemical elements from which they are made. He tests this in countless series of experiments, particularly on various salts. Berzelius uses the recently invented galvanic battery to conduct electrochemical experiments with salts, from which he obtains insights into the importance of electricity for the bonding of chemical elements. While examining minerals and also waste products of the emerging chemical industry, Berzelius discovers new elements, such as thorium. His *Textbook of Chemistry* becomes the standard reference work in Europe and profoundly influences an entire generation of chemists.

Marie Curie

Marie Curie (1867–1934), the fifth child of a Warsaw teacher's family, begins to study science in Paris in 1891. She overcomes a great deal of opposition in a male-dominated

field. In her doctoral dissertation, she examines the rays emitted by uranium crystals – discovered shortly before by Becquerel – using an ionization chamber in place of the photographic plates used previously. While looking for other elements that emit radiation, Curie first investigates thorium; then she examines pitchblende, a waste product of uranium mining, and discovers polonium and, ultimately, radium. For her pioneering work on radioactivity, a term she introduces in 1898, Marie Curie wins the Nobel Prize for Physics in 1903 together with her husband Pierre Curie and Henri Becquerel. In 1911, she is awarded the Nobel Prize for Chemistry for her groundbreaking work on radium and its compounds.

Laboratory Equipment

1 (Iron Mortar and Pestle
16th – 18th century
From the Middle Ages on smelting works
start securing the success of their production processes by "assaying" which rocks
are best suited for processing. The smelting process is simulated on a small scale in
the laboratory using crucibles, live coals,
and bellows. Prior to this the ores and all
auxiliary materials are pulverized in a
mortar, the most important utensil for
crushing and powdering hard and brittle
substances.

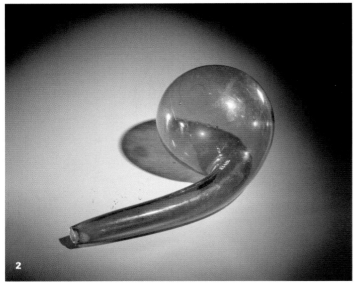

2 (Retort Made of Green Glass
18th–19th century
By the mid-sixteenth century at the latest
the retort is introduced into Europe as a
universal implement in alchemy. With the
help of a "receiver" it enables the extraction of substances by distillation, sublima-
tion, or gasification achieving far higher
quality levels than ever before. Liquid or
solid materials are heated in the retort.
The expelled substances precipitate in the
cool neck of the retort and are collected
in the receiver.

3 (Florentine Bottle for "aqua regia"
17th–18th century
The Florentine bottle is used in alchemy
to separate non-mixable liquids and as a
dropper for carefully administering minute
quantities of a valuable liquid. The inscrip-
tion "aqua regia" indicates a special substance: it is a mixture of acids that dis-
solves metals, including gold. A stroke of
gold on a porcelain plate and a drop of
"aqua regia" reveal the quality of the gold.

4 (Blowpipe Instruments
1907
Blowpipe analysis is a quick method of investigating minerals by means of heating or melting and is especially useful on site in the field. The method, perfected in the nineteenth century, derives from the minia-turization of ore analysis techniques used in the laboratory. By blowing air through the blowpipe into the flame of a burner it is possible to produce temperatures hot enough to melt, oxidize, or reduce a small grain of the mineral.

4

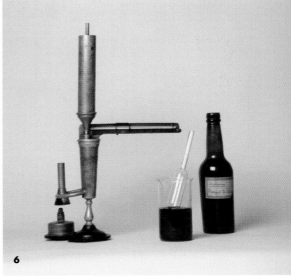

6

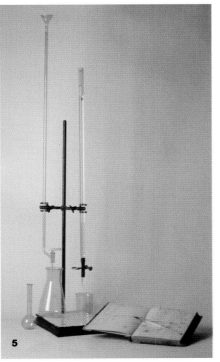

5

5 (Burette
Second half of the 19th century
The burette is a calibrated glass tube with an aperture and stopcock used to deliver small controlled quantities of a reactant agent into a reaction vessel. The end of the reaction is then signaled by the changing color of an indicator. The burette is the most important instrument in volumetry invented toward the end of the eighteenth century. The volume of the added reactant agent indicates the amount of reacting substance in the reaction vessel.

6 (Ebullioscope
1890
The ebullioscope is an instrument for determining the alcohol content of wine and beer by measuring the decrease of the boiling point of water induced by the alcohol. From the mid-nineteenth century onwards traveling salesmen are equipped with such gauging instruments for quality testing during their journeys. After calibrating the instrument with boiling water it is possible to determine at constant air pressure the alcohol content to within 0.1% precision.

Light, Heat, Magnetism, Electricity

Making light and heat with fire is, like the use of mechanical forces, fundamental to human civilization. In contrast, magnetism, known as a property of the mineral magnetite, and the electricity generated by rubbing amber remain of no interest for a long time.

Many of the early attempts to find explanations prove to be of little historical endurance. In the case of light, for example, there are two competing basic assumptions about vision that are unreconcilable. According to one assumption, the eye emits rays that scan objects, whereas in the other, objects emit something that encounters the eye.

Major progress in understanding the nature of physical phenomena begins in modern times as a result of investigations such as the study of the refraction of light, of thermal dilatation, and of the polarity of magnetism and electricity. However, there are still fundamental questions to which there are no answers. Is light a wave that propagates within a medium, or does it consist of smallest particles that travel in straight lines through space? Does light move with finite or infinite velocity? Do colors originate through transformation of light, or are they different kinds of light? Is heat weightless matter or is it the motion of smallest particles of matter? Is electricity a manifestation of the elasticity of an "aether," that is, of a substance that permeates all bodies, or does a specific kind of electrical matter exist, which a body may contain in greater or lesser amounts? Is magnetism an innate force of the matter of a magnet, or is it the result of a vortex of particles that streams through the magnet and the metal it attracts?

Such questions partly find answers within the framework of classical nineteenth-century physics. In this respect, in-

Nicola Tesla experimenting with spark discharges, around 1900

creasing mastery of electricity plays a key role. The development begins with the construction of machines, which can generate large amounts of electricity through charging by friction, and with the construction of devices like the Leyden jar for storing it. The invention of the chemical battery makes it possible to study the flowing electrical current. The discovery that an electrical current generates a magnetic field around the conductor, and vice versa, that an electrical current is generated in a conductor by a changing magnetic field, sheds light on the connection between magnetism and electricity. Experiments to create electrical currents through a vacuum and through rarefied gases using discharge tubes generate electrical particle rays, that is, electricity separated from any material carrier. Strange light phenomena are observed when these rays hit atoms, including the invisible light, known as x-rays. The discovery of electromagnetic waves brings order to many of the new discoveries. The colors of light, heat radiation, x-rays, and the gamma component of radioactivity, which was discovered at the end of the century, are interpreted as electromagnetic waves of different wavelengths. On the other hand, other kinds of radiation, such as the other components of radioactivity, prove to be particle rays. Using such particle rays, the first nuclear reactions are achieved. This opens the gates to the era of Einstein.

○ HISTORICAL CONTEXT

Science and Superstition

At the beginning of the twentieth century, rays stimulate the imagination of science and the public. Cathode rays can be deflected by a magnet. X-rays penetrate opaque bodies. Nothing seems impossible. Belief in supernatural phenomena

Dowser with divining rod, around 1900

○ PERSONALIA

Alessandro Volta

Alessandro Volta (1745–1827) studies electrical phenomena starting in his boyhood, and particularly phenomena associated with electricity generated by friction. After four years as a physics teacher at a gymnasium and as director of the

○ FOUNDATIONS

Rays

From the mid-nineteenth century, new experimental technologies reveal the existence of a hitherto unknown world of rays, from cathode rays to x-rays and radioactivity. The interpretation of these rays with the aid of familiar concepts, such

Alessandro Volta, around 1820

also receives backing from science. With strong academic participation, spiritualist and parapsychological associations are formed. Occultism is ubiquitous.

In this situation, even prominent scientists no longer see any strict division between science and superstition. In 1930, Albert Einstein attends parapsychological meetings with clairvoyants and appears deeply impressed. In a radio broadcast in 1932, Max Planck recommends the foundation of an institute of the Kaiser Wilhelm Society to investigate the putative "Earth rays" and the alleged achievements of diviners.

public school in his native town of Como, Volta is appointed professor of experimental physics at the University of Pavia, where he researches on electricity, magnetism, heat, pneumatics, acoustics, meteorology, and optics. As director of the physics cabinet, from 1778–1819 Volta is responsible for inventing numerous instruments, for example, the electrophorus (1775) and the voltaic pile or electric battery (1799). The voltaic pile becomes the most useful source of electricity for it guarantees continuously flowing electric current. Seventy years after Volta's death, the unit of electromotive force is named in his honor: the volt.

as waves and particles, remains controversial for a long time, until quantum theory calls such interpretations into question. This development begins with the investigation of how electricity travels through diluted gases, and leads to the discovery of rays that Goldstein in 1881 names cathode rays. In 1897 Thomson is first able to prove that they behave like electrically charged particles ("electrons"). In 1895 Röntgen discovers by chance a new type of ray, which he calls "x-rays." For many years the nature of these rays also remains unclear. Encouraged by Röntgen, in 1896 Becquerel while experimenting with radium salts discovers a further phenomenon involving rays – radioactivity.

Mysterious Effects

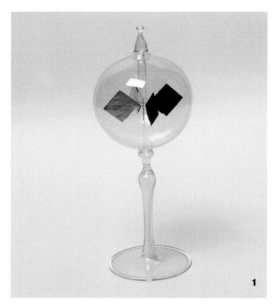

1

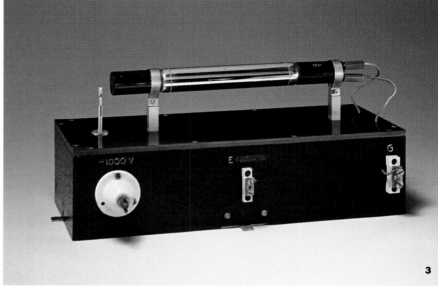

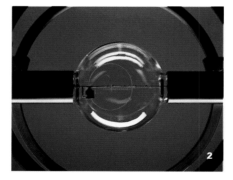

3

1 (Light Mill

Phywe Company, Göttingen

In 1875 William Crookes constructs a "light mill" based on observations of the mechanical effects of light and heat. An impeller wheel with suitably coated vanes encased in a glass bulb is set in motion solely by the entering light. Cookes first describes this mysterious phenomenon as the mechanical effect of the "aether waves" of light. But he soon notices that the gas in the bulb also plays a role. This role is still not completely understood in Einstein's day.

2 (Cathode Ray Tube with Helmholtz Coils

Phywe Company, Göttingen

Beams of electrons are deflected by magnetic fields. In the tube a strongly accelerated and bundled beam of electrons collides with the atoms of an inert gas causing them to emit light. As a result of this the beam of electrons becomes visible. The homogeneous magnetic field of Helmholtz coils can force this beam into an orbit.

3 (Geiger-Müller Tube

Otto Preissler Company, Leipzig, 1938

In 1928 Hans Geiger and Walther Müller invent a detector for radioactive rays. A metal cylinder contains a rarified inert gas and an electrode at high voltage against the cylinder. Each particle of radiation that hits an atom of the inert gas triggers an avalanche of released electrons which causes a short electric pulse. This signal can be amplified so it can be heard. It can also be electronically counted to serve as a measure of the strength of the radiation.

Heat, Light, Radiation

4 (Steam Engine

Henry Maudslay, 19th century

The invention of the steam engine to transform heat into mechanical energy is one of the most significant preconditions of the industrial revolution.

During the nineteenth century Henry Maudslay's constructions of compact and efficient steam engines help to spread such technology to smaller factories and workshops in many different sectors of production.

5 (Flint Glass Prism

Joseph Fraunhofer, beginning of the 19th century

The "spectral analysis" of light sources gains increasing importance following Newton's observation that white light can be split into the colors of the spectrum with the help of a prism. One of the pioneers in this development is Joseph Fraunhofer. He perfects the production of lenses and prisms and makes numerous discoveries, especially that of the absorption lines in the solar spectrum (Fraunhofer lines).

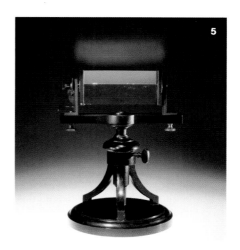

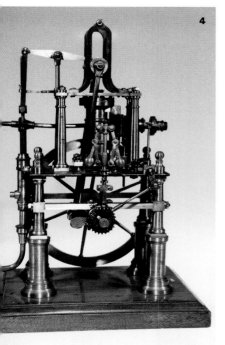

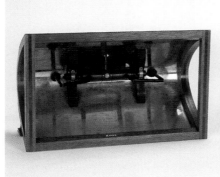 

6 (Parabolic Mirror for "Hertzian waves"

Around 1900

In an experiment in 1886 Heinrich Hertz succeeds in generating the electromagnetic waves predicted by Maxwell. He also proves that these waves are reflected in a similar way to light waves. This leads to the construction of parabolic mirrors with which electromagnetic waves can be bundled and transmitted like light waves.

Magnetism, Electricity, Induction

1 (Volta's Electric Pile
Alessandro Volta, 1800
In 1800 Volta invents the first fully-functioning chemical battery based on his discovery of contact electricity. He utilizes the differences in tension between two metals in a liquid conductor. His column consists of alternating layers of zinc and copper plates separated from each other by pieces of felt soaked in diluted acid. All subsequent wet cell batteries are based on this invention.

2 (Electromagnetic Motor
C. A. Grüel, Berlin, around 1858
After the discovery of electromagnetism mechanical practitioners try to convert electricity into mechanical movement for practical purposes. Their efforts sometimes result in weird and wonderful contraptions. Grüel's motor has an iron anchor which is alternately tilted from side to side by an electromagnet. This movement is transferred to a flywheel which in turn regulates the switching of the electromagnet in time with the motion.

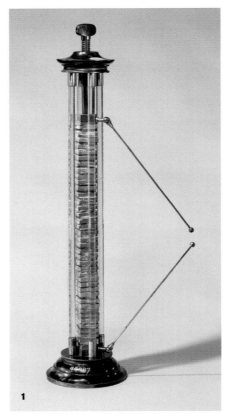

1

3 (The Spherical Gyrocompass by Anschütz-Kaempfe and Einstein
Hermann Anschütz-Kaempfe, around 1926
In many years of collaboration with Einstein, the manufacturer Anschütz-Kaempfe develops a gyrocompass freely floating in a liquid. It is Einstein's idea to center the compass by means of an electromagnet. After countless attempts to improve the stability of the compass Anschütz-Kaempfe finally manages to get it ready for production.

3

2

Demons and Ghosts

4 (Letter from Einstein to Zangger with Accompanying Letter from His Son "Tete"
Einstein's letter, undated; accompanying letter dated 12 March 1930
In his letter Tete writes: "At the moment clairvoyants are popular in Berlin. Almost every family has a clairvoyant, a psychometer, or at least a graphologist. Papa also visited a clairvoyant and now believes in them." Einstein comments: "Tete writes to you about 'clairvoyance.' This is a strange thing. A little woman of 55 sat there. Items of jewelry, pencils, pocket watches were given to her. She takes an object and feels it. The person in question then raises his or her hand. 'You have had gas poisoning (inhaled poison). You work in a big company and are dreaded by your subordinates.' And so it goes on with great accuracy. Here fact, there reason, both in hopeless conflict."

5 (Max Planck in a Radio Conversation
Broadcast in "Gelehrtenfunk"
(The Scholars' Program), end of 1932
"You know that certain questions such as those concerning Earth rays and the divining rod make people's feelings run high. [...] The Kaiser Wilhelm Society should establish a research institute for such matters [...]. Admittedly, some people in scientific circles do have their doubts about addressing such things at all [...]. But I think differently, and here as well I would like to support research in a universal way."

6

6 (Automatic Spring Finder
Adolf Schmidt, Bern, around 1911
The device produced by one of Einstein's contemporaries is a kind of "scientific divining rod." It is supposed to be able to locate subterranean watercourses by measuring horizontal variations in the Earth's magnetism. Its actual efficacy is questionable.

Labyrinth of Microworlds

When we perceive something by seeing or hearing it, the sources of the impressions we hear or see are at a distance from our sensory organs. How is this distance bridged? Are there processes in the world of seemingly infinitely small things, in a microworld, that produce this connection?

It takes a long time before instruments are invented that make it possible to explore this world directly. The diverse and often fantastic ideas about microworlds remain speculative.

Two ideas, which are essentially irreconcilable, lead to alternative explanations for bridging the space between source and appearance. Either the source of the appearance emits something that traverses space and evokes the appearance in a distant place, or some kind of material medium between source and appearance is changed locally into a different state from which it can only return to its original state if it transfers its changed state to the adjacent medium. In this way, the changed state propagates and spreads out in the medium without making it necessary for the medium itself to move from the source to the appearance.

In modern times, these two alternative notions found separate theoretical traditions. The first notion is closely connected with atomism. Matter is composed of smallest particles in space that is otherwise empty. A radiating body emits small particles, which fly through the empty space and create as "particle rays" the perceived appearance in a distant place. The second notion leads to the model of a wave, which propagates in a medium, like a wave in the sea. Therefore, the space that is bridged cannot be entirely empty.

In the seventeenth and eighteenth centuries, numerous theories are developed, which, in one way or another, attribute the propagation and the effects of sound, light, gravity, heat, magnetism, and electricity to mechanical processes within complex structures of microworlds

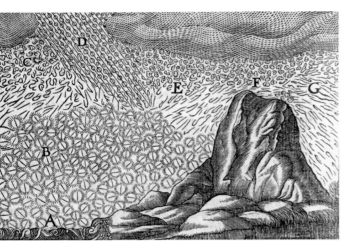

The fine structure of matter according to René Descartes, 1650

and attempt to explain them in this way. However, as such effects exhibit similarities, but on closer scrutiny unmistakably also irreconcilable differences, the explanations advanced become increasingly differentiated and complicated.

The ringing of a bell placed in a bell jar cannot be heard if the air has been pumped out. It is thus fair to interpret sound as a wave that propagates through the medium of air. This supposition is specified and verified by deriving and checking from assumptions about the change of state in the microworld caused by the sound source phenomena in the macroworld of how sound propagates.

Light, however, penetrates the vacuum in the bell jar unhindered. Therefore, the propagation of light cannot be explained as a wave through the medium of air. Is light a stream of particles, which are emitted by a light source? But then why does the light source not decrease in weight? Is light a wave in a medium called "aether" which is still present in the evacuated bell jar? If so, why is this aether weightless? Was it originally in the bell jar together with the air? How else could it enter the jar? At the end of the nineteenth century such questions are still unanswered and assumptions about processes in the microworld are controversial.

○ FOUNDATIONS

Descartes' Theory of Light

According to Descartes (1596–1650), light propagates without any time lag through a solid medium, which fills the whole of space. The propagation of light is a direct mechanical transmission of motion. Descartes thinks that the medium through which light propagates consists of matter that is composed of particles of different sizes, which are only defined by their extension. The medium is not elastic and cannot be compressed. Its existence cannot be proven. According to Descartes' third law, which describes the motion of particles of matter, all moving bodies strive to move in a straight line. However, as it is not possible to move in a straight line in space that is completely filled, the particles take on circular movements. As a result of this process, space is full of vortices, which carry the celestial bodies around their orbits. There are three kinds of particle. The smallest form the stars, because their circular motion drives them toward the central point of the vortex.

Because the smallest particles are moving through filled space, they continually press up against neighboring particles. Light is simply this pressure. Light propagates in straight lines because, unlike particles of matter, it can obey Descartes' third law, which prescribes a tendency to move in straight lines.

Huygens' Theory of Light

Christiaan Huygens (1629–1695) is convinced that light does not propagate instantaneously but with finite velocity. In fact, while Huygens is still pursuing his studies on the nature of light, the speed of light is actually measured for the first time. As Huygens rules out that bodies are able to travel at such high velocity, he rejects a corpuscular model of light as postulated by Gassendi and Newton. Instead, for his explanation of light Huygens takes up Descartes' ideas, where space is completely filled with an aether that is composed of smallest particles. These particles pass on impulses from a light source in a straight direction like a rigid stick and in this way transport light without delay from source to destination.

To avoid assuming that the speed of light is infinite, Huygens replaces Descartes' rigid particles with elastic ones, thus postulating an aether in which light pulses propagate as wave fronts.

Huygens formulates a principle according to which each point on these wave fronts can be considered the starting point of a new, secondary wave. These secondary waves add up to a new wave front. With the aid of this principle, Huygens succeeds in deducing the laws of reflection and refraction, that is, the deflection of light upon entering a medium. When Huygens finally finds an explanation for the unusual phenomenon of birefringence, he is ultemately persuaded that his wave theory of light is correct.

Newton's Theory of Light

Newton sees light as a stream of particles that is emitted by a light source and travels at a finite speed in a straight line. The particles are of different sizes; the largest ones,

when they strike the eye's retina, produce a red color sensation, whereas with decreasing size the smaller ones produce sensations of orange, yellow, green, and blue. White light is composed of rays with different sized particles. Newton's arguments in favor of his theory of light are based on his experiments in which a ray of white sunlight passes through a prism and disperses into the colors of the spectrum. With his corpuscular theory of light he attempts to explain phenomena that we understand today on the basis of the idea of light as a wave. These include the diffraction of light, when light appears to be deflected from the edges of

solid bodies, and interference, the pattern of light and dark areas produced when light from two sources overlaps. Newton speculates about an interpretation of diffraction as deflection of the particle stream due to an attractive force between edge of body and light particles. However, he does not find any satisfactory explanation for the phenomenon of interference.

On the basis of Newton's corpuscular theory of light, his successors calculate the bending of a ray of light that passes the Sun and is modified by its gravity. The value calculated is half of the one calculated with Einstein's general theory of relativity, which is confirmed by observations.

Pelagio Palagi:
Newton discovers
the refraction of
light – by observ-
ing a soap bubble,
1827

The Vacuum

1 (Experimenting with the Vacuum
Workshop on "Nothingness," Berlin-Neukölln, 2003
Drita (7 years) and Granit (8 years) are experimenting with a vacuum bell jar. An alarm clock is placed under a bell jar and the air is then pumped out. The alarm clock is now in a vacuum. When it rings, it can hardly be heard. The air is let in again. When the alarm clock rings this time, it can be heard very clearly.
The children give their explanations to the interviewer, Katja Bödeker:

Katja: Why does it sound loud now, but quiet before?
Drita: You made the air smaller.
Granit: You sucked the air in!
Drita: You made the sound quieter with the machine.
Katja: How can I make the sound quieter with the machine?
Drita: You can take it away with the air. The air pulls it, doesn't it. The air goes down and the noise gets smaller and smaller.
Katja: How does the noise get smaller?
Drita: The wind is down at the bottom. It's all here [in the pump], isn't it. And then the wind takes the sound and then it gets smaller and smaller.
Granit: The sound is sucked in.
Katja: The sound is sucked in?
Drita (laughing): Yes.
Granit: The sound is sucked in and then the double sound is pumped into the clock… When we put the clock inside, it wasn't as loud, was it? And then the vacuum pump sucked in the sounds and pumped twice as much into the alarm clock.

Katja: It pumped in the sound or the air?
Granit: The sound, but double the loudness.
Katja: Really, and the loudness, does that have something to do with the air being sucked out?
Granit: Yes.
Katja: How is the air connected to the loudness?
Granit: For example, if there weren't any air, then we wouldn't be able to talk either. Our vocal chords would just dry up.
Katja: … But if we were to switch the radio on now and there were no air, would we still be able to hear it?
Granit: No.
Katja: Why not?
Granit: Because we need the air for electricity as well for our voices. You wouldn't be able to sing. And then the CD would crumble into dust.

3 (Lucretius

*De rerum natura, second edition, London,
1683, frontispiece*

The atomist Lucretius (97–55 BCE) sees
light and sound equally as the effect of
minutest particles. The Sun projects the
tiniest of particles which travel fast
through vast spaces. We see objects because images in the shape of thin skins
constantly detach themselves from the
objects' surfaces and strike our eyes.
They always move in a straight direction
and are thus unable to penetrate opaque
objects, because they break apart when
striking them. Similarly, we only perceive
sound when its particles strike our ears.
Sound can disperse in all directions and
thus also penetrate the winding pathways
of opaque bodies.

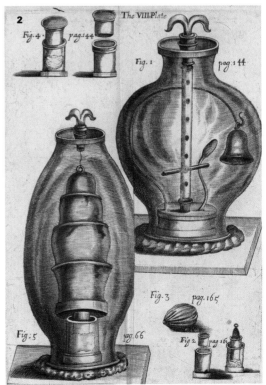

2 (The Experiment "de vacuo"

*Boyle, A Continuation of New Experiments
Physico-Mechanical, Oxford, 1669*

Ever since the invention of the air pump by
Otto von Guericke in the seventeenth century scientists experiment with the vacuum. One of them is Robert Boyle who investigates the propagation of sound in a
vacuum. Inside an evacuated glass jar he
suspends a bell the hammer of which can
be operated by a magnet outside the jar.
He also places a large, loud mechanical
clock inside a glass jar and pumps out all
the air. By means of a valve he varies the
amount of air in the jar. Boyle notices that
the sound volume changes. He concludes
"that whether or no the Air be the onely,
it is at least, the principal medium of
Sounds."

Models of the Cosmos

The starting point for constructing any worldview that encompasses heaven and Earth is terrestrial experience. Active conquest of the Earth's realm offers manifold possibilities of making such experiences. Geographical explorations providing growing amounts of new information determine and change the role of the individual, the group, and culture in the cosmos. Progressive expansion of the geographic sphere of experience in the course of the enlargement of local cultures remains for a long time the sole changing element in a worldview that can hardly be altered at all by individual experience.

A typical example of such a model of the cosmos is the Ebstorf world map. It embeds, like other medieval world maps, geographic information in a Christian perspective on the world: Jerusalem is the oversized center of a disk surrounded by water, which Jesus Christ holds in his open arms. The map shows the known places and countries, which are depicted with progressive inaccuracy and more fantasy the nearer they get to the edges.

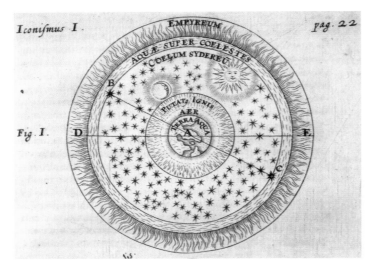

Aristotelian-Ptolemaic world system, in: Athanasius Kircher, *Iter extaticum coeleste,* 1660

The mathematical description of the Earth, explorers' travels and voyages around the world, and natural scientists' expeditions gradually substantiate and stabilize the image of the Earth. Looking at the sky becomes increasingly more important for the change of ideas about the cosmos.

Similar to on Earth, the dimensions in the sky are underestimated the farther they are from the immediate sphere of human experience. Even geocentric armillary spheres of early modern times still have a model Earth at their center that is far too large compared with the spheres of Sun, Moon, and planets that encircle it and the final sphere of the fixed stars which follows virtually without any space in between.

Such models make evident that, at the time of the great controversies about the Copernican system, there is still no conception of the immense distances that separate us even from the nearest fixed star.

The advance toward the infinite space of the cosmos, which commences in early modern times, is driven by the continual invention of new instruments – from Galileo's telescope to contemporary radiotelescopes and gravitational wave detectors. The planetary system suddenly appears rather small compared to the fixed stars of the Milky Way, which lend the starry sky its seemingly eternal appearance. The discovery of more and more nebulas supplement a view that at first is determined primarily by stars. However, even at the beginning of the twentieth century, nobody knows that millions of these nebulas are Milky Ways, "galaxies" that are infinitely more distant than our stars.

Advancing toward the depths of space with the aid of new observation instruments does not mark the end of developing classic models of the cosmos, for initially the traditional conceptions of space and time remain unaffected. When Einstein revolutionizes these concepts with his theory of general relativity, the empirical consequences are hardly different than for classical physics. Why should one throw Newton's theory overboard because of such subtle differences?

Einstein's theory celebrates its real triumph only after the discovery of the deep space beyond our Milky Way. Today his theory is the only useful basis for the models of the expanding universe of modern cosmology, with its galaxies, quasars, black holes, and dark matter.

Mythical Cosmologies

The task of mythical cosmologies is to bring together knowledge about humankind and nature in an overall picture. Thus, in creation myths, religious, social, and cosmological notions of order are combined. This

is also due to the ancient role of astronomical observations in planning the agricultural calendar. In ancient empires such myths support the cohesion of society. The Babylonian creation myth describes the development of spatial and temporal order from chaos. After his victory over chaos Marduk, the deity of the city of Babylon, divides the year into twelve months and names the constellations of the stars. To the Egyptians, the world is a disk surrounded by water and spanned by the vault of the heavens, the star-covered body of the goddess Nut. According to the Chinese doctrine of elements, human affairs obey the same laws as natural processes.

Top left:
Egyptian
cosmology,
around
1000 BCE

The Hierarchical Structure of the Modern Cosmos

Matter in space is not scattered without order, it exhibits a variety of structures, depending on the size of the part of space one is observing. The "glue" that keeps the

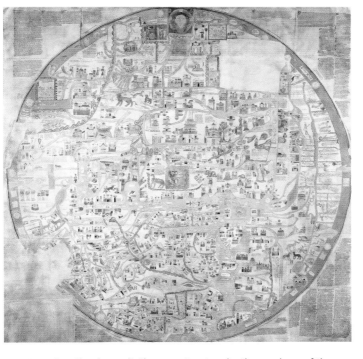

structures together is gravitation, the mutual attraction of all matter. If one considers a space through which light takes only a few seconds to travel, one sees the Earth and the Moon revolving around their common center of gravity. If one considers a space through which light takes a few hours to travel, one sees the solar system: nine planets, including Earth, circle around the system's center of gravity, which lies within the Sun. If one goes on to consider a space, through which light takes a hundred thousand years to travel, one sees a hundred billion stars – the Sun is one of them – that form a

disk with spiral-formed arms, the Milky Way, the center of which they orbit.

If one considers a space through which light needs many millions of years to travel, one sees an entire cluster of galaxies that mutually attract each other, and one of them is the Milky Way.

If one considers a space through which light takes more than a hundred million years to travel, one sees that the clusters of galaxies in turn form clusters, so-called "superclusters." Often, the force of gravity is not strong enough to keep a supercluster together against the expansion of the universe.

A space, which light needs many billions of years to travel through, looks like a gigantic sponge: fibers and walls formed by superclusters enclose vast, empty spaces.

The Hubble Constant

The Hubble constant is a fundamental term of modern cosmology. It describes the relation between the distance of a galaxy and the speed at which it is moving away from us. This velocity is calculated from a change in the color of the light emitted from the galaxies towards the long-wave, red end of the spectrum. Through the discovery of this approximately valid relation by the American astronomer Edwin Hubble in 1929 came the insight that our universe is expanding, and,

as later measurements and their interpretations have shown, the universe developed from a "big bang" around fourteen billion years ago.

Center:
Ebstorf world
map, 1293

Right:
Spiral nebula
and galaxies,
photographed by
the Hubble Space
Telescope, 1996

Mechanical and Electronic Models

1 (Mechanical Model of the Ptolemaic
System
Michael Sendtner Company, 1912

The human view of the heavens is subject to a natural limitation: they are seen from the perspective of an observer who is restricted to a particular point on the Earth, and who, without noticing it, accompanies every movement of the Earth. As a result, all of the movements of the celestial bodies are superposed with virtual movements that are in fact caused by the movement of the Earth itself. This results in complex movements of the Sun, the Moon, and the planets against the background of a sphere with the other stars, seemingly rotating around a celestial axis.

It is not easy to free oneself from this observer's perspective and to relate these apparent movements to an observer who is detached from the movements of the Earth. This difficulty accounts for the particular attraction of models that, based on certain assumptions, produce mechanical representations of the real movements of the stars. The observer of the model is then outside the moving celestial bodies, as if in a position independent of their movements.

Models of this kind gain in popularity in the wake of the discussions surrounding Copernicanism in the seventeenth century. In the following centuries precision mechanics workshops produce innumerable telluriums, planetariums, and cosmographic clocks either for representative purposes or as teaching aids in astronomy. In keeping with this tradition Oskar von Miller, the first director of the *Deutsches Museum* in Munich, commissions two highly representative models for the newly founded museum in 1903. The idea is to create a complex system of gear wheels that reflects the movements of the planets as accurately as possible, with one model based on the Ptolemaic system and the other on the Copernican system.

The task is given to the Michael Sendtner precision mechanics company in Munich. Each planetarium contains a special mechanism for moving celestial bodies based on the assumptions of each of the systems. The mechanisms are set in motion by means of a crank.

To make the two planetariums more appealing, the mechanism is encased in a transparent glass sphere composed of eight segments. The fixed star constellations are etched onto the surface of the glass spheres. A special firm in Dresden is responsible for producing these special spheres. The mechanisms differ in each planetarium depending on the system they are designed to represent. In the case of the Ptolemaic system the crank sets the celestial bodies surrounding the Earth in motion and simultaneously rotates the glass sphere with the fixed star constellations by means of a sprocket.

The construction of the two planetariums takes almost six years, and they are finally installed in the museum in 1911 and 1912. They remain in operation until World War II. In December 1944 both of the glass spheres are destroyed during a bombing raid. In 1947 they are reproduced again in a simplified version with the constellations painted instead of etched on the glass. The two models then remain in the museum's permanent exhibition until 1991 when the astronomy section is redesigned.

2 (Celestia, an Electronic Model of the Cosmos
Chris Laurel, 2001–2005

2

Mechanical models of the cosmos are essentially models of the solar system surrounded by an immobile sphere with the fixed star constellations. But with the increase in knowledge about objects outside the solar system these types of mechanical models reach their limits as an appropriate representation medium. During the 1920s projection planetariums developed by Zeiss replace the old mechanical models, but even this technical improvement fails to provide a fundamental solution to the problem. Traditional techniques in model construction are focused on objects and movements in the sky visible to a human being on Earth. The far greater complexity of cosmic structures in contrast to those of the planetary system, and the vast differences in the orders of magnitude of astronomical objects in modern cosmology, make these representation techniques obsolete.

The situation changes dramatically with the advent of computer technology. As the technology matures, it enables people to program new electronic planetariums that vastly surpass even the best mechanical and optical models. Moreover, computer technology even offers new possibilities which otherwise could never have been realized. For instance, theoretical models of developments in the cosmos as described in theories about the birth of stars and their development, or theories on the dynamics of galaxies, are visualized by complex representations of developmental processes.

One of the most original programs that uses the new possibilities of computer technology is *Celestia* which allows for virtual space travel that even extend into extragalactic space. The program, free to all on the Internet, is an extendable "open source" development. The first version of the program is published in 2001 and results in the formation of a group numbering over a thousand people who participate in the development and expansion of the program, including detailed visual depictions of astronomical objects.

http://www.shatters.net/celestia/

Models of the Earth

Even today cultures exist whose members never leave a narrowly limited territory and rarely posses any geographical information extending beyond the boundaries of this territory. Their worldview corresponds to this space of experience. An Earth structured by mountains and waters is covered by the dome of a faraway and unapproachable sky. Since the Earth is far larger than actual geographical experience, the transition between heaven and Earth lies out of reach beyond the horizon. Myths, often creation stories or notions about the hereafter, fill the information gap.

The idea that the Earth might be a sphere would not be meaningful to a culture that possesses only a limited radius of activity for this idea would contradict any real experiences. However far one travels, the newly entered territory only expands the area that is familiar. The assumption that the Earth is geometrically flat, to use a modern expression, cannot be shaken solely by expanding the geographical scope of experience. The insight that the Earth is a sphere is based on other kinds of experience.

Wanderers at the end of the world, 19th-century engraving

In antiquity, seafaring and astronomy provide the first indications of the spherical shape of the Earth. On leaving the coastline, a sharp eye realizes how land appears to sink gradually into the sea and can deduce that the surface of the water is curved. Observing the starry sky at various geographical locations confirms indirectly the curvature because the elevation of the stars above the horizon changes with the geographical latitude. On the basis of such observations, Greek philosophers develop a model of the Earth as a sphere divided mathematically by lines of latitude and longitude, and surrounded by concentric spheres of the celestial bodies.

Initially, geographic knowledge about the Earth is very incomplete and its size cannot be precisely determined. For centuries the correct positioning of the known landmasses and seas poses a major problem.

The spherical model of the Earth raises not only geometrical questions but also physical ones. Is the sphere moving or is it standing still? Which forces are behind the way land and water is distributed over the sphere's surface?

For a long time, the physics of Aristotle influenced the answers to such questions. The four elements, earth, water, air, and fire, determine the natural place of material things. Correspondingly, the structure of the world is seen as a sphere with the Earth at the center surrounded by the waters of the oceans and the enveloping atmosphere as well as fire that strives in an upward direction. The difficulty with this model – that not all continents are submerged beneath the waters of the oceans – is explained mechanically in the Middle Ages. If the Earth is not perfectly round, then its center of gravity is not precisely in its center and parts of the Earth must protrude out of the oceans as firm ground. In the age of voyages of discovery, this explanation becomes problematic because then the side of the globe opposite the continents would consist of oceans. Physics and astronomy of the modern era change the model of the Earth and pose new physical questions. What effect on the Earth's form do the forces have that arise from its rotation? Which forces keep the moving Earth and the planets in their paths? What causes the Earth's magnetic field that makes a compass needle point north? Einstein and even present-day scientists are still engaged in clarifying these questions.

○ HISTORICAL CONTEXT

Measuring the Earth

Since ancient times the Earth is
mapped using lines of latitude and
longitude, which makes it easier to
explore its surface and size. Well
into the eighteenth century, longitude is calculated most exactly
with the aid of eclipses of the Moon
observed from two locations that
lie apart in an east-west direction.
From the time difference between
the beginning of the eclipse at the

○ PERSONALIA

Christopher Columbus

As of 1477 Christopher Columbus
(1451–1506), probably the son of
a family of woolweavers and mer-
chants, works as a cartographer
and draughtsman together with his
brother Bartolomeo in Lisbon.
Columbus' four voyages of discovery between 1492 and 1504 aim to
open up a western sea passage to
India. While seeking the support
necessary for his first expedition,

◉ FOUNDATIONS

The Magnetic Field of the Earth

The Earth behaves like a gigantic
magnet. For this reason it is possible to find one's direction with the
aid of a compass. Since the early
modern era, scientists carry out
experiments to understand the
secret of magnetism and its role
as a characteristic of our planet.
Models based on technical experience and laboratory experiments
lead to ever new insights. Today,

one assumes that the Earth operates like a sort of dynamo. The
largest part of the Earth's magnetic
field is generated by the outer
liquid core, which is composed
mainly of iron. The Earth's magnetic field protects humankind from
dangerous radiation from space.

Left: Galileo and
Viviani discussing
models of the
Earth, 19th-century
painting

Right: Allegory of
the crisis in
geography in the
modern era, in:
Heinrich Scherer,
*Critica quadri-
partita*, 1710

two observations points, the distance between their longitudes is
calculated.
From the fifteenth century, the
pre-calculated daily positions of
celestial bodies are written down
in so-called ephemerides which accompany sailors all over the world.
An eclipse of the Moon, the occurrence of which has been calculated
for a specific place, is observed
elsewhere. From the perceived time
difference, one can determine one's
present position. In this way, an
ever more precise picture of the
Earth and its surface develops.

Columbus experiences many
rebuffs. One reason is his exaggerated estimate of the size of the
Eurasian continent and, consequently, his underestimation of the
dimensions of the ocean: in the
view of his critics, his calculation
of a journey to India is far too short.
They prove to be right. In 1492,
under the Spanish flag, Columbus
reaches the North American
continent. But until his dying day,
Columbus is convinced he has
discovered a route to the "East
Indies."

Maps, Globes, Magnets

1 (Columbus Predicts a Lunar Eclipse
on Jamaica
*Washington Irving, Life and Voyages of
Christopher Columbus, New York, 1892*
Aristotle already argues that during a lunar
eclipse the round Earth casts a circular
shadow on the Moon. However, the naïve
idea that some strange spirit of darkness is
threatening to devour the Moon is still
widespread in many cultures. Columbus
uses the successful prediction of a lunar
eclipse to intimidate hostile natives, threatening punishment by his God and forcing
them to surrender.

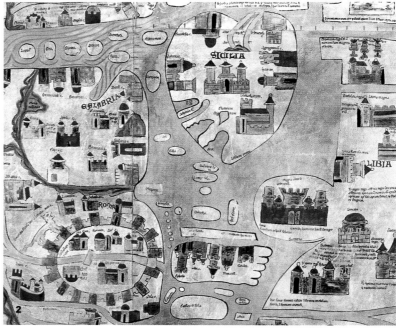

2 (The Ebstorf World Map
*13th century, destroyed during World
War II, reconstruction*
Medieval European world maps often sub-
ordinate geography to a Christian worldview. Over a thousand picture notes make
the Ebstorf world map into a chronic of the
medieval world. It depicts the world as a
place of God's salvation work. On the other
hand the map lacks the precision in geographical representations that has already
been possible for a long time. This precision is lacking even in the representation
of well-known parts of the world such as
the area of the Roman Empire.

3 (Globe of the Earth after Martin Behaim
1492
Reproduction
Without knowledge of the existence of the
American continent Martin Behaim tries to
distribute the known continents and
oceans on a sphere. His globe documents
the spirit of enterprise in the early modern
age with its great voyages of discovery. Behaim wants to show that it is also possible
to reach the east by traveling to the west.
On his globe the geographical distance between Europe and eastern Asia to the west
is far too small.

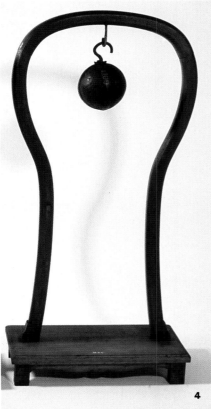

5 (Barlow's Sphere
C. Dell'Acqua, Milan, around 1840
The comparison of the Earth with a magnet is criticized by Peter Barlow because, after the discovery of induction, it becomes clear that an electric current also produces a magnetic field. Barlow's own model of the Earth is a spherical coil with an electrical current flowing through it that exhibits similar characteristics to Gilbert's "Terrella." Consequently, in Barlow's opinion "magnetism [...] has no real existence."

6 (Einstein to Schrödinger
Princeton, 10 September 1942
Einstein asks Schrödinger to examine empirical data in order to check whether the Earth's magnetic field is symmetrical. He assumes that it is induced by electrical cur-

rents. In the Earth's crust such currents would produce an asymmetrical field due to the high conductivity of the oceans, while the field would be symmetrical at the Earth's core.
"Let's assume that the [magnetic] field is produced by electrical currents, [...] in this case the currents would presumably be stronger at places where the electrical resistance is weaker. It occurs to me that the southern hemisphere is mostly covered by seas which conduct far better than the firm crust of the Earth. This would result in a certain asymmetry in the system of currents, and in particular it would mean that the vertical components of the field in the southern hemisphere would be greater than in the northern."

4

4 (Gilbert's "Terrella"
End of the 18th century
Around 1600 William Gilbert founds modern geomagnetism by comparing a spherical magnet with the Earth. His "little Earth" (Terrella) is a round magnet "just like the Earth" which can be used to demonstrate the behavior of a magnetic compass needle. Geographic designations, such as equator, parallel circle, or pole acquire a new, geomagnetic definition.

5

Advancing towards Infinity

What we know about the heavens is dependent on the means of observation. Looking at the sky with the naked eye familiarizes us with the constellations of fixed stars, the daily rotation of the firmament, the motions of the planets, whose positions change continually in relation to the fixed stars, and rare but unsettling phenomena such as eclipses and comets.

Yet it is almost impossible to determine even the movements of the planets without the aid of instruments to establish their relative positions more precisely and without long-term records of the results. Thus, the primary purpose of early observation instruments is to determine the precise positions of astronomical bodies on the celestial sphere.

The knowledge gained in this way raises the question of what determines this complex system of daily, monthly, annual, and irregular motions. Does the Earth rotate once a day on its axis or do the heavens rotate? Does the Earth orbit the Moon once a month, or does the Moon go round the Earth? Does the Earth orbit the Sun once a year, or does the Sun go round the Earth? Why do some planets have orbits with strange loops? From the beginning of astronomical records, therefore, observations are interwoven with mythical, mechanical, and physical interpretations.

Observing the heavens with a diopter, 13th-century illustration

This linking of observation and interpretation triggers a dynamic that changes conceptions of the world, because each time technical progress is made in observation instruments, this has the power to call into question existing interpretations. During the course of history, three technical innovations above all others bring about a dramatic extension of knowledge and trigger corresponding changes in world conceptions: the invention of the telescope, the application of spectral analysis to the light emitted by astronomical objects, and – after the discovery that astronomical objects emit not only visible light but also other types of radiation – detectors for these other types of radiation such as radio waves, x-rays, and particle radiation.

With each new observation instrument we advance deeper into space. After the telescope is invented, it becomes apparent that the stars we can see with the naked eye represent only a tiny fraction of the astronomical objects in the heavens, and their number increases with each improvement of the telescopes pointing skywards. Additionally, after the invention of the telescope, ever new objects are discovered: new planets, moons of the planets, asteroids, double stars, variable stars of various kinds, and nebulae, more and more nebulae, until they vastly outnumber the stars.

During Einstein's lifetime it becomes clear that most nebulae are much further away from us than the most distant star. They are islands of stars in empty space, galaxies like the Milky Way to which our planetary system belongs.

The application of spectral analysis not only provides for the first time information about the composition of the stars, it also leads to the discovery of a displacement of the wavelength of spectral lines from galaxies toward the red end, which increases with distance. Radio telescopes and other detectors of invisible types of radiation provide information about hitherto unknown astronomical objects like quasars and gamma-ray bursts which continue to puzzle astronomers and astrophysicists today.

Giordano Bruno

After studying theology, which ends in a break with his religious order, Giordano Bruno (1548–1600) leads the life of a wanderer, residing in many places in Europe where he repeatedly comes into conflict with ecclesiastical and secular authorities. In 1591 he is arrested and, after nine years in prison, is condemned as a heretic by the Inquisi-

Herschel's forty-foot telescope, 1795

tion. In 1600 Bruno, who refuses to recant, is burned alive at the stake. The exact reasons why Bruno was condemned remain unknown because the trial documents are missing. His worldview definitely plays a major role: Bruno not only adheres to the Copernican doctrine, he also believes that every celestial body moves by itself in an infinite universe, with no center and no limits, and that this universe is filled with innumerable inhabited worlds. Bruno, who thus anticipates hypotheses proposed by modern science, has still not been rehabilitated by the Catholic Church.

Edwin P. Hubble

At first, Hubble (1889–1953) is undecided whether to become a boxer or to study astronomy. He ultimately chooses the latter. At the Yerke Observatory he works on his dissertation on photographic studies of faint nebulae, which he completes in 1917. Afterwards, he is fortunate in being able to conduct research using the world's largest tele-

scopes: first at Mount Wilson, later at the Palomar Observatory. His work opens up an entirely new view of our universe. Using Cepheids stars, he calculates the distance to the Andromeda nebula. Hubble demonstrates that the Milky Way is merely one among many star systems, and that the galaxies are moving away from us, which leads to the conclusion that the universe is expanding. Hubble's observations confirm the mathematician Friedmann's cosmological interpretation of Einstein's general theory of relativity.

Distance Determination within the Cosmos

How can one determine the distance to an object that one can see but cannot reach? Determining astronomical distances requires different methods for increasing distances.

If one measures the angles at which one sees an object from the two end points of a baseline of known

Cassini-Huygens spacecraft, Kennedy Space Center, 1997

length, one can determine the distance to the object. By taking the diameter of the Earth's orbit around the Sun as the baseline, one can determine the distances of many thousands of neighboring stars. For objects that are further away, different methods are required.

If the luminosity of an object is known, then one can determine its distance using the object's visible brightness. One of the various objects to which this method is applied is Cepheids, stars whose brightness undergoes a periodic fluctuation, which is correlated with

the luminosity. From the length of the period one can determine the luminosity, and from this and the observed brightness one can determine the distance. Cepheids can still be observed in neighboring galaxies, and so they permit us to calculate the distance to these galaxies.

On their way to us, light waves from far-off objects are elongated due to the expansion of the universe. The light from these objects is thus shifted toward the long-wave red end of the spectrum. The greater the distance to the object, the more the light is shifted towards the red end of the spectrum. Thus, the redshift is a measure of distance and with its help we can determine the distance to objects whose light takes many billions of years to reach us.

Sights, Telescopes, Spectrographs

1 (Astrolabe by Regiomontanus
Nuremberg, 15th century
An astrolabe serves to determine the part of the sky visible on a given date, at a given time, and at a specific place. The instru-

ment is based on the projection of the sky onto a circular disc mounted with an additional pivoted dial specifying the visible horizon. It is typically combined with a device for measuring angles used to determine the height of a celestial body above the horizon.

2 (Quadrant
From the Andechs Benedictine monastery collection, 16th century
The quadrant serves to determine the height of celestial bodies above the

horizon. Before the invention of the telescope the quadrant was the most important measuring instrument for astronomical observations. It is also often integrated as an essential part into more complicated instruments.

3 (Sights with Celestial Globe
*Johann Gabriel Doppelmayr, 1730
(globe only)*
The sights for determining the position of the stars are combined with a celestial globe manufactured by Doppelmayr. The globe bears the coordinates and star constellations used for orientation since antiquity. The celestial globe also includes newer star constellations of the more recently investigated southern hemisphere.

3

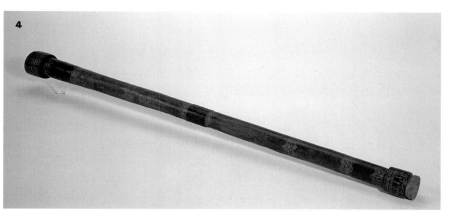

4 (Galileo's Telescope
Galileo Galilei, around 1620, reproduction
The telescope is invented around 1609.
Galileo is among the first to produce and
improve the construction of telescopes
in their workshops.
Galileo points the telescope to the sky.
He discovers mountains on the Moon, and
four moons orbiting Jupiter. Kepler enthusi-
astically interprets Galileo's publication
Sidereus Nuncius as a manifesto of
Copernicanism. The telescope becomes
the symbol of the heliocentric astronomy
condemned by the Catholic Church.

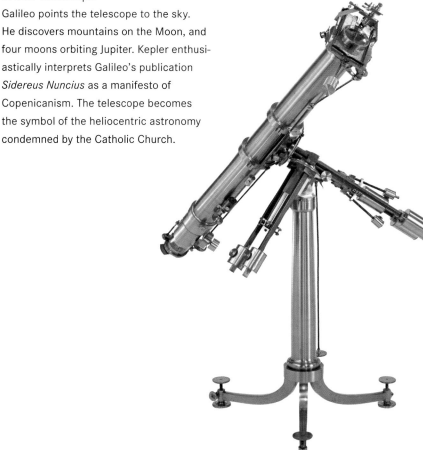

5 (Heliometer
Utzschneider and Fraunhofer, 1815
The heliometer is a special telescope containing a divided objective with two sliding
halves. This instrument is designed to
measure the angle between two adjacent
celestial bodies with great precision.
As this angle changes depending on the
observation point on Earth or the Earth's
movement along its orbit, the heliometer
makes it possible to measure planetary
distances and even the distance to close
fixed stars.

6 (Spectrograph for Decomposing the
Light from Stars
*Erich Neubauer Company, Berlin-Karow,
around 1940*
The spectrograph operates with a prism
that decomposes light into its constituent
colors. The instrument opens a new win-
dow to the sky.
As the spectrum of a luminous body, or a
body penetrated by light, depends on the
body's composition, the spectrograph is
able to provide information about the
chemical composition of the stars.

Models of the Heavens

The most conspicuous events in the sky are the daily course of the Sun, the rotation of the stars in the night sky, the movements and phases of the Moon, and the changing positions of the planets. The Moon and the planets revolve in the night sky but change their position nightly. The Moon does not seem quite able to keep pace with the rotation of the night sky and lags behind a little. Some of the planets move even more erratically: they move at different speeds and sometimes even backwards and forwards.

Attempts to comprehend these movements with the help of instruments and to document them in writing go back as far as Ancient Egypt and Babylon. The observations are used to make calendars and predictions about astronomical events. Additionally, in classical antiquity models of the heavens are proposed which are designed to explain the complex movements of the celestial bodies. Historically, the most influential model is the Ptolemaic system. The Sun, the Moon, and the planets revolve around the Earth which is stationary at the center. The explanation for the curious paths of the planets is that a planet not only moves in its orbit around the Earth but also describes a smaller circle around a point in its orbit. The dominance of the Ptolemaic system is only broken in the modern era by the Copernican system in

Geocentric world system, in: Hartmann Schedel, *Register des Buchs der Croniken* [The Nuremberg Chronicle], 1493

which the planets do not revolve around the Earth, but around the Sun. The paths of the planets are not perfectly circular but elliptical, as Johannes Kepler proves with the aid of more exact observations.

After a long struggle with the Roman Catholic Church which sees the astronomical worldview as unchangeable through the revelation of the Bible, the Copernican system is finally generally accepted. The points of contention in this struggle are many and various. They concern not only the question of whether the Earth or the Sun moves but involve the entire conception of the heavens. However, in the long run, neither the execution of Giordano Bruno nor the condemnation of Galileo Galilei by the Inquisition can halt the success of the Copernican system.

To avoid deciding whether the Sun or the Earth moves, René Descartes defines all motion as relative. For him, whether the Sun revolves around the Earth or the Earth around the Sun is a question of the framework of reference. Instead, Descartes thinks about a larger connection between structure of matter and the entire cosmos in which to see our planetary system. Descartes' cosmos consists of many worlds through which comets wander and sometimes appear in our planetary system.

With Newton's theory of gravity and his inference of elliptical orbits of the planets in absolute space, Descartes' relativist and cosmological questions recede for the time being into the background. At the latest with Einstein's theory of general relativity, however, these questions again come to the fore. Einstein's theory, in which all frames of reference are equally valid, becomes the basis for cosmic models, which cover far greater spaces and other objects than those of our planetary system.

○ HISTORICAL CONTEXT

World Views and the Church

Copernicus' new astronomy of 1543, which places the Sun instead of the Earth at the center of the solar system, initially concerns only scholars who use it as a simplified method for determining the constellations of planets. However, rapidly advancing knowledge about nature in the early modern era leads inevitably to conflict with the all-embracing worldview of the

Nicolaus Copernicus, 16th-century painting

Catholic Church, which arises from the inclusion of Aristotelian theory of nature into Christian dogma. At the beginning of the sixteenth century, through scholars like Bruno and Galileo, the Copernican doctrine becomes the controversial foundation of a new world system. With the aid of telescopic observations of the Moon and the planets, Galileo provides an impressive confirmation of this system. In 1616, the Catholic Church bans the teachings of Copernicus, and in 1633 Galilei is condemned for embracing Copernicanism.

○ PERSONALIA

René Descartes

Today, René Descartes (1596–1650) is best known for his contributions in the field of philosophy. Descartes believes that through the method of doubt it is possible to arrive at certain knowledge, such as the certainty of one's own existence, expressed in the phrase "I think therefore I am."

In the course of the revival of classical atomism, Descartes under-

takes the attempt to construct a world view that traces the motions of celestial bodies back to mechanical motions and impacts of minute particles. He conceives of the universe as filled with particles that form vortices upon which particularly stable aggregations of particles, the planets, are carried around the Sun. All other physical phenomena, such as light, he explains by the motion and impacts of these particles. In this way he constructs a mechanical world view, which particularly through its clearness exerts an enormous influence on the development of physics.

◉ FOUNDATIONS

Absolute and Relative Motion

To make judgments about the movements of any body, one needs a point of reference. Does an absolutely stationary point of reference exist? In Aristotelian understanding of the world such a point of reference exists at the center of the universe in which the Earth

rests. According to Copernicus, however, the Earth revolves around the Sun. Why is it that we do not notice this motion? Galileo's answer is to distinguish between absolute and relative motion: in relation to the Sun, we move rapidly through space, but in relation to the Earth we are stationary. The laws of terrestrial physics remain

almost unaffected by our joint movement with the Earth around the Sun. This insight leads to the relative principle of classical physics. Is it possible to extend this principle to accelerated motion? This question is the starting point of Einstein's general theory of relativity.

Copernicus' heliocentric world system, in: Andreas Cellarius, *Harmonia macrocosmica*, 1708

Representations of Celestial Movements

1 (Armillary Sphere

17th century

The armillary sphere is a mechanical model
of the geocentric world system. A sphere
at the center represents the Earth. It is surrounded by pivoted systems of rings representing celestial coordinate systems. The

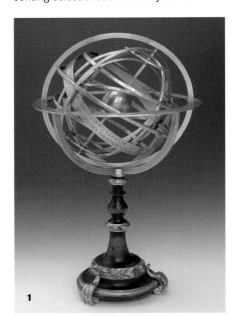

1

external rings connected with the stand
represent the horizontal coordinate system. Inside it a coordinate system represents the firmament that rotates around
the Earth's axis every twenty-four hours.
This system comprises the celestial equator, the two tropics, and both polar circles.
The ecliptic is attached to it at an oblique
angle representing the apparent annual
orbit of the Sun in the sky. It bears the
names of the signs of the zodiac. Two additional pivoted coordinate systems probably
serve to locate positions of the Earth, the
Sun, or the planets in relationship to the
sky and consequently to adapt the model
to times and places of observations.

2

2 (Galileo's "Dialogo"

Second Italian edition, Florence, 1710

Galileo's *Dialogue Concerning the Two
Chief World Systems* compares the Ptolemaic system in which the Earth rests at the
center of the cosmos, with the Copernican
system in which the Earth moves around
the Sun. The publication leads to his persecution by the Inquisition. Galileo is forced
to renounce his theory and is kept under
house arrest until his death.

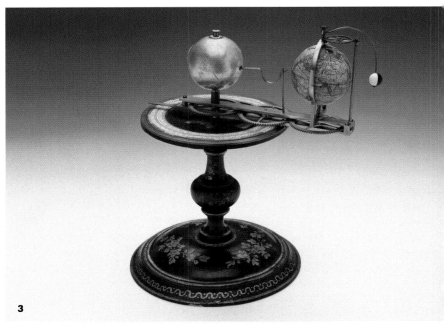

3

3 (Tellurium

*Charles François Delamarche, around
1800, reproduction*

A tellurium is an astronomic model that
represents solely the movements of the
Earth, Moon, and Sun. Such models are a
popular means of visualizing the astronomical explanations of the seasons and the
phases of the Moon.

4 (Herschel's Planetarium

William and Samuel Jones, London,
around 1800

The planetarium made by the Jones
brothers is a model of the planetary

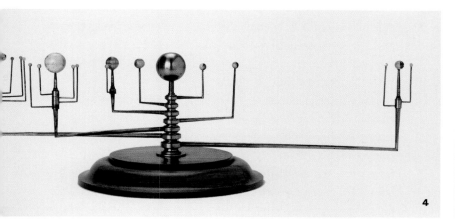

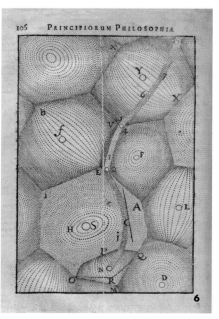

5 (Pamphlet on the Comet Appearing in
1664

Christian Theophilus, Cometen Propheten,
Nuremberg, 1665, frontispiece

The reappearance of comets remains unde-
tected for a long time and their paths can
not be calculated. Comets seem to depart
from the normal course of nature. Their appearance evokes dread and horror. Interpreted as a warning it promotes beliefs in
prophecies. The appearance of the comet
in 1664 still results in the publication of numerous pamphlets in which astronomical
observations are linked with such warnings
and prophecies. It is not until later times
that systematic observations uncover the
true nature of comets so that their appearance becomes a normal natural phenomenon.

6 (Path of a Comet through the Cosmos
into the Solar System

René Descartes, Principia philosophiae,
Amsterdam, 1644

Following the Church's condemnation of
Galileo, René Descartes hesitantly publishes his main work which focuses on a mechanical theory of the whole cosmos. All
physical phenomena are supposedly explained by the movements of matter which
fills the whole of space. Once the matter
is set in motion its movements are transferred by pressure and impact whilst the
total mass of the movement is eternally
conserved.

system which contains not only the planets
Mercury, Venus, the Earth, Jupiter, and
Saturn with their moons orbiting the Sun,
but also the planet Uranus with two moons.
It belonged to Wilhelm Friedrich Herschel
who discovered this planet in 1781 and two
of its five moons in 1787.

Curved Spaces

Our conceptions of space are influenced strongly by one particular work and the proofs it contains: Euclid's *Elements*. His proof that the sum of the angles in any triangle is 180° renders it inconceivable that measuring the angles of a real triangle could disprove such a theorem.

Doubts about the universal validity of Euclidean geometry first arise in the nineteenth century and lead to the spectacular construction of "non-Euclidean" geometries. Analytical methods whereby plane geometry is generalized to the geometry of curved planes provide the basis for this development. In a similar way, three-dimensional space is generalized to space of multiple dimensions in which – analogous to curved planes – curved three-dimensional spaces entailing laws other than those of Euclidean geometry can be defined.

Einstein curving space

With the mathematical possibility of non-Euclidean geometry the question arises as to whether the space in which we live is really Euclidean. As the sum of all angles of a triangle is not 180° in curved spaces, in principle the question can be settled by measuring triangles.

However, as we know, all sums of the angles of triangles that deviate from 180° result from inexact drawings with compass and ruler. If empirical space is in fact curved, then its curvature must be so minute that even with very precise measuring instruments no deviation from the Euclidean sum of the angles can be detected.

In order to track down such miniscule curvature, larger triangles have to be measured. Friedrich Gauss is one of the creators of the geometry of curved spaces. It is said of Gauss that he long questioned the truth of the Euclidean nature of empirical space and, during his work as a surveyor, secretly checked the sum of the angles of surveying triangles.

The astronomer Karl Schwarzschild considers using much larger astronomical triangles. In 1900 he publishes a work in which he analyzes the question as to how strongly space can be curved without astronomers detecting any effect. Schwarzschild discusses the consequences of a possible curvature with reference to a triangle consisting of a fixed star and two points opposite on the Earth's orbit around the Sun from which the star is visible from a slightly different direction. If the star is not too far from the Earth, then this difference can be measured and the distance of the star estimated – under the provision that space is Euclidean. However, if space is curved and the sum of the triangle's angles deviates from 180°, then the estimation of the distance will yield a false result, which, if the curvature is sufficient, must inevitably lead to contradictions with astronomical observations.

Einstein's theory of general relativity demonstrates to us today that neither the triangles in land surveying nor Schwarzschild's astronomical triangles are large enough to ascertain the actual curvature of space. Nevertheless, Schwarzschild's ideas are on the right track and he is practically the only astronomer who recognizes immediately the importance of Einstein's general relativity for astronomy and collaborates on the elaboration and testing of this theory.

○ PERSONALIA

Carl Friedrich Gauss

Carl Friedrich Gauss (1777–1855) is raised in humble circumstances and from an early age shows an exceptional gift for mathematics. He contributes pioneering work to nearly all areas of mathematics, also to astronomy and physics. Achievements in number theory and the calculation of the orbit of the small planet Ceres make him famous throughout Europe. At the

age of thirty, Gauss is appointed director of the observatory in Göttingen, where he is active for more than forty years. There, he develops his ideas on non-Euclidean geometry, prompted by doubts about a possible proof of the parallel axiom of the Euclidean geometry, which states that parallel lines never cross. Around 1820, Gauss starts work on a large-scale geodesic survey. In connection with this project he works on a variety of mathematical problems. In particular, Gauss develops the basis of differential geometry of curved surfaces; a starting point for Einstein's insight into the curvature of space and time.

Bernhard Riemann

Bernard Riemann (1826–1866) is one of the most important mathematicians of the nineteenth century. In addition to making fundamental contributions to many areas of

mathematics, he also investigates physical problems, in particular those concerning the theory of heat, electricity, and magnetism. In 1850 he enrols at the mathematical-physical seminar of the University of Göttingen and becomes assistant to the physicist Wilhelm Weber. His habilitation lecture, *On the Hypotheses That Lie at the Foundations of Geometry*, is of seminal importance for the development of multidimensional non-Euclidean geometry, the mathematical language of the general theory of relativity. Additionally, his work on quadratic differential forms, which ensued from the investigation of questions concerning heat conduction, provides fundamental prerequisites for Einstein's theory.

Left: Carl Friedrich Gauss, 1828

Center: Bernhard Riemann, 1863

Karl Schwarzschild

The astronomer Karl Schwarzschild (1873–1916), son of a Jewish businessman in Frankfurt am Main, is famous for a multitude of pioneering scientific achievements in various fields of research.

Schwarzschild makes improvements to observation methods and plays a leading role in introducing photographical techniques to astronomy. He applies methods and findings in physics and chemistry to astronomy thus becoming the pioneer of German astrophysics. He investigates fundamental problems in astronomy and neighboring disciplines, which result in his principal contributions to physics. In particular, he discovers the first exact solutions to the equations that describe the interaction of the gravitational field and matter in Einstein's general theory of relativity.

In early 1916, Schwarzschild completes this work while on service at the Russian front. There, he contracts a skin disease which leads to his death a few months later.

Right: Karl Schwarzschild, around 1910

○ FOUNDATIONS

Non-Euclidean Geometry

The sum of the angles of a triangle which one draws on a sheet of paper is always the same – 180° – regardless of the size or form of the triangle. This does not change if the paper is rolled up into a cylinder or a cone. The geometry on a sheet of paper is known as Euclidean geometry. If one tries to curve the surface of the paper to a sphere, one soon realizes that this is not possible: the geometry of this surface is non-Euclidean. Here, triangles have a sum of angles that is greater than 180°, and the sum is also dependent on the size of the triangle. If one looks at a triangle where one side is formed by one quarter of the sphere's equator and the others by lines running up to the North Pole, one sees that the sum of the angles is 90°+90°+90°=270°. In addition to this "spherical" geometry, there is also "hyperbolic" geometry where the sum of the angles of a triangle is always less than 180°. Analogous to the surface of a sphere, one can imagine the hyperbolic surface as the surface of a saddle.

All this also applies – with modifications – to spaces with three or more dimensions. Although such spaces surpass the powers of human imagination, mathematically it is possible to show that they can be consistently constructed. According to Einstein's general theory of relativity, physical spacetime is in fact non-Euclidean. Whereas the spaces cited above exhibit a constant curvature, in Einstein's theory the curvature of space can vary from point to point.

Geometry and Experience

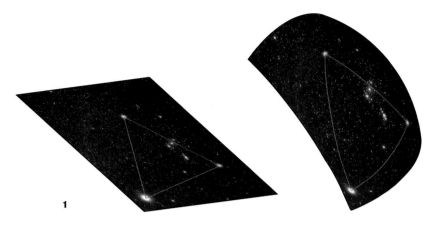

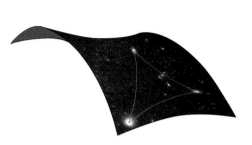

1 (Models of Curved Spaces

The development of mathematics in the nineteenth century produces definitions of geometrical objects that seem to contradict all common-sense perception and at best can only be made plausible by models. An example of this is the definition of curves that change their direction infinitely often in any arbitrarily small interval, so that no direction can be given for any single point of these curves. Einstein shows that such "mathematical monsters" can well correspond to physical realities.

He proves that in the Brownian motion of a particle floating in liquid, the particle changes direction and speed so often that it is impossible to reasonably attribute a velocity to its movement.

The case of curved spacetime in Einstein's theory of relativity is equally irritating. For instance, it is simple to imagine a curved surface, but it seems meaningless to transfer the idea of curvature to the whole of space, let alone to the four dimensional space of spacetime with one dimension of time and three dimensions of space. But in this case, too, mathematical methods make it easy to transfer calculations with the three coordinates of a three-dimensional space to calculations with four coordinates of a four-dimensional space. Curved two-dimensional surfaces in a three-dimensional space are defined by certain equations of three variable coordinates, and curved three-dimensional spaces are defined in a four-dimensional space by analogue equations of four coordinates. This analogy can be used to illustrate the properties of three-dimensional curved spaces by two-dimensional models. Projecting the starry sky onto a flat surface, an elliptically curved surface, and a hyperbolically curved surface provides an illustrative visualization of how a curvature of space affects geometry. For instance, it can be seen that the sum of the angles in a triangle is changed by the curvature. In fact the larger the triangle, the larger the change is.

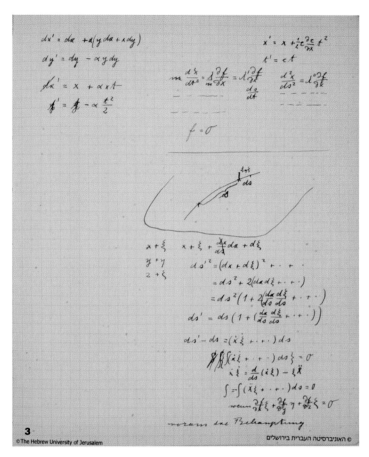

3

3 (Einstein's Initial Work on a Physics in Curved Spaces
Zurich notebook, 1912/13
Einstein's Zurich notebook is a key document in the development of the general theory of relativity. One of the pages of the booklet contains a short note that shows how Einstein imagines the possibility of a physics in curved spaces by recapitulating what he once learned about the geometry of curved space in a lecture by his teacher Carl Friedrich Geiser. The sketch of a curved surface serves him as a model of a curved space. Einstein notes the Newtonian law of force for a motion in a curved surface and proves that under this restriction an inert body, upon which no forces are exerted, moves on a geodetic of the surface, that is, on the shortest line connecting two points.

2 (Einstein on Geometry and Experience
Berlin, 1921
Up to the nineteenth century is the prevailing opinion that the theorems of mathematics, and especially those of Euclidean geometry, necessarily apply to reality. For instance, the philosopher Immanual Kant maintains that the theorems of geometry provide conclusive evidence for the existence of a priori judgments which are valid for experience, although they do not derive from it. He asks how this can be possible, and replies that space and time are forms of intuition which are necessary preconditions of our intellect to experience objects. The assumption that Euclidean geometry is universally valid in reality is finally rejected at the latest with the advent of Einstein's theory of relativity. Looking back during a public lecture Einstein again asks Kant's question and replies in the opposite way: "How is it possible that mathematics, a product of the human mind completely independent of all experiences, fit the objects of reality so perfectly? Can human reason fathom properties of real things simply through thought? In my opinion the short answer to this is: As far as the theorems of mathematics refer to reality, they are not certain; and as far as they are certain, they do not refer to reality."

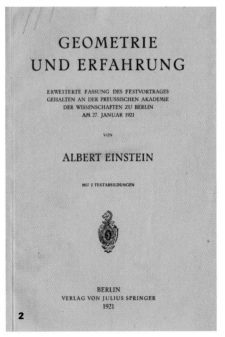

2

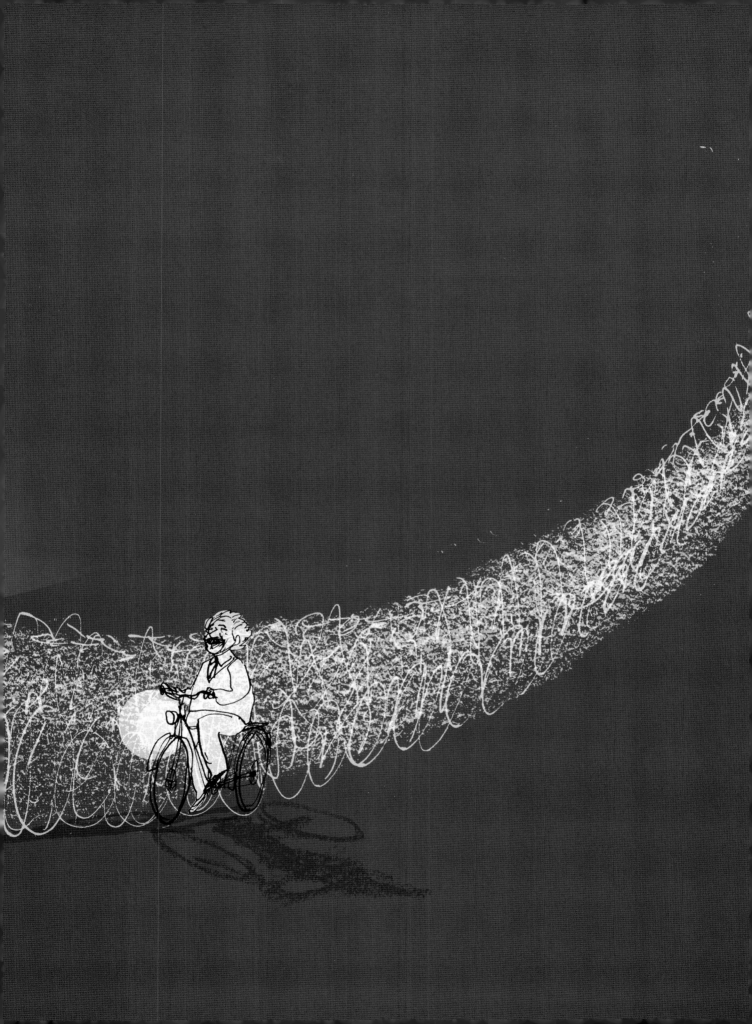

EINSTEIN –
HIS LIFE'S PATH

**"As punishment for my contempt
for authority, fate has made me
an authority myself."**

Einstein's life begins in an epoch characterized by industrial and artistic renewal, and ends within the shadows of confrontation between superpowers on the threshold of nuclear war. His pathway was at the beginning shaped by the industrial and intellectual upheaval of the ninteenth century with its tensions between bohemian and bourgeoise worlds. Towards the end of his pathway he becomes a cultural symbol of an architect of the world of the twentieth century.

His path takes him through many different kinds of worlds, beginning in the middle class world of the family electrical business, confronted by the emergence of a large-scale electrical industry; through to the world of the Swiss Bohème, where he is torn between student pretension and established science. To the world of the Prussian elite, torn between violent nationalism and pacifistic protest; through to the world of Judaism caught between anti-Semitism and new Zionistic beginnings. Finally, to the world of emigration in the tense atmosphere of fortified democracy and political mistrust.

Borderline Problems in Classical Physics

Planck: Gentlemen, physics is approaching its completion.

Boltzmann: It could also be a first rate funeral, my dear colleague Planck.

Lorentz: I think we should avoid both exaggerated optimism and exaggerated pessimism, my dear Boltzmann,

Boltzmann: Always trying to create harmony, our colleague Lorentz! What would we do without him!

Planck: Then we would have no electrodynamics, one of the pillars of our physics alongside mechanics and thermodynamics.

Lorentz: You're exaggerating, my dear Planck. Electrodynamics was the work of Maxwell. I merely tried to free it of a few problems. And it didn't take much to do that. As I understand things physics is basically simple; it consists merely of aether and atoms. The aether is a continuous medium that fills all of space. It is at rest and bears the waves of light like the sea bears the waves of water. On the other hand matter is composed of atoms which move in the waves of this sea like tiny ships. The sea of aether is ruled by the laws of electrodynamics and the little ships are ruled by the laws of mechanics.

Boltzmann: But there are also borderline problems. Wouldn't we be bound to sense at least a slight perceptible breeze from the aether as we move through it with the Earth?

Lorentz: If we assume that bodies shrink in the direction of their movement in relationship to the aether, that would explain why we do not perceive this aether wind.

Boltzmann: Then wouldn't it be simpler to just do away with it?

Lorentz: In that case which medium would bear the electromagnetic waves? After all, waves of water are inconceivable without water!

Planck: At any rate, colleague Lorentz's aether seems more trustworthy to me than your atoms, Mr. Boltzmann! Admittedly, I can imagine that heat phenomena may be attributed to the movement of atoms. But it's a different question whether this image also corresponds with reality. In that case heat would merely be a statistical problem, a dance of atoms.

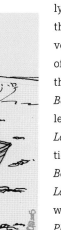

Dialogue in the sea of aether

Boltzmann: Oh really, my dear Planck, you see everything purely from the point of view of thermodynamics. You have two principles: the conservation of energy and the increase of entropy, and that's all there is to it, your physics. A bit unexciting don't you think? In my school we have more fun. You'd certainly learn something there, for instance that entropy, seen from the point of view of mechanics, is a measure of the disorder of atoms and molecules. And this disorder increases for statistical reasons, as I observe daily by the state of my desk.

Planck: But what happens when you let time run backwards? Then your disorder will decrease again. And according to the laws of mechanics such a reversal should be possible, but not according to the laws of thermodynamics. So you see that these are not attributable to mechanics. My tendency is to place my hopes rather in a combination of thermodynamics and electrodynamics. Radiation in a state of thermal equilibrium, a borderline problem in the theory of heat and electromagnetism, that could lead to a keystone in classical physics!

Boltzmann: Or the beginning of the end. Anyway, without statistics you're hardly likely to discover how energy is distributed throughout the individual radiation waves. It should in fact function in a similar way to the distribution of kinetic energy throughout the individual molecules of a gas. Unfortunately, my theory is not directly applicable to this particular case.

Planck: You reckon? Well, I'm in favor of any method, so long as it gets me to the goal.

○ PERSONALIA

Ludwig Boltzmann

"Yet I believe I can say with confidence of molecules: but they do move!" In light of the great successes of atomic and nuclear physics, this final sentence of a

Ludwig Boltzmann, around 1890

publication by Ludwig Boltzmann (1844–1906), echoing Galileo's legendary words, sounds rather anachronistic to us. But when Boltzmann writes this in 1897 atomism is not yet part of the established core of contemporary physics, indeed it is contested – after all, one cannot see atoms. Boltzmann deserves a special place among the scientists who vigorously propagate the recognition of atomism in physics at the end of the nineteenth century. He contributes significantly to the introduction of statistical methods in physics. His kinetic theory makes it possible to derive numerous properties of matter, particularly gases, from basic atomistic premises.

Hendrik Antoon Lorentz

The Dutchman Hendrik Antoon Lorentz (1853–1928) is one of the most significant theoretical physicists of the modern age. His theory of electrons – a logical extension of Maxwell's electrodynamics based on the hypothesis of atomistically structured matter in a stationary aether – resolves a number of open problems in late nineteenth-century physics. Equally, it marks the culmination of classical physics and lays the foundations for Einstein's formulation of the special theory of relativity in 1905. Despite his re-

luctance to follow Einstein in abandoning the concept of aether, Lorentz becomes a fatherly friend to him. In particular during and after World War I he uses his outstanding reputation as a scientist to mediate and promote the cultivation of international scientific relations.

Max Planck

Max Planck's (1858–1947) still-valid formula for the distribution of energy over the various colors of heat radiation make him an unwitting pioneer of the quantum revolution. Planck's formula represents the first indication of a fundamental crisis in classical physics. His hypothesis that the emission and absorption of radiated energy occurs

Max Planck, 1919

not continuously but discontinuously in tiny units or quanta proves to be a sharp break with the concepts of classical physics. Planck is also one of the first to recognize the significance of the special theory of relativity and its novel understanding of space and time, and this makes him an early supporter of the young Einstein. Planck is not just one of the most important, but also one of the most influential, German physicists in the first half of the twentieth century. He holds a number of organizational positions within scientific institutions during the Wilhelminian era and the Weimar Republic as well as under National Socialism.

Center: Hendrik Antoon Lorentz, 1902

Classical Physics in Theory and Practice

During the nineteenth century three branches of physics are firmly established: mechanics describes the motion of heavy bodies in space and time, thermodynamics explains phenomena associated with heat, and electrodynamics deals with electricity, magnetism, and electromagnetic waves, for example, light.

These three disciplines also have their secure place in industrial praxis. Mechanics describes motion processes and power transmission within increasingly complicated machines, thermodynamics provides the theory for steam engines, and the electrical engineering industry is grounded on the findings of electrodynamics.

Towards the end of the century, most physicists are convinced that physics has thus reached completion and only minor marginal problems remain to be solved.

2 (Balancer after Franz Reuleaux
Berlin, around 1880, model
Franz Reuleaux, professor for mechanical engineering at the Technical University in Berlin, wants to provide a scientific basis for mechanical engineering. His goal is the analysis and the classification of mechanical motion. He speaks of "pair elements," for instance pivots and bearings, that can be put together in "kinematic chains," out of which a machine is created.
To illustrate the concept he builds more than 800 models representing the connection of pair elements into kinetic chains.

3 (Ice Calorimeter after Antoine Laurent de Lavoisier and Pierre Simon de Laplace
E. Leybold's Nachfolger, Cologne, 1911
The calorimeter is used for measuring thermal energy. It is developed in this form by the French physicists Lavoisier and Laplace. It is made of a series of containers nested in one another. The innermost container holds the test sample and the outermost is filled with crushed ice. The melt water is caught in the drain and its volume provides a measure for the amount of heat released by the sample.

1 (Atwood's Fall Machine
Milan, around 1865
With the Atwood's fall machine, Newton's second law (force equals mass times acceleration) can be proven to a high degree of accuracy. No physicist in the nineteenth century seriously doubts the truth of this law. Thus, this machine is more likely to be used for demonstrations in physics lectures than for research.

4 (Hot-Air Engine
Goetze Company, Leipzig, around 1925
The Scottish engineer Robert Stirling constructs the first hot-air engine in 1816. The
piston is moved by the enclosed volume
of air that is being alternately heated and

5 (Galvanometer with Astatic Needle Pair
A researcher's construction, around 1830
In 1820 Hans Christian Ørsted describes
the deflection of a magnetized needle
caused by an electric current. This effect
can be used to measure such currents.
Two oppositely poled needles counteract
the disturbing influence of the Earth's magnetic field, thus the name "astatic needle
pair." This example is an instrument built
for laboratory purposes. The serial production of galvanometers first begins in
the 1870s.

cooled. Heating occurs through a burner,
the cooling with water. Despite their simplicity and efficiency hot-air engines do
not establish themselves – the closed
circulation puts far too high demands on
the material and mechanics.

6 (Flat Ring Dynamo
Schuckert & Co., Nuremberg, around 1893
In 1866 Werner von Siemens builds the
first functional dynamo, thus establishing
the foundations for the development of
electrotechnology. The distinctive feature
of this dynamo is its armature: it is fixed on
the drive shaft and is constructed in the

shape of a disk that spins between four
solenoids. Siegmund Schuckert, founder of
the electrotechnical company *Schuckert &
Co.* in Nuremberg, first builds this machine
in 1874 and presents it at the Paris Electricity Exhibition in 1881.

Irritating Experiments

Around 1900 scientists are convinced that classical physics is approaching its completion. The foundations for the branches of mechanics, electromagnetics, and thermodynamics are settled and allow for the understanding of most known phenomena. On the other hand, experimental research in laboratories as well as industrial experience is constantly creating new challenges to theoretical understanding. Furthermore, there are questions as to the connections between the various branches. Can all of physics be traced back to mechanics? Or must mechanical ideas be modified in an electromagnetically based physics? Can the thermodynamic idea of energy be the starting point of a unification of physics? All of these alternatives have their supporters and opponents, not to mention the irritating experiments that do not appear to fit into the worldview of a complete physics.

Early on, Einstein considers such an experiment examining the interaction between a magnet and a conductor moving relative to one another. The result of this relative movement is that a current is created in the conductor. It thus stands to reason that the relativity principle is also true for electrodynamic phenomena. However, classical electrodynamics based on the idea of aether calls for the strict separation of the case when the conductor is motionless with respect to the aether and the magnet moves, versus the case in which the magnet is stationary and the conductor moves. This means, if the magnet moves it generates an electrical field that produces a current in the conductor. But if it is the conductor that moves, a similar electrical field is not generated. Instead, an electromotive force is generated that likewise leads to an electrical current.

Brownian motion, the random movement of microscopically small particles in a fluid, is also irritating for classical physics. According to thermodynamics, such particles should reach a thermal equilibrium with the fluid surrounding them and come to rest. According to the kinetic theory of heat, in which heat is a result of molecular movement, they should instead behave like large molecules and move with a velocity that can be calculated from the theory. Efforts to measure this velocity, however, show results that are not concordant with the predictions. As Einstein finally recognizes, these small particles have no measurable instantaneous velocity!

The conception of classical electrodynamics that the energy of light is continuously distributed in space, has problems when it comes to explain the photoelectric effect or the thermodynamic equilibrium of radiation. In fact, experiments conducted by Philipp Lenard suggested that the energy of electrons released from an irradiated metal surface do not depend on the intensity of the light, as is to be expected from the wave theory of light, but instead depend upon its color. Precise measurements of the distribution of energy across the various colors of thermal radiation lead Planck to establish a formula that, as Einstein realizes, cannot be reconciled with the classical idea of a uniform distribution of energy.

Laboratory for radiation measurement at the *Physikalisch-Technische Reichsanstalt* (National Institute for Physics and Technical Standards) in Berlin, around 1900

○ HISTORICAL CONTEXT

Laboratory and Seminar

In the mid-nineteenth century, the basic principles of scientific instruction change. The romantic philosophy of nature is increasingly replaced by a conception of sciences more interested in experiments and technical implementation. Students are becoming familiarized with the current research of their teachers. Little

of ideas and research results. In the last third of the nineteenth century a strong differentiation between the disciplines sets in, each of which develops its own technical terminology and qualifications profile for its students. This is where the academic natural and technical sciences are encountered with the practical demands of industrial applications.

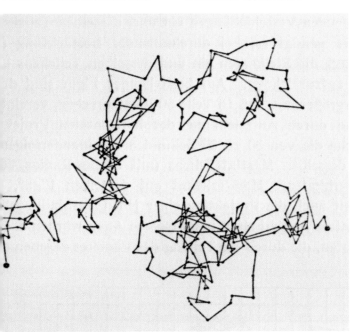

by little, professors manage to purchase instruments at the universities' expense. With their laboratories, Justus Liebig and Gustav Magnus create facilities that serve as models that others emulate. Parallel to the institutionalization of scientific communication in societies, colloquia and seminars take on increasing importance for the free exchange

Representation of Brownian motion, 1914

○ PERSONALIA

Gustav Kirchhoff

Together with Robert Bunsen, Gustav Robert Kirchhoff (1824–1887) is seen as the founder of spectral analysis. He significantly influences the development of theoretical physics in Germany. In 1860, in the context of his spectroscopic experimentation, he formulates the so-called Kirchhoff's law of thermal radiation.

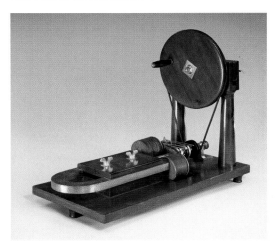

According to this law, the relationship between emission and absorption capacities is a function equal for all bodies and depending solely upon temperature and wavelength. In this context, he also develops the idea of a "black body": a body which absorbs all radiation such that its emission curve is identical with that of the universal law of thermal radiation. Determining this body's emission curve as a function of its temperature and the radiation wavelength is thus for him "a task of greatest importance." Yet it takes more than three decades to physically realize such a black body and measure its emissions.

Wilhelm Wien

Wilhelm Wien (1864–1928) belongs to a group of physicists who still have an equal mastery of both experimental and theoretical physics. His most important contributions are in the field of thermal radiation. His formulation of laws of thermal radiation and the design, developed with Otto Lummer in 1895, for the realization of an actual black body

Early induction device developed by Joseph Saxton in 1833

blaze new trails for investigations in this field. This work gains him the Nobel Prize in 1911. Wien provides the essential precondition for Planck's radiation law as well as for Einstein's work about the light quantum hypothesis of 1905.

At the Boderline between Mechanics and Electrodynamics: Electromagnetic Induction

The relative motion between a magnet and a coil generates an electric current within the coil. This phenomenon is called induction. It is immaterial whether the magnet or the coil is moved, a difference is not observed. In Maxwell's theory of electromagnetism, however, there are two entirely different explanations given for this, depending on whether the coil or the magnet is moved. Einstein points out this assymmetry at the beginning of his paper *On the Electrodynamics of Moving Bodies*, published in 1905, the work in which he develops his special theory of relativity.

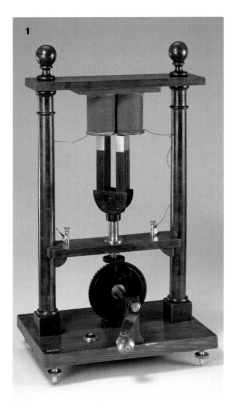

At the Borderline between Electrodynamics and Thermodynamics: Black Bodies and the Photoeffect

Around 1900, the wave theory of light appears unshakeable. However, the radiation of black bodies and the photoeffect elude satisfactory explanation on its basis.

A heated black body emits light. In the 1890s, the spectrum of this light is measured precisely at the *Physikalisch-Technische Reichsanstalt* (National Institute for Physics and Technical Standards) in Berlin. However, it is not possible to reconcile the results of these measurements with classical physics.

When a metal plate is irradiated with light, it receives a positive charge through the emission of electrons. This effect is observed by Heinrich Hertz in 1887, and consequently becomes the subject of intensive research. It turns out that the electrons' energy depends not on the intensity of the light, as expected, but, inexplicably, on its color.

1 (Induction Machine after Hippolyte Pixii

Demonstration Model, 1908

A magnet is moved beneath a fixed coils by means of a crank, thus causing electromagnetic induction. Hippolyte Pixii, son of a Parisian instrument maker, constructs this device in 1832. It is the first machine for generating alternating current.

At the Borderline between Mechanics and the Theory of Heat: Brownian Motion

In 1857, Rudolf Clausius shows that one can understand the heat of a body as the rapid motions of its invisible atoms or molecules. Microscopically small bodies suspended in liquid also exhibit an irregular motion, which is called "Brownian motion." It seems plausible to assume that this irregular motion is caused by the bodies colliding with the liquid's atoms or molecules. Attempts at explaining Brownian motion with the help of kinetic theory of heat, however, produce problematic results.

2 (Black Body
Physikalisch-Technische Bundesanstalt, Berlin, around 1950
In 1860 Gustav Robert Kirchhoff shows that a body only emits radiation of the same wavelength it can also absorb. A body that absorbs all light, that is a black body, emits light in all colors when heated. A black body is thus an ideal light source as the emitted light spectrum merely depends on the body's temperature. In 1899, after years of effort, the *Physikalisch-Technische Reichsanstalt* (National Institute for Physics and Technical Standards) finally manages to construct a black body that can also be used at high temperatures.

3 (Brown's Communication of his Discovery
Notizen aus dem Gebiet der Natur- und Heilkunde, October 1828
In 1827 the English botanist Robert Brown examines under the microscope the random motion of small particles floating in a fluid. He systematically tests various substances, from plant pollen to minerals and even samples from an Egyptian mummy. Brown shows that "life force" cannot be the reason for the motion, as other researchers have presumed. However, Brown is also unable to explain how this motion occurs.

Milieu of a Childhood

Albert Einstein is born in Ulm, Germany, on 14 March 1879. The world inhabited by his parents, Pauline and Hermann, is one marked by industrial change and the emancipation of the Jewish middle class, but also by the growing self-confidence of the second culture of science and technology. At the instigation of Einstein's Uncle Jakob, an innovative engineer, the family moves to Munich in 1880 to found an electrical company. The firm participates in the electrification of Schwabing in 1888, but soon runs into difficulties, mainly because of growing concentration in the electrotechnical sector. The factory moves to Pavia in Italy in 1894, with the family at first settling in Milan. Young Albert stays in Munich to finish his secondary education. The military drill and the prospect of soon having to don a real uniform, however, poisons his schooldays, and that same year he decides to quit school and join his parents in Italy.

Einstein discovers early on the pleasure of independent activity, from puzzles to solving geometry problems. The family's electrotechnical firm also offers occasions for reflection, from the boy's wonder at the invisible forces that make compasses work to the question of the state assumed by aether in a magnetic field, about which he writes an essay at the age of sixteen.

The electro-technical factory *J. Einstein & Comp.* in Munich, around 1890

His surroundings offer other stimulation as well, though. While his immediate family are freethinkers, the schoolboy receives religious instruction from a relative. It awakens religious feelings and aspirations in him which are later replaced by a philosophical way of thinking marked by a strict commitment to conscience. His encounter with Max Talmey (Talmud), a Jewish medical student from Poland for whom the Jewish community arranges free meals with Einstein's family, proves important for his interest in philosophy and natural sciences. Through him, Einstein becomes acquainted with the core texts of the second culture – from Humboldt's *Cosmos* and Büchner's *Force and Matter* to Bernstein's *Popular Scientific Books*. Through reading and discussions, Einstein thus receives an introduction to the world of the natural sciences, which leads him to see the search for scientific knowledge as an adventure, a striving for an all-encompassing view of the world, and a human endeavor that helps to bridge national and political differences. Einstein's later rejection of overspecialization and nationalist prejudice owes much to the broad knowledge he gains in this period, which serves to provide him with orientation.

Einstein's open-minded and democratic stance is strengthened by attendance at the Canton School in Aarau. He attends this school on the advice of the director of the Swiss Federal Institute of Technology in Zurich after his ambitious plan to be admitted there without first passing the final school examination fails in 1895. In Aarau he finds a home away from home with Jost Winteler, a politically interested and democratically minded teacher and linguist, but also stimulation for his study of physics. It is here that he begins to think for the first time about what a light ray would look like if approached at the speed of light. Einstein passes the Matura examination in 1896, after which he begins his study of physics at the Federal Institute of Technology.

○ HISTORICAL CONTEXT

Industrialization, Modernization, and Emancipation

In the Enlightenment period, many Jews in Germany move towards assimilation, away from the confined world of their traditional communi-

ties. The reforms introduced in various German states in the wake of the Napoleonic occupation awaken hopes of legal equality. However, the German middle classes turn to nationalism during the Restoration and the period leading up to the revolution of 1848, and new social barriers are created.

After the revolution of 1848 industrialization offers new occupational opportunities and chances for social mobility. Education becomes the central intellectual and social mediating institution between Jews and non-Jews, and with it the emerging natural and technical sciences. For a start, study in these fields promises better professional prospects in the expanding industrial sector, especially when access

to a university career is closed. At the same time such studies also gain increasing social recognition. Decisive for the final third of the nineteenth century is the fact that the rapid spread of technology and

science start breaking down traditional and discriminatory social structures. In this way, along with the growing significance of theoretical or interdisciplinary approaches, talented Jewish academics who have previously worked in thematic niches or on the margins now have the chance to gain the recognition of a broader scientific audience.

○ PERSONALIA

Maria (Maja) Einstein

Maja Winteler-Einstein (1881–1951) is not simply Einstein's younger sister, but also his confidante since childhood. In 1910 she marries Paul Winteler, son of Einstein's Swiss

host family, a year after completing her doctorate in Romance languages. She frequently visits her brother in Berlin, where she had also studied for several semesters as one of the first women students at the university. Shortly before the outbreak of World War II she follows him into exile in Princeton. A stroke prevents her return to Europe after the end of the war and leaves her bedridden. "In her last years I read to her every evening from the finest works of ancient and modern literature," Einstein informs an acquaintance after his sister's death, noting that he misses her more than anyone could imagine.

Left:
Albert Einstein and his sister Maja, around 1893

Right:
Einstein's graduating class at the Canton School in Aarau, 1896. Albert Einstein is seated first from the left

Electrotechnology in the Nineteenth Century – The Rise of Industry and Challenges to Science

The nineteenth-century world of electrotechnology strongly influences Einstein's youth.
Quite early on, it provides him with stimulation for independent reflection and research.

1 (The First Electrical Lighting of the Oktoberfest
Postcard, Munich, 1885

2 (Invoice from J. Einstein & Comp.
Munich, 28 February 1889

Albert Einstein's uncle, the engineer Jakob Einstein, founds a "Factory for Water Conveyance and Central Heating" in Munich in 1876. In 1879 Albert's father Hermann becomes a partner in the business. Soon, the company shifts its production entirely to electrotechnical equipment.
The factory's business progresses well at first. In 1885 the company installs the first electrical lighting for the Oktoberfest; in 1889 it provides the fifty-five arc lamps and sixty-six incandescent lamps that illuminate the Munich Fairgrounds during the German Gymnastics Festival and installs the public lighting system for the township of Schwabing.

Competition from the larger companies *AEG* and *Siemens & Schuckert* make life increasingly difficult for *J. Einstein & Comp.* In 1894 the brothers move production to Pavia, Italy, but the business there fails after only two years. Hermann tries to start again with money from his Cousin Rudolf. In 1898 and 1900 he erects two power stations in towns in northern Italy, before dying suddenly in 1902.

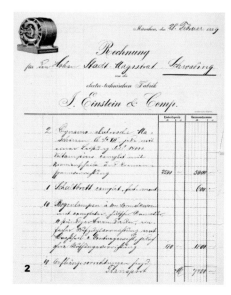

3 (Generator for Manual Operation
C. & E. Fein, Stuttgart 1881

The Electrotechnical Exhibitions

The first International Electricity Exhibition opens in Paris in 1881. In the next two decades it is followed by countless others in cities around Europe. These exhibitions function both as personal meeting places and as discussion forums for a specialist public. The wide-ranging audience can learn about electrotechnology in the most entertaining of ways and stare in wonder at the latest inventions. Electrotechnology generates the hope of increasing levels of affluence for all and lasting economic growth through technical progress.

The transmission of high voltage electricity over large distances succeeds for the first time at the International Electricity Exhibition of 1891 in Frankfurt on Main. This clears the way for further expansion of the still young electrotechnical industry. Also *J. Einstein & Comp.* from Munich presents its products at this exhibition. Its generators provide the electricity to light a beer hall, a café, the labyrinth, and the shooting gallery.

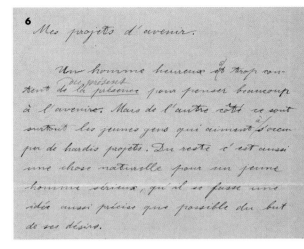

4 (Compass
19th century
At the age of four or five, Albert receives a compass as a gift from his father. The simple instrument makes a deep impression on the young lad. He wonders why the needle moves although he cannot see any cause for it. This "wondering" will remain particularly important for Einstein through-

out his life, as the point of departure for reflections and thus of science.

5 (On the Investigation of the State of the Aether in a Magnetic Field
Einstein's first scientific essay, 1895
At the age of sixteen, Einstein writes his first physics essay, which he sends to his uncle Caesar Koch. Consistent with the physics of the time, he assumes the existence of an elastic substance that fills all of space, the aether. Einstein investigates the issue of the influence of magnetic fields on the aether. Such fields cause a mechanical deformation of the aether affecting the velocity of propagation of the electromagnetic waves within it. The problems at the borderline between mechanics and electrodynamics still occupy him in 1905 as well, and they form the departure point for the special theory of relativity — the consequence being the abandonment of the aether concept.

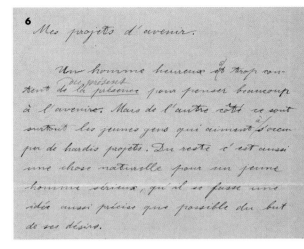

6 (Mes projets d'avenir
Einstein's Secondary-School Final Essay, 1896
In 1896 Einstein takes his secondary-school leaving examination at the canton school of Aarau. The topic of his essay for the French exam is *My Plans for the Future*. In this essay he writes that he wants to study mathematics and physics at the *Politechnikum* in Zurich (Federal Institute of Technology). He would like to focus especially on the theoretical parts of these sciences, as he believes that he has no practical talent

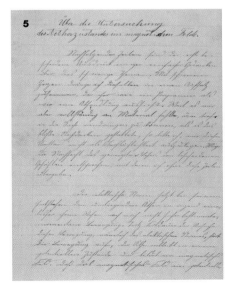

Milieu of a Revolution

Einstein: A consecrated silence prevails among us. To me it seems almost like a sinful desecration to interrupt it with some meaningless babble.

Marić: Didn't you want to tell us something scientific, my dear?

Habicht: Well, if Einstein's got nothing, Paul and I have something cobbled together according to your scheme.

Einstein: At last, you miserable bum. And when are you going to bring me your dissertation? Don't you know that I am one of the one-and-a-half fellows who would be interested in, or even enjoy, reading it through? I promise you four works for your one. Herr Boltzmann will be amazed.

Marić: Haven't we always criticized him because of his narrow-minded notions of atoms?

Einstein: He's never seriously considered that the motion of atoms can perhaps be observed, as is the case with the smallest of particles suspended in a liquid. I was convinced, by the way, that even the problems of electromagnetism can be solved if light as well is ascribed atomic properties.

Habicht: Meaning that light is made up of particles, just as I read back in Newton? That's terribly revolutionary!

Besso: It is held that Newton's theories are obsolete since the discovery of light's wave properties.

Habicht: But Dr. Albert Frankenstein brings the dead Sir Isaac back to life with electricity!

Einstein: But it just didn't work. With the notion of particles, you can get tantalizingly close to Planck's formulas for radiation, but it isn't completely correct.

Besso: And Planck's formula has proved unassailable due to the experimental results.

The "Olympia Academy": Conrad Habicht, Maurice Solovine, Albert Einstein, around 1903

Einstein: Exactly! That's why I had to rethink it, especially as it was impossible to develop a full-blown theory from the assumption of light particles. Lorentz's electrodynamics was unbeatable, as monstrous as it seemed to me with its contraction of moving bodies.

Habicht: You've not become a converted advocate of aether because of that, I hope – after all our discussions about the problematic character of such spooky concepts.

Einstein: There's no going back to the aether, not least due to the problem of thermal radiation. If it is composed of aether waves, there is no way it can reach thermal equilibrium. Imagine that each wave receives the same portion of energy...

Besso: How many aether waves are there actually in thermal radiation?

Einstein: Infinitely many, and they all sould get an equal share of energy.

Habicht: Well, that definitely settles it. So, the situation was desperate. Aether had been abolished, but Lorentz also got it right somehow. What did you two do?

Marić: We discussed everything again. What remained was the paradox that the speed of light is always the same, the same whether observed from a moving or stationary frame of reference.

Einstein: I was just about to give up. Then I went to Besso. It was he who posed the decisive question about time. How do you prove that two events occur at the same time? Is it not possible that events appearing to occur simultaneously to a stationary observer on a railroad embankment can, for a conductor on a moving train, appear to occur at different times? Common sense speaks for the absolute character of time. Lorentz's theory speaks for the absolute character of the speed of light, and its validity is unassailable. What we need is hence a new paradigm of space and time, if only to explain contraction.

○ HISTORICAL CONTEXT

Natural History Societies

In the second half of the nineteenth century, the development of science and technology and the growing importance of their achievements for everyday life are linked with hopes for extensive social progress. After the revolution of 1848, a newly self-confident middle class increasingly demands participation in all aspects of public affairs, including science and technology. Against this background, many scientists feel it is their duty not just to communicate among themselves, but also to make their knowledge accessible to the wider public.

The natural history societies founded in many towns become meeting places between scholars and laypeople interested in science. Members discuss scientific matters in a convivial atmosphere and go on nature study walks together. Einstein joins the *Naturforschende Gesellschaft* (Natural History Society) of Bern in 1903 and regularly attends its evening meetings at the *Storchen* inn together with a colleague from the patent office. That same year he gives his very first lecture on *The Theory of Electromagnetic Waves*. Together with two teachers, Einstein and his older friend Michele Besso frequently perform experiments in the school laboratory.

Even after becoming a famous scientist, Einstein still attends meetings at various scientific societies and lectures to the general public.

Olympia Academy

Soon after his arrival in Bern, Einstein founds the "Olympia Academy" with his friends Maurice Solovine and Conrad Habicht. They debate literature, philosophy, and science and make fun of the academic world's narrow-mindedness and infatuation with authority. The models for their critical reflections on the conceptual and methodological problems of contemporary science are the works of Mach and Poincaré, but they also read Spinoza, Hume, Cervantes, Dickens, Dedekind, and Helmholtz. Einstein finds inspiration and an open ear for his own revolutionary ideas

Left: Anna and Michele Besso, around 1900

Right: Mileva and Albert Einstein with their son Hans-Albert, 1904

among the members of the Olympia Academy. The convivial meetings of these Swiss bohemians produce life-long friendships and cooperation. Einstein still returns to them evenafter becoming a member of more illustrious academies where, however, his specialist colleagues' lack of open-mindedness often disappoints him.

○ PERSONALIA

Michele Besso

Michele Besso (1873–1955) is one of Einstein's closest friends, and he stays in touch with him throughout his life. They meet as students in Zurich, where Besso studies mechanical engineering. The two become colleagues at the Swiss Patent Office in Bern and continue their intensive discussions of the foundations of physics.

In particular, these conversations are the key source of inspiration for Einstein's work on the special theory of relativity. He refers to no previous literature there, but thanks his "friend and colleague M. Besso" instead.

Besso also supports Einstein's work on the general theory of relativity. He helps him not just to calculate the motion of the perihelion of Mercury, but also to grasp conceptual problems.

Mileva Marić

Mileva Marić (1875–1948) comes from Serbia, and begins her study of mathematics at the Swiss Federal Institute of Technology in 1896 as the only woman in her class. There she meets Einstein with whom she has intense discussions of their common interest in physics and their problems as students. Their friendship soon develops into a romance, with the prospect of becoming a work and life partnership,

though Einstein's parents disapprove of the this. The birth of their illegitimate daughter Lieserl in 1902 represents a problem in light of Einstein's precarious professional prospects, which are probably the reason why they decide to give her up for adoption. After their marriage they have two sons, Hans-Albert in 1904 and Eduard in 1910. The couple separate immediately after moving to Berlin, and divorce in 1919. Mileva lives in Zurich for the rest of her life. Her later years are marked by frequent illness.

Search for Clues

1 (Einstein's Advertisement for
Private Tutoring
Anzeiger für die Stadt Bern,
19 February 1902
After completing his studies, Einstein
searches in vain for a suitable position.
For a short time he works as an assistant
teacher at schools in Winterthur and
Schaffhausen. In 1902, hoping for a position at the patent office, he moves to Bern
where he initially attempts to earn a living
by giving private lessons.

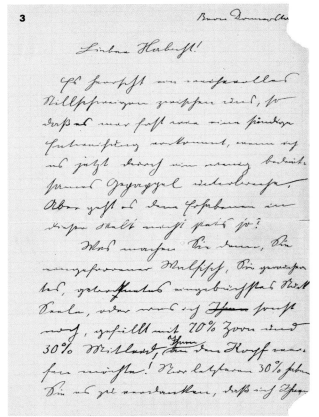

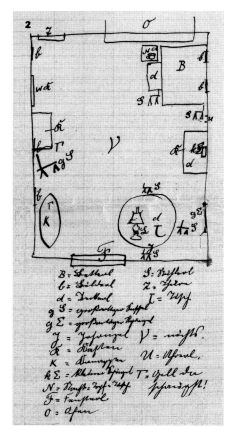

2 (Einstein's First Room in Bern,
Gerechtigkeitsgasse 32
From a letter by Einstein to Mileva Marić,
Bern, 4 February 1902
In his first letter from Bern to his girlfriend
Mileva, Einstein describes his new room
and includes a small sketch of it. Mileva
is currently living with her parents after
giving birth to their daughter Lieserl.

3 (Einstein to Conrad Habicht
Bern, May 1905
Einstein announces four papers in this
letter to his friend. The first paper "deals
with radiation and the energy properties of
light and is very revolutionary." He is talk-
ing about his work on the light quantum
hypothesis.

"The second paper is a determination of
the true size of atoms from diffusion and
internal friction." This is Einstein's doctoral
dissertation.
"The third proves that, on the assumption
of the molecular theory of heat, bodies
on the order of magnitude 1/1000 mm,
suspended in liquids, must already perform
an observable random motion." Here he
means his work on Brownian motion.
"The fourth paper is only a rough draft at
this point, and is an electrodynamics of
moving bodies which employs a modification of the theory of space and time." In
this work Einstein formulates his special
theory of relativity.

4 (Potential Multiplier

Paul Habicht, Schaffhausen, around 1920

Einstein, together with the Habicht brothers, develops this measurement device in the years around 1908. He often refers to it in his letters as "our little machine." In order to be able to measure extremely small electrical currents, the potential multiplier amplifies the voltage by a factor of 10,000.

Around 1920, Walther Gerlach measures the contact voltages of metals with this device. This study serves to confirm experimentally Einstein's paper on the photoeffect.

5 (Einstein's First Lecture to the Natural History Society

Naturforschende Gesellschaft Bern, minutes, vol. 10

On 5 December 1903 Einstein gives his very first lecture. It is on *The Theory of Electromagnetic Waves* ant is held at the

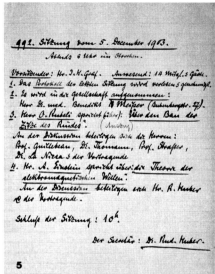

Natural History Society of Bern. The minutes note a subsequent discussion between Einstein and Rudolf Huber, a professor at the *Freies Gymnasium* in Bern and chair of the society. Huber often gives lectures himself, for example about electrons, and the minutes record Einstein's lively participation in the following discussions.

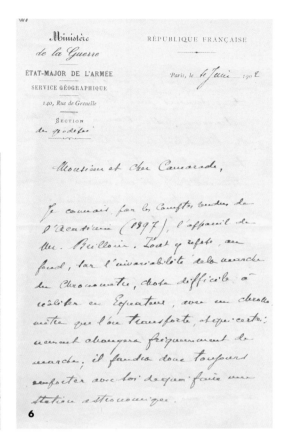

6 (Robert Bourgeois to Henri Poincaré

Quito, 4 June 1902

On assignment for the French government, the mathematician Poincaré is dealing with the problem of determining longitude. To this end, he intensively studies the problems of time measurement, including the telegraphic comparison of two distant clocks. In this letter the military geographer Bourgeois answers Poincaré's suggestion to measure gravity by means of a clock. Gravity influences the period of the pendulum and thus the running of the clock. Clearly, Einstein is not the only person dealing with the problems of simultaneity.

The *annus mirabilis* 1905

In 1905 Einstein publishes a total of twenty-six scientific articles, including his dissertation on molecular dimensions. This is an astonishing amount of work, even if the majority are just brief reviews with which he earns a little extra income. Three of his works, though, about the light quantum hypothesis, the Brownian motion, and the electrodynamics of moving bodies each unleash a scientific revolution. In yet another work this year, he derives a relationship between the energy and mass of a body. Its representation is arguably the most famous equation in the world, $E = mc^2$. These works and their repercussions dramatically alter the understanding of space and time, of matter and radiation. The year 1905 marks the transition from classical to modern physics. Today's modern physics is still characterized by the tension between quantum theory and relativity theory. Both have their genesis in Einstein's revolutionary works.

The revolutionary

How is it possible to achieve so many scientific breakthroughs with such far reaching consequences in so many different fields, fields that, at first glance, have very little to do with one another? An answer to this question must take into account both the history of science as well as Einstein's exceptional perspective on this body of knowledge. The tensions inherent in the classical physics that Einstein learns in his early studies come to a head at the turn of the century. They emerge in problems lying at the borderlines between mechanics, electrodynamics, and thermodynamics. Einstein's breakthrough comes with his novel interpretation of the solutions proposed by the masters of classical physics for these borderline problems. His perspective on the borderline problems is initially characterized by his search for a conceptual unity in physics, which he hopes to find on the basis of an atomic theory. In order to broaden the scope of the atomic hypothesis and find proof for the reality of atoms, Einstein develops statistical mechanics based on Boltzmann's kinetic theory of gases between 1902 and 1904. This provides clues about observable fluctuations and eventually leads Einstein to interpret the long-known Brownian motion as such a fluctuation phenomenon. Based on considerations of statistical physics, he can further show that Planck's formula for thermal radiation's energy distribution is not concordant with the classical idea of radiation as a wave phenomenon in a continuous aether. With the assumption of atomic light quanta Einstein manages to find a new, albeit only partial interpretation of Planck's formula as well as explain peculiarities in the interaction of radiation and matter.

In contrast, the attempts to transfer the relativity principle of mechanics to electrodynamics with the aid of an atomic theory of radiation go awry. Einstein undertakes such attempts, among other reason, to explain the interaction between a magnet and a coil, independent of assuming which of the two moves in relation to the aether. Einstein is caught between having to accept the validity of Lorentz's electrodynamics and being unable to accept its assumption of aether. As an outcome of this predicament Einstein formulates new conceptions of space and time in his theory of relativity.

⊙ FOUNDATIONS

Space and Time

IIn our daily experience, measures of time and space are absolute. Regardless of where observers stand and how they move, the passing of an hour or a distance of one meter always have the same meaning. Measurement of the velocity of a moving body, however, does not possess this absolute character, as it depends on the observer's velocity.

Surprisingly enough, this idea does not hold true for the speed of light, which is the same for all observers. Einstein raises this observation to a general principle. He states that not only the laws of mechanics, but also those of electrodynamics are independent of the state of motion of the observer. On the basis of these two postulates he develop his special theory of relativity which contradicts our every-day observations: time intervals and length are no longer independent from the observer's motion. Moreover, two observers moving relative to each other come to different conclusions about the simultaneity of events.

In the special theory of relativity space and time are no longer independent of one another – they are unified in four-dimensional spacetime. The perception of how this is divided into time and space depends on the observer's motion – much the same as the perception of what is an object's width and depth depends on the angle from which it is observed.

Matter and Energy

As a consequence of the special theory of relativity, a body's mass increases with its velocity. This provides the first chance for an experimental verification of the theory. The increase in mass is proportional to a body's energy. In a short essay published just a few short months after his paper on the theory of relativity, Einstein generalizes this conclusion: in general, every mass should correspond to an energy and, vice-versa, every energy should correspond to a mass. This

$$E = \frac{mc^2}{\sqrt{1 - \frac{q^2}{c^2}}}$$

is what Einstein's famous formula says: $E = mc^2$, energy equals mass times the speed of light squared. Because of light's enormous velocity, this means that even small masses correspond to vast amounts of energy.

Does this equivalence of mass and energy really mean that it is possible to turn mass into energy? Einstein sees one case in which this conversion could be observed, in radioactive decay. There, such a large amount of energy is released that there must be a measurable difference between the mass of the original radioactive material and the material into which it decays. It isnt't until decades later that a far more dramatic application of this formula is found – the splitting of an atomic nucleus. Einstein's theoretical postulate becomes the symbol of the atomic age.

Waves and Particles

The wave theory of light is well established during the nineteenth century as part of classical physics. It explains light's refraction, diffraction, and polarization. Maxwell is

able to move beyond this, explaining light waves as electromagnetic waves.

Yet Einstein questions this generally accepted theoretical framework in 1905 and suggests that in certain situations light acts as if it were made up of individual particles. This helps him to partially explain the puzzling formula that Planck published a few years before about the spectral distribution of light from glowing bodies. It also helps clarify other irritating experimental results, like the photoelectric effect and the unusual properties of x-rays. But even Einstein

does not suggest that light is only made up of small particles, as the wave properties of light are too well confirmed.

But how can light simultaneously have characteristics of both parti-

Right:
Albert Einstein at the patent office, Bern, around 1905

Left:
Equivalence of mass and energy from Einstein's manuscript on the special theory of relativity, 1912

cles and waves? Even after years of struggling with the problem, Einstein himself is unable to find an answer. As it turns out, the problem is not just restricted to light. In 1924, Louis de Broglie conversely postulates that particles of matter also have wave properties. The dualism of waves and particles becomes the central feature of quantum mechanics. Whether we see a particle or a wave depends on how it is measured. Einstein, however, rejects this interpretation, as he is not ready to abandon the question of the true nature of light or matter.

Four Works Make History

The patent office clerk from Bern turns classical physics on its head in 1905 with four works. Afterwards, the concepts of space, time, energy, mass, radiation, and atoms all take on new meanings.

1 (The Special Theory of Relativity
On the Electrodynamics of Moving Bodies.
Annalen der Physik, vol. 17 (1905)
Since Galileo, the notion of absolute rest has been meaningless in mechanics. In two frames of reference that move relative to each other in uniform rectilinear motion, the same mechanical laws hold, regardless which frame of reference is considered to be at rest and which in motion. This is the relativity principle of mechanics.

At the beginning of his work Einstein points out that in the aether theories of electrodynamics and optics there is a difference between frames of reference that are moving and those at rest, although this difference is not measurable in the phenomena. He thus postulates that the relativity principle must also be true for electrodynamics and optics. Together with the postulate of the constancy of the velocity of light, this leads to the conclusion that space and time depend on the state of motion of the frame of reference.

2 ($E = mc^2$
Does the Inertia of a Body Depend on its Energy Content? Annalen der Physik, vol. 17 (1905)
In September of 1905, Einstein publishes a short paper in which he draws an important conclusion from his special theory of relativity – mass and energy are equivalent. Mass is thus a measure of the amount of energy contained in a body. Conversely, light also transmits mass. However, Einstein introduces the well-known form of his formula $E = mc^2$ only later.

3 (Light Quanta

On a Heuristic Point of View Concerning the Production and Transformation of Light Annalen der Physik, vol. 17 (1905)
In the nineteenth century, the idea of light as a wave is firmly established through countless experiments and practical applications. No "serious" physicist believes that this theory will ever be called into question. Einstein shows, however, that it is inconsistent with the idea of the thermo-dynamic equilibrium of radiation. He makes the daring assumption that in certain

4 (Atoms

On the Movement of Small Particles Suspended in Stationary Liquids Required by the Molecular-Kinetic Theory of Heat. Annalen der Physik, vol. 17 (1905)
Around 1900 physicists are still debating about whether or not atoms actually exist or if they are merely a useful theoretical model. Also the cause of Brownian motion, the random movement of microscopically small particles suspended in a fluid, is still unknown. Einstein shows that the move-ment can be statistically described and can

situations light acts as if it were made up of small particles. With the help of his light quantum hypothesis, he is able to explain some previously puzzling experimental results.
Most physicists, including Planck and Lorentz, hold this idea for untenable for quite some time. In 1922 Einstein receives the 1921 Nobel Prize for Physics for his contribution to the understanding of the photoelectric effect contained in this article.

be understood as the elementary process that causes diffusion. He deduces a law for the average displacement of the small particles. In the framework of the mech-anical theory of heat it is thus possible to infer the size of the atoms from the ob-served displacements. In this way Einstein delivers a crucial argument for the accep-tance of the atom hypothesis.

Academic Career

Einstein's breakthroughs in 1905 do not remain unnoticed, however they do not lead immediately to an academic career. In 1908, although he receives the permission to lecture as a *Privatdozent* at the University of Bern, he also applies for a job as a secondary school teacher. It is not until 1909, at the instigation of his former professor Alfred Kleiner, that he is appointed associate professor of theoretical physics at

the University of Zurich. This starts his brisk climb up the academic ladder. Later that year, he is invited to speak to the Society of German Natural Scientists and Physicians. One year after that the German University of Prague makes him an offer which he finally accepts. An invitation to the Solvay Conference in 1911, one of the first international physicists' congresses, marks his transition to the league of most outstanding scientists of the time. He receives his first Nobel Prize nomination and is appointed as ordinary professor at the Swiss Federal Institute of Technology in Zurich. Two years later, in the summer of 1913, Planck and Nernst visit Ein-

The Natural Science Institute of the German University, Prague, around 1910.

stein and suggest a move to Berlin. The offer includes membership in the Prussian Academy of Sciences, as well as a directorship at the Kaiser Wilhelm Institute for Theoretical Physics, which is in planning.

Is his ascent into the academic Olympus a direct result of his revolutionary work in the *annus mirabilis*? This would be too simple an explanation, in view of the misunderstanding, skepticism, and resistance Einstein's challenges to classical physics encounter – despite all the recognition he receives. At first the change in mass depending on velocity, as predicted by the theory of relativity, cannot be confirmed by measuring accelerated electrons. Bucherer's experiments indicate confirmation while Kaufmann's do not. Einstein's statistical analysis of Brownian motion, in which he states that the particles do not have velocity in the usual sense of the word, meets with bewilderment at first. In his proposal for Einstein's election to the Academy, Planck feels he has to apologize for Einstein's boldness in his light quantum hypothesis, which he thinks is a bit over the mark. In 1907 Einstein also does little to increase his standing among his established colleagues: he starts questioning the foundations of his freshly formulated theory of relativity, in order to avoid conflicts between his ideas and Newton's law of gravitation which has been verified time and again.

The secret behind Einstein's unforeseeable rise lies rather in a different work, also published in 1907. As with the previous interdisciplinary speculations he pursued as a student at university, he transfers the hypothesis of energy quanta to a different branch of physics – the thermal behavior of solids. He draws on the quantum hypothesis to help explain their strange thermal behavior at low temperatures, and Nernst provides impressive experimental confirmation of this hypothesis in 1910. Einstein's appointment to Berlin is primarily combined with hopes that he can develop a new theory of matter with far-reaching consequences that will extend as far as physical chemistry and its practical applications. Einstein's international reputation is also heightened by the spectacular confirmation of his theoretical predictions about Brownian motion delivered by Jean Perrin in Paris from 1908 on. This confirmation leads to the ultimate acceptance of the atom hypothesis.

⌣ HISTORICAL CONTEXT

**The German Empire as
an Aspiring Great Power**

After the failed revolution of 1848, Prussia's military victories over Austria and France bring the long hoped-for national unification in the form of the *kleindeutsch* (lesser German) solution. The success of Bismarck's policies unites the economic bourgeoisie and the educated middle classes, while old-school liberalism wanes. The rising middle classes, together with the aristocracy, become loyal supporters of the new powerful state. Industry, at first plagued by crisis after 1871, experiences an enormous upswing. Science refines production and Germany becomes an outstanding exporter of high-quality goods. However, success brings arrogance. Imperial Germany fails to find its place in Europe. On the eve of World War I aspirations to become a great power, social tensions, and signs of decadence are eroding the fragile civil façade of Wilhelminian society.

The toughness of international competition brings with it greater social and political interest in the sciences. The founding of the Kaiser Wilhelm Society for the Advancement of the Sciences in 1911 is not least a matter of national pride in the struggle for often invoked intellectual supremacy. The splendidly equipped institutes are intended to supplement university research at disciplinary borderlines, undisturbed by teaching duties and under the leadership of renowned directors.

Fritz Haber and Gas Warfare

Fritz Haber's research at the Kaiser Wilhelm Institute for Physical Chemistry create the preconditions for modern chemical warfare in Germany. Haber hopes that new weapons can end the war more quickly. In early 1915, at Haber's suggestion, the German military decide to test and deploy chlorine gas on the warfront. The first

deployment on 22 April 1915 at Ypres in Flanders is supervised by Haber personally. The future Nobel Prize winners Otto Hahn, James Franck, and Gustav Hertz also participate in this work. After the deployment Haber is given the rank of captain. His institute is placed under military command in 1916 and expands substantially. Until 1918 Haber continues to believe that his weapon, the deployment of which violates the Hague Convention of 1907, can determine the outcome of the war.

Despite his pacifist convictions and his friendship with Haber, Einstein never addresses Haber's role in the development of such weapons of mass destruction.

◯ PERSONALIA

Elsa Löwenthal-Einstein

Elsa Einstein (1876–1936) is Einstein's cousin and they have known each other since childhood.
In 1896 she marries Max Löwenthal, a textile entrepreneur from Swabia. Their marriage, which ends in divorce in 1908, produces two

daughters, Ilse (born 1897) and Margot (born 1899). During Einstein's first visit to Berlin in the spring of 1912, Elsa and Albert renew their old acquaintance, which soon develops into a romance. When Einstein falls seriously ill in 1917/18 Elsa cares for him devotedly. They marry in 1919, and Elsa enjoys life at the side of her famous husband as well as in Berlin society. Emigration to America represents a sharp break in her life. Her final years are overshadowed by severe illness.

Left:
Fritz Haber in
captain's uniform,
1916

Right: Elsa and
Albert Einstein,
1921

Heinrich Zangger

Heinrich Zangger (1874–1957), professor at the University of Zurich and one of the leading Swiss forensic medical specialists, comes into contact with Einstein because of their common interest in Brownian motion. They become close friends, who also remain in constant communication on scientific, political, academic, and personal topics during Einstein's time in Berlin. Zangger not only advises Einstein on issues of academic appointments but on personal matters, too. In particular he acts as a mediator after Einstein's separation from his wife Mileva, and sometimes takes care of Einstein's sons.

Academic Rise and Experimental Confirmations

With his works of 1905 Einstein lays the cornerstone for his academic career, which reaches an initial highpoint in 1913 when he is elected into the Academy of Sciences in Berlin. The fact that many of the predictions in his 1905 works are tested experimentally and found to be correct naturally contributes to his rise as a scientist.

1 (Einstein Applies for Reimbursement of Removal Costs
Einstein to the Education Authority of the Canton Zurich, 1 November 1909
On 7 May 1909 Einstein is appointed extraordinary professor of theoretical physics at the Swiss Federal Institute of Technology, and moves from Bern to Zurich in October. He applies for a subsidy from the Education Authority to help pay his removal costs of 225 franks. The application is granted, and Einstein is reimbursed in full.

2 (Petition from Zurich Students
Letter to the Education Authority of the Canton Zurich, 23 June 1910
When it becomes known that Einstein is to be appointed to the vacant chair of theoretical physics at the German University of Prague, 15 students submit a petition to the Education Authority: "The undersigned students attending Prof. Einstein's lectures hereby request that you do all you possibly can to keep this outstanding scientist and teacher at our university."

3 (Ballot Ball of the Berlin Academy of Sciences
On 3 July 1913 the physical-mathematical class of the Berlin Academy meets to vote on Planck's proposal that Einstein be elected a full member. According to the traditional ritual, black and white (silver) balls are used in a secret ballot. The "balling" results in 21 white and only one negative, black ball. Planck and Nernst then travel to Zurich to present Einstein with the Academy's offer.

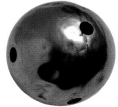

3

1

2

4 (Part of an Experimental Setup by
Walter Kaufmann
*For studying changes in the mass of fast
electrons, 1906*
It follows from the theory of relativity that
the mass of an electron depends upon its
velocity. Other theories of the electron also

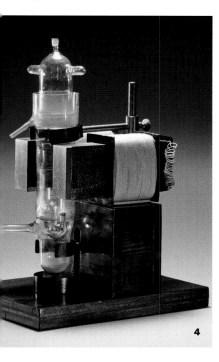

predict a change in mass, but of a different
magnitude.
Walter Kaufmann accelerates electrons in
a tube and deflects them with a magnet.
He can calculate an electron's mass from
the curve of its path. His initial results
seem to contradict Einstein's theory, while
later experiments confirm it.

5 (Ultramicroscope
Carl Zeiss, around 1910
The ultramicroscope makes it possible to
see structures that remain hidden to normal microscopes. An object is diagonally
flooded with light, so that the direct path
of the light past the lens and the image is
generated only by light that is scattered at
the object. The method is developed in
1903 by the chemist Richard Zsigmondy
in cooperation with Zeiss. Jean Perrin uses
an ultramicroscope to study the Brownian
motion of minute spheres and confirms
the results Einstein predicted in 1905.

6 (Spark Chamber for the Detection of
Muons
*Deutsches Elektronen-Synchrotron (DESY),
1990*
A muon is a negatively charged particle
that is generated by cosmic radiation in the
upper atmosphere. Muons are very short-
lived: their decay time is just 0.00002

seconds. Although it is very fast, its life
span means it can only travel a very short
distance and therefore should not reach
the Earth's surface. But according to Einstein's theory of relativity, moving clocks
tick more slowly, which also holds true for
the muon's "biological clock." That is why
on the Earth's surface we can see traces
of numerous muons in spark chambers.

War and Revolution

Einstein returns to Germany in April 1914, on the eve of World War I. He has been in Berlin just four months when Germany invades Belgium. He does not share the general enthusiasm for the war. But although Einstein looks forward to his "Berlinization" with skepticism precisely because of his rejection of Prussian militarism, he does not arrive in Berlin as a politically active scientist. It is his opposition to the broad societal consensus that first provokes his pacifist and democratic engagement. Einstein's political activity is triggered by the appeal *An die Kulturwelt!* (To the Civilized World!) of October 1914, an attempt by the political leadership to draw scholars and artists directly into the political conflict. In the appeal ninety-three signatories, including almost all the notable personalities in the arts and sciences of Wilhelminian Germany, justify the war and German militarism as a supposed necessity to protect German culture. The "Appeal of the 93" is an expression of the instrumentalization of culture and science for German war propaganda. It is intended to counteract international protests against German war crimes in Belgium by appealing to the same cultural values as the protesters.

The Berlin physiologist Georg Friedrich Nicolai immediately writes an opposing *Aufruf an die Europäer* (Appeal to the Europeans), propagating a "common world culture" of peace and mutual understanding among nations. This appeal, which is not published until 1917, is signed by four people, among them Einstein, whose anti-war commitment soon isolates him from his colleagues.

On the other hand, his stance brings him into contact with other war opponents and leads to his participation in pacifist organizations and activities. In March 1915 Einstein joins the *Bund Neues Vaterland* (New Fatherland League), a group of middle-class democratic opponents to the war with quite varied party political and religious affiliations. Although Einstein publicly expresses positions far removed from radical pacifism, his correspondence reveals an uncompromising attitude towards Germany's war

Enthusiasm for war in Berlin, August 1914

Right: Einstein's lecture notes, 1918: "9.XI. canceled because of revolution."

policies, which he expressly hopes will end in defeat. In the course of 1915 the military authorities paralyze the organized pacifist movement with arrests, censorship, and all manner of repressive measures, virtually forcing it into the underground. In this situation, Einstein limits his political statements mainly to the private sphere.

The situation only changes in the autumn of 1918 with the end of the war and the beginning of revolution. Einstein reenters the political stage and supports the establishment of democratic organizations, but does not join a political party. In an atmosphere of hope and confidence he also resumes his work in the *Bund Neues Vaterland* with renewed vigor. He speaks at public events and even participates in unofficial German missions on the question of the conditions of peace. In contrast to the wartime years, there is no longer any contradiction between Einstein's public activities and his opinions expressed in private.

○ HISTORICAL CONTEXT

The November Revolution

The sailors' uprising in Kiel in November 1918 unleashes unrest throughout Germany. After the emperor's abdication Friedrich Ebert is appointed German chancellor and the Republic is declared. The Social Democratic government secures the support of the military in order to keep control of political events. This deepens the split be-

Bund Neues Vaterland

The Bund Neues Vaterland (New Fatherland League) is the first political association Einstein joins. Founded in November 1914, the *Bund* pursues the objectives of a rapid peace and a system of European states that would prevent future conflicts.

Members include such diverse individuals as the journalist and later

○ PERSONALIA

Romain Rolland

The French author Romain Rolland (1866–1944), son of middle-class Protestant parents, receives the Nobel Prize for literature in 1915. After studying philosophy and history in Paris and Rome he teaches history of art and music in Paris from 1891 to 1912. Then he devotes himself full-time to his writing, living in Switzerland from 1914

Georg F. Nicolai

The physician Georg Friedrich Nicolai (1874–1964) comes from a middle-class Berlin Jewish family, but is baptized a Lutheran. He studies in Königsberg, Berlin, Paris, and Heidelberg and gains his doctorate in Leipzig in 1901. After that he works as a ships doctor for one year and is then an assistant in physiology in Halle, Berlin, Leyden, and St. Petersburg. He habilitates for lecturing in 1907 and in 1909 becomes full professor and assistant medical director at the Second Medical Clinic of the *Charité* Hospital in Berlin. In 1910, together with his superior Friedrich Kraus, he publishes a fundamental textbook of electrocardiography. He is one of the physicians to the empress. From the outbreak of the World War I he works in a military hospital. Nicolai condemns the war and is committed to a European cultural community. In October 1914 he writes the *Appeal to the Europeans* in response to the pamphlet *To the Civilized World!* Apart from Einstein the only other signatories he is able to acquire are the astronomer Wilhelm Foerster and the private scholar Otto Buek. He speaks out against the war in lectures and publishes the anti-war book *The Biology of War* in 1917. These activities bring him a court-martial and demotion in rank. After the war he is prevented from resuming his teaching duties at the *Charité*, and in 1920 the university revokes his permission to lecture. For this reason he emigrates to Argentina in 1922, where he is welcomed as a "great European." In 1936 he is appointed professor of physiology at the College of Veterinary Medicine in Santiago de Chile.

Revolutionary demonstration, Berlin, Unter den Linden, 1918

tween the Social Democrats and the radical left. Violent unrest continues well into 1919. The tensions between the extremes of the political spectrum, which culminate in the murders of Rosa Luxemburg and Karl Liebknecht by members of the *Freikorps* (Volunteer Military Corps), dog the Republic until the end.

Most of the professors remaining loyal to the emperor feel deeply unsettled and personally offended by these developments. The Republic is unpopular and, with the handling of the consequences of war, constant unrest, and economic crises, seems ill-fated. Albert Einstein is one of the few academics who actively oppose nationalism and militarism, and who welcome the end of the Wilhelminian regime as a liberation.

governing mayor of West Berlin Ernst Reuter and the banker Hugo Simon.

The organization formulates not only petitions to the government but also helps persecuted and jailed pacifists – an activity that is not without personal risk. The *Bund Neues Vaterland* is banned in February 1916, but resumes work after the war and is renamed the German League for Human Rights in 1922.

to 1937. In the 1920s he studies Indian religion and philosophy and is committed to socialist developments in the Soviet Union.

Rolland sees his own writing as dedicated to a humanistic view of mankind in the Enlightenment tradition, and many of his works champion international understanding and pacifism. He denounces the widespread war hysteria at the beginning of World War I, and is later active in the struggle against fascism. Einstein starts corresponding with Rolland in March 1915 in order to support his pacifist efforts, and meets him for the first time in September of that year in Switzerland.

Appeals

Sebastian Merkle,
Professor der kath. Theologie,
Würzburg.

Friedrich Naumann,
Berlin.

Wilhelm Ostwald,
Professor der Chemie, Leipzig.

Albert Plehn,
Professor der Medizin, Berlin.

Alois Riehl,
Professor der Philosophie,
Berlin.

Max Rubner,
Professor der Medizin, Berlin.

August Schmidlin,
Professor der Kirchengeschichte,
Münster.

Martin Spahn,
Professor der Geschichte, Straßburg.

Hans Thoma,
Karlsruhe.

Richard Voß,
Berchtesgaden.

Wilhelm Waldeyer,
Professor der Anatomie, Berlin.

Theodor Wiegand,
Museumsdirektor, Berlin.

Richard Willstätter,
Professor der Chemie, Berlin.

Eduard Meyer,
Professor der Geschichte, Berlin.

Albert Neisser,
Professor der Medizin, Breslau.

Bruno Paul,
Direktor der Kunstgewerbeschule,
Berlin.

Georg Reicke,
Berlin.

Karl Robert,
Professor der Archäologie, Halle.

Fritz Schaper,
Berlin.

Gustav von Schmoller,
Exz.,
Professor der Nationalökonomie,
Berlin.

Franz von Stuck,
München.

Wilhelm Trübner,
Karlsruhe.

Karl Voßler,
Professor
der romanischen Philologie,
München.

August von Wassermann,
Professor der Medizin, Berlin.

Wilhelm Wien,
Professor der Physik, Würzburg.

Wilhelm Windelband,
Professor der Philosophie,
Heidelberg.

Heinrich Morf,
Professor
der romanischen Philologie, Berlin.

Walter Nernst,
Professor der Physik, Berlin.

Max Planck,
Professor der Physik, Berlin.

Prof. Max Reinhardt,
Direktor des Deutschen Theaters,
Berlin.

Wilhelm Röntgen, Exz.,
Professor der Physik, München.

Adolf von Schlatter,
Professor der protest. Theologie,
Tübingen.

Reinhold Seeberg,
Professor der protest. Theologie,
Berlin.

Hermann Sudermann,
Berlin.

Karl Vollmöller,
Stuttgart.

Siegfried Wagner,
Bayreuth.

Felix von Weingartner.

Ulrich von Wilamowitz-
Moellendorff, Exz.,
Professor der Philologie, Berlin.

Wilhelm Wundt, Exz.,
Professor der Philosophie,
Leipzig.

1

1 (To the Civilized World!
4 Oktober 1914
The appeal is initiated by the writers Ludwig Fulda and Hermann Sudermann and signed by ninety-three scholars and intellectuals, including Einstein's colleague Max Planck. The pamphlet has serious consequences for international academic relations. In it the authors deny German war atrocities in Belgium and identify themselves unreservedly with the German military. Intellectual frontlines develop parallel to the military frontlines, and natural scientists take up position alongside other members of national educated elites. Some of the signatories later distance themselves from the appeal.

Denn der heute tobende Kampf wird kaum einen Sieger, sondern wahrscheinlich nur Besiegte zurücklassen. Darum scheint es nicht nur gut, sondern bitter nötig, daß gebildete Männer aller Staaten ihren Einfluß dahin aufbieten, daß — wie auch der heute noch ungewisse Ausgang des Krieges sein mag — die Bedingungen des Friedens nicht die Quelle künftiger Kriege werden, daß vielmehr die Tatsache, daß durch diesen Krieg alle europäischen Verhältnisse in einen gleichsam labilen und plastischen Zustand geraten sind, dazu benutzt werde, um aus Europa eine organische Einheit zu schaffen. — Die technischen und intellektuellen Bedingungen dafür sind gegeben.

2

2 (Appeal to the Europeans
Mid October 1914
This text, written by Georg Friedrich Nicolai and Albert Einstein, is intended as a counter-manifesto to the appeal *To the Civilized World!* Since nobody else is prepared to sign it, it becomes known only through Nicolai's book *The Biology of War* published in 1917. In his book Nicolai notes, "Although we received much friendly assent when we privately circulated the petition by post, most did not wish to sign it. One man found the passage on Greece historically not quite correct, while another said that such an appeal came too late, and a third that it came too early, while yet another considered it inopportune for scientists to meddle in world affairs, etc."
"For the fight that rages today will leave behind hardly a victor, but probably only vanquished. For that reason it seems not just good but also highly necessary that educated men of all nations exert their influence to ensure that – whatever the still uncertain outcome of the war may be – [firstly] the conditions of the peace will not become the source of future wars, and [secondly] that the fact at this war has placed all European circumstances, as it were, into a fragile and malleable state, will be used instead to make Europe an organic whole. – The necessary technological and intellectual preconditions exist."

3 (My Opinion about the War

The Berlin Goethebund:

The Land of Goethe 1914/1916

In March 1900, Theodor Mommsen, Georg Reiche, Hermann Sudermann, and others found the Berlin *Goethebund* (Goethe League), which later changes its name to "Cultural League of German Scholars and Artists." It comes out of the resistance to the so-called *Lex Heinze* of 1899, a planned restriction of artistic and press freedom in the guise of a new obscenity law.

During World War I the organization contributes to German war propaganda with a lavish anthology bringing together authors who, according to the editors, "represent the glory of the nation in these great and difficult times, and whom we look to with all the more pride the more enemy countries exhaust themselves in impotent attempts to tear down our culture." It is the editors' stated aim here "to provide a review of Germany's intellectual and moral leaders. [...] We find here, in words, music, and pictures, an all-encompassing declaration of the German intellectual world, testimony to the unshakeable belief in the victory of German arms and the German soul."

More than 120 politicians, scholars, and artists provide contributions, which are however quite diverse and sometimes amount to no more than moderate greetings. Einstein's contribution to this celebratory coffee-table book overflowing with florid nationalism stands out from the rest. His piece, which is toned down for publication, is sarcastic, and at the same time strongly warns against the brutality of war.

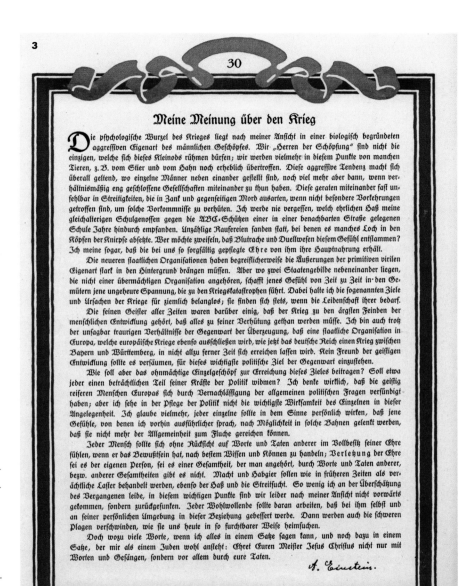

Propaganda and Critique

1 (Propaganda Postcards

a) Heroism and religious pathos

b) Chauvinist rhyming couplets

c) Faces Boches – French political cartoon

World War I enflames nationalist sentiment across Europe. The various governments systematically rely on the effects of aggressive propaganda. Not just cultural arrogance and nationalist pathos, but also clearly racist opinions spread. After the end of the war the wounds can hardly be healed. National clichés, suspicion, and prejudice long mark the public consciousness in the enemy states. In Germany, a mentality of bitter resentment unites with a desire for revenge.

2 ("Shrine of Patriotism"
Manuscript of the contribution to the anthology The Land of Goethe, 1915
Einstein is repelled by the combination of self-righteousness and petty-minded mean-ness with which sections of the German middle classes enthusiastically welcome the war in 1914. His acid commentaries are often cutting, but frequently also leave a

back door open for future understanding. One passage struck from the piece for the *Goethebund* reads,
"When I look into the mind of a decent, normal citizen, I see a moderately illumina-ted cozy room. In one corner stands a well-tended shrine, of which the master of the house is mightily proud, and which he loudly points out to every visitor. Inscribed on it in large letters is the word "Patrio-tism." To open this cabinet is generally frowned upon, however. Indeed, the master of the house barely knows, or does not know at all, that his cabinet holds the mo-ral requisites of bestial hatred and mass murder, which he obediently takes out in case of war in order to use them. You will not find this shrine in my little parlor, dear reader, and I would be happy if you would adopt the opinion that in that same corner of your little parlor a piano or a little book-shelf would be a more fitting piece of furni-ture than the one you only find tolerable because you have been used to it since your youth."

The Unfinished Quantum Revolution

The quantum revolution starts at the beginning of the twentieth century with Planck's formula for the energy distribution of thermal radiation, reaching its first conclusion in the mid-1920s with the formulation of quantum mechanics. Only in retrospect can it be seen as an all-encompassing, fundamental revision of the classical ideas of matter and radiation. At the time, it seems more plausible to search for quantum properties in the previously unknown microworld of atoms and molecules, while at the same time limiting them to these dimensions. However, Einstein is the first to realize not only that Planck's formula is a break with classical theories of radiation, but also that the introduction of quanta cannot be restricted to just one branch of physics.

Einstein's first contribution to broadening the quantum hypothesis is his work about the specific heat of solid bodies in 1907. He takes conclusions drawn from Planck's theory and transfers them to the atomic structures responsible for the thermal motion of solids. So he gives up the idea that atomic motion is sub-

Niels Bohr (right) and Arnold Sommerfeld (left) in Lund, 1919

ject to the same laws as bodies in our normal range of perception. Einstein's theoretical conclusions are later proven experimentally by Nernst and others. "This quantum question is so terribly important and difficult that everyone's endeavors should be focused on it," writes Einstein to a colleague in 1909. Yet initially, its study remains the purview of a few outsiders. Only slowly do interested scientists begin networking – their exchanges and cooperation allow for the gradual understanding of the true dimensions of the problems being faced.

Einstein's step in the direction of a non-classical physics to describe atoms opens up the perspective that other problems of the atomic concept of matter can also be solved in novel ways. Speculations about atoms as the building blocks of matter started with Democritus in ancient Greece. Atoms were traditionally assigned properties based on experiences with objects of the world accessible to our senses. Through assumptions about their shape, it is hoped that atoms will help explain certain properties of the bodies made up of them such as heat and compressibility.

Starting in the mid-nineteenth century, many branches of physics and chemistry develop more differentiated atomic models – based, for example, on newly won knowledge about electricity. These models, however, do not coalesce into a unified whole. The spectral analysis of chemical compounds, especially the multitude of spectral line patterns seen in a burning sample, can be interpreted as clues to an extremely complex internal structure of atoms and molecules. Such a structure, however, cannot be reconciled with knowledge gleaned from classical physics, especially about the thermal properties of solid bodies.

In 1913, Bohr constructs an atomic model based on the astronomical paradigm of a planetary system and consistent with experiments suggesting that matter at the atomic level is rather empty. In order to explain the spectral line patterns, Bohr's model uses the idea of quantized energy and assumes that the electrons spinning around the atom's nucleus make quantum jumps between the orbits of the atomic planetary system. By refining the analogy of solar systems and inner structure of the atom, Sommerfeld and his students are able to interpret the fine structure of the spectral lines as well. This combination of classical and quantum building blocks results in a sort of half-classical atomic theory that later becomes the springboard for the completion of quantum theory.

○ PERSONALIA

Walther Nernst

Walther Hermann Nernst (1864–1941) studies in Zurich, Berlin, and Graz. He gains his doctorate in Würzburg in 1887. After working as assistant to the physical chemist Wilhelm Ostwald in Leipzig, he is appointed professor of physical chemistry in Göttingen in 1891. In 1905 he moves to Berlin University, and is full professor of physics there from 1924 to 1933. From 1922 to 1924 he also serves as president of the National Institute of Physics and Technical Standards. In 1897 he invents an efficient incandescent lamp, the Nernst lamp.

His 1905 formulation of the third law of thermodynamics earns him the Nobel Prize for chemistry in 1920. During experiments to demonstrate this theorem it becomes evident that at lower temperatures, the temperature curves of specific heats correspond to Einstein's quantum theoretical predictions of 1907.

Nernst is a dedicated scientific organizer. He is co-founder of the Electrochemical Society and a promoter of the Kaiser Wilhelm Society. The influential Solvay Conferences are attributable to his initiative.

Arnold Sommerfeld

Arnold Sommerfeld (1868–1951) is considered one of the founders of modern theoretical physics and trains generations of important physicists. He studies in Königsberg, where he gains his doctorate in mathematics in 1891. He is influenced above all by his subsequent time as assistant to Felix Klein in Göttingen. In 1900 he is appointed professor of technical mechanics at the Technical University of Aachen, and full professor of theoretical physics at the University of Munich in 1906. He starts incorporating Einstein's special theory of relativity into his lectures as early as 1907, and writes important articles on refining its mathematical formalism. He is among the early supporters of the quantum hypothesis and further develops the Bohr model of the atom.

Sommerfeld's fundamental works include his 1919 treatise *Atomic Structure and Spectral Lines*, that is regarded as the "bible" of atomic physicists at that time.

First Solvay-Conferenz, Brüssel 1911. Standing, from left to right: Robert Goldschmidt, Max Planck, Heinrich Rubens, Arnold Sommerfeld, Friedrich Lindemann, Louis de Broglie, Martin Knudsen, Fritz Hasenöhrl, Georges Hostelet, Édouard Herzen, James H. Jeans, Ernest Rutherford, Heike Kamerlingh Onnes, Albert Einstein, Paul Langevin. Seated, from left to right: Walther Nernst, Léon Brillouin, Ernest Solvay, Hendrik Antoon Lorentz, Emil Warburg, Jean Perrin, Wilhelm Wien, Marie Curie, Henri Poincaré.

Atoms and Quanta

1 (Model of the Atom According to Democritus

According to Democritus there is an infinite number of atoms. Atoms are differentiated by their form and, due to their innate force, are always moving through empty

space where they collide with one another. These collisions form aggregations of atoms that make up natural things and the entire cosmos. The motion of atoms has no purpose, but follows a necessity intrinsic to matter. Its regularity enables the identification of causes of recurring natural processes.

2 (Bohr-Sommerfeld Atom Model
Reconstruction

At the threshold of the twentieth century, the study of cathode rays and the spectrum leads to the construction of the first models of atomic structure. The discoverer

of the electron, Joseph John Thompson, presents one of the first models in 1902. In his model the positive charge is distributed across a sphere the size of an atom. The electrons are embedded in this positive field like raisins in a cake. In 1910, after experimenting with alpha particles scattered on gold foil, Ernest Rutherford concludes that an atom consists of an almost point-like concentration of positive charge, an atomic nucleus, surrounded with a spherically formed distribution of negative charge equal to the positive charge of the nucleus.

In 1912 / 13 Niels Bohr takes up Rutherford's idea and explains by means of quantum theory under what conditions atoms remain stable and how they absorb and emit light. His model of the atom is similar to the planetary model – electrons travel in defined circular orbits around the positively charged nucleus. When an electron changes orbit it undergoes either a loss or a gain of energy.

In Munich, Arnold Sommerfeld refines Bohr's atom model by introducing eliptical electron orbits. This improves the explanation of the qualitative structures found in the spectral lines.

3 (The Franck-Hertz Experiment
Phywe Company, Göttingen

In 1913/14 the Berlin physicists James Franck and Gustav Hertz attempt to verify the fundamental laws of electrical discharge processes. For that purpose, they

4 (Nernst's Hydrogen Liquefier
Berlin, around 1910

In 1905 Walther Nernst formulates the third law of thermodynamics, describing the behavior of matter as temperatures approach absolute zero, –273.16 degrees

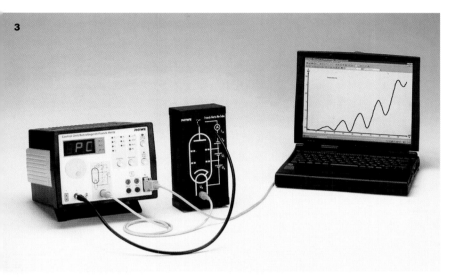

examine collisions between electrons and gas molecules. They show that collisions with electrons below a certain threshold energy E are elastic, meaning the molecules do not absorb any energy from the electrons. If the electrons have a higher energy, the collision is inelastic, whereby the energy transfer is a multiple of E. Simultaneously, the atoms radiate light at a particular frequency.

Franck and Hertz link their experimental results with Planck's quantum theory and show that the radiation emitted follows exactly the formula $E = h\nu$, where the proportionality factor is Planck's constant h. Only months later they realize they have also delivered a splendid confirmation of Bohr's model of the atom.

Celsius. To verify it, it is necessary to know specific heat values at very low temperatures. Specific heat is the amount of energy required to raise the temperature of a particular substance by a certain amount. Very little is known about very low temperatures at the start of the twentieth century. Therefore, Nernst develops a large-scale research program at his institute in Berlin, where reaching these very low temperatures takes top priority. In 1910, together with Hönow, the institute's mechanic, he develops his own hydrogen liquefier, with which it is possible to reach temperatures as low as –250 degrees Celsius. The contraption is temperamental and needs Hönow's "magic touch" to function properly. Nernst's experimental findings on the

drop of specific heat at low temperatures correspond well with the results Einstein predicts as a consequence of the quantum hypothesis in 1907. Impressed by this agreement, Nernst becomes an adherent of the quantum hypothesis and a sponsor of the young Einstein.

The Riddle of Gravitation

Until the beginning of the twentieth century, Newton's law of gravitation is considered unassailable – its validity having been consistently proved by experience. But because this law assumes an interaction taking place without a time lag, it is repeatedly cast into doubt on theoretical grounds. How is such an interaction transmitted? Does it not in fact need time in order to propagate? These sorts of questions are prompted especially by the advent of aether-based electrodynamics. With Einstein's theory of relativity of 1905 these questions cannot be ignored any longer – according to this theory, an effect that propagates faster than light is impossible.

Relativity's view of space and time thus requires a revision of Newton's theory, and along with Einstein various researchers address the task. But such a revision also includes a challenge – to either give up the well-established insights of mechanics, or once again to question the new understanding of space and time.

Dancing with gravity

Indeed it seems that a relativistic theory of gravitation is incompatible with the idea of all bodies falling with the same acceleration. In 1907, however, Einstein recognizes that a theory of gravitation meets these requirements if one assumes that gravitation and the inertial forces acting in an accelerated frame of reference are essentially the same. This idea becomes the starting point for his famous equivalence principle. According to it, one may assume that an accelerated frame of reference is at rest while exhibiting a gravitational field. This creates a connection between force of gravity and the inertial forces occuring in an accelerated frame of reference.

The equivalence principle suggests, at the same time, a generalization of the relativity principle for accelerated motion. Already Mach has wondered whether a rotating bucket could be interpreted as being at rest, while the centrifugal forces appearing during its rotation could be interpreted as the effect of the cosmic masses rotating around the bucket.

Recourse to Mach's interpreation solves a central problem of a relativistic theory of gravitation. Attempts to describe gravitation as the effect of a field, in analogy to the electromagnetic field, face the problem of finding an equivalent to the magnetic forces created by electrically charged bodies in motion. The relationship between gravitational and inertial forces now makes it plausible to think of the centrifugal forces created in the rotating bucket as an equivalent of the magnetic forces, and to include them in a field theory of gravitation.

How does a body move in such a field? Simulating gravitational fields with accelerated motion makes for an easy answer to this question. Einstein merely has to translate the inertial motion along a straight line into a language appropriate for an accelerated frame of reference. This translation has similarities to the problem of translating the description of a straight line in an orthogonal coordinate system into a description using a curvilinear coordinate system, as is the case in the theory of curved surfaces. In 1912 this analogy inspires Einstein to conceive of gravitation as the curvature of spacetime.

○ PERSONALIA

Marcel Grossmann

Marcel Grossmann (1878–1936) is Einstein's friend and fellow student at Zurich Federal Institute of Technology, as well as his collaborator in research concerning the theory

Marcel Grossmann

of general relativity. He studies mathematics and writes his dissertation on non-Euclidean geometry. He lends Einstein his lecture notebooks, becoming his friend's savior for the exams. Later he helps him get a position at the Swiss Patent Office. In summer 1912, when Einstein realizes that the general theory of relativity is related to non-Euclidean geometry he turns to his mathematician friend seeking aid. Together they search for the field equations of gravity. With Grossmann's help Einstein comes across the correct solution, but he does not yet accept it. Instead, he and Grossmann publish a preliminary theory in 1913, the so-called *Entwurf* (outline) theory. In 1915 Einstein returns to the first, correct solution. Grossmann, a professor at the Swiss Federal Institute of Technology since 1907, falls ill with multiple sclerosis in 1920, and retires in 1927.

◎ FOUNDATIONS

Four-Dimensional Spacetime

Events in space can be described in four coordinates, three for location on the axes of physical space and one for time. The special theory of relativity is the first to connect these four coordinates and suggests an extension of the three-dimensional geometry of space to a four-dimensional geometry of spacetime. According to classical physics, changing the frames of reference only changes the spatial coordinates while time remains the same. But, according to the theory of relativity, changing from one frame of reference to another that is moving in reference to the first, also changes the temporal coordinates. In 1908, the mathematician Hermann Minkowski realizes that this melts space and time into a new unity, which he describes as a four-dimensional geometry. In 1912 Einstein concludes that a relativistic theory of gravitation requires the assumption of a curved spacetime.

On the way to the equivalence principle

The Equivalence Principle and its Consequences

Newton's law of gravitation, which states that the gravity force spreads without any time lag, is not compatible with the special theory of relativity, which allows no propagation of physical forces faster than the speed of light. In 1907, attempts to adapt Newton's law to meet this requirement lead Einstein to see that gravitation and inertial forces in an accelerated system are equivalent, represented by the fact that gravitational mass and inertial mass are the same. Light rays are deflected in an accelerated system, meaning they are deflected by gravity as well – an astounding consequence of this "equivalence principle."

1 (Eötvös Balance
Budapest, 1902
Around 1900, the Hungarian physicist Lorand Eötvös develops a torsion balance in order to prove the equivalence of heavy and inertial masses. Heavy mass is the cause of gravitational force, while inertial

mass resists its acceleration. Eötvös places two different substances on each end of the balance. Gravitation exerts its force on the heavy masses and the inertial masses are affected by the centrifugal force of the rotating Earth. If these two types of mass were not equal, the horizontal beam would start rotating. But such movement cannot be observed.

2 (Drop Tower
University of Bremen
According to Einstein in a lecture of 1922, he was sitting on his chair in 1907 in the Bern patent office when he had an astounding idea: in free fall you cannot feel your own weight. This little thought experiment is the first decisive step on the path to the general theory of relativity. A drop tower enables an experimental verification of the idea. Objects in free fall actually do behave as if they were weightless.

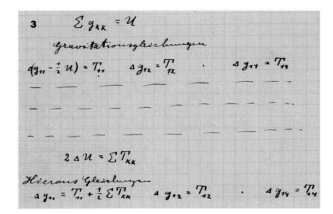

4 (Einstein to Ernst Mach

25 June 1913

Einstein sends the publication of the *Entwurf* theory to Ernst Mach. In the accompanying letter he emphasizes the role of Mach's "ingenious studies about the foundations of mechanics" and suggests the possibility of verifying the equivalence principle by observing the deflection of light rays.

3 (Einstein's Zurich Notebook

Winter 1912/13

In the winter of 1912/13, Einstein is searching for the field equations of gravitation with the help of the mathematician Marcel Grossmann. Einstein's calculations are documented in his Zurich notebook. He writes in a letter to Sommerfeld on 29 October 1912:

"I am working exclusively on the problem of gravitation at the moment and, with the help of a mathematician friend here, I believe I will master all difficulties. But one thing is certain: never before in my life have I toiled this much, and I have been instilled with great reverence for mathematics, whose more subtle parts I used to regard, in my ignorance, as pure luxury! Compared with this problem, the original theory of relativity is child's play!"

Einstein finds the right approach for the field equations but rejects it, as it does not correspond to his physical expectations. Instead, in spring 1913, he and Grossmann publish a theory that is untenable from today's perspective, the so-called *Entwurf* (outline) theory. Nevertheless, its mathematical elaboration creates the preconditions for returning to the correct formulation at the end of 1915.

The Second Revolution

The special theory of relativity of 1905 puts the development of a field theory of gravitation, analogous to the field theory of electromagnetism, on the agenda in physics. Minkowski's formulation of the theory of relativity as a geometry of four-dimensional spacetime delivers a framework in which this problem can be solved. But Einstein's special perspective on contemporary physics leads him, unlike his colleagues, to perceive the challenge as more than just a routine exercise. Indeed, against the backdrop of Mach's philosophical critique of mechanics, Einstein is willing to question yet again the recent changes to the concepts of space and time brought about by the special theory of relativity. His understanding of the problem of gravitation as a borderline problem between field theory and mechanics leads Einstein to de-

velop Minkowski's four-dimensional geometry into the concept of a curved spacetime, the properties of which are determined by the gravitational field. Distances in this curved spacetime can be calculated with the help of a "metric tensor," which also represents the gravitational potential.

A formula describing motion of matter in such a potential follows directly from the analogy of the new theory of gravitation with the geometry of curved surfaces. The biggest challenge in formulating such a theory, on the other hand, is the search for an equation describing how the gravitational potential is in turn created by the distribution of matter in the universe. With the help of his mathematician friend Grossmann, Einstein finds the mathematically correct point of departure for solving the problem in the winter of 1912/13, but soon discards it on physical reasons. Instead he works out a theory that meets his criteria, which are still grounded in classical physics. This so-called *Entwurf* (outline)

Albert Einstein at his desk, Berlin, 1927

theory integrates previously won heuristic insights, such as the deflection of light and the redshift in a gravitational field, under one theoretical roof. Together with his friend Besso, Einstein also searches on this basis albeit in vain for an explanation of the precession of Mercury's orbit, one of the astronomical puzzles of the time. In 1915, the mathematician David Hilbert takes up Einstein's results to formulate a theory of gravitation and electromagnetism. This theory, however, just as little as Einstein's *Entwurf* theory, initially does not fulfill the requirements of a generalized principle of relativity.

It is only through the elaboration of the *Entwurf* theory, in an interplay of mathematical formalism and physical interpretation, that the necessary conditions are created for the completion of Einstein's second revolution of space and time, just as the elaboration of Lorentz's theory of aether created the conditions for the first relativity revolution. In a dramatic succession of revisions to his theory, Einstein finally succeeds in formulating the general theory of relativity in November 1915. It remains the basis for our understanding of gravity to this day.

○ PERSONALIA

David Hilbert

David Hilbert (1862–1943), one of
the most significant mathematicians of the twentieth century
comes from Königsberg (today
Kaliningrad), where he finishes his
studies in 1884. In 1895, he is appointed as professor of mathematics at the University of Göttingen,
where he remains until his retirement in 1930. During his career
Hilbert changes the focal point of

◎ FOUNDATIONS

Generalization of the Relativity Principle

The unaccelerated motion of a
closed box in a gravity-free space is
indistinguishable from it being at
rest, seen by an observer within the
box. This indistinguishability is the
essence of the relativity principle
of classical physics. Can this principle be generalized to accelerated
motion as well? The accelerated
motion of a closed box in free fall

Fields

The concept of a field describes
the effect of a body on the space
around it. Other bodies exposed to
this effect experience a force. This
force depends on the field's type,
direction and strength, as well as
on the composition of the body
experiencing the force. There are
different types of fields although
relationships between them can
exist. Thus, a magnet creates a

Remark in
Einstein's note-
book: Grossmann
helps Einstein,
1912.

his research activities many times
and provides new impulses for
mathematics in many branches, including the theory of invariants, al-
gebraic number theory, geometry,
the theory of integral equations,
logic, and the foundations of mathematics. At a congress in 1900 he
presents a vision for mathematics
in the twentieth century, putting
forth twenty-three fundamental
problems waiting for solution, including the axiomatization of
physic. Hilbert attempts such an
axiomatization of the foundations
of physics in 1915, using the work
of Gustav Mie on electrodynamics
and Einstein's theory of gravitation.

in a uniform gravitational field also
remains unnoticed by an observer
within the box. The inertial forces
caused by the acceleration counter-
balance the gravitational forces in
such a way that the observer in the
box cannot decide if he is in free
fall in a gravitational field or if the
box is floating in a gravity-free
space. This indistinguishability is
the central thought in Einstein's
equivalence principle and the
starting point on his search for
a generalization of the relativity
principle to accelerated motions.

magnetic field that attracts iron.
Electrically charged bodies create
an electrical field. Changing electrical fields produce a magnetic field
as well. Heavy bodies cause a
gravitational field that effects
other heavy bodies. Temporal
changes in a field are transmitted
as waves through space. Light is a
wave phenomenon of the electromagnetic field. Gravitational waves
are created by temporal changes in
a gravitational field. A field theory
incorporates both a field equation
and an equation of motion. The
field equation describes the generation of the field while the equation
of motion describes its effect.

The Completion of the Relativity Revolution

In November 1915, Einstein's work on the general theory of relativity reaches a dramatic conclusion. He discards the *Entwurf* theory, to which he has clung for almost three years, and on 4 November at the Prussian Academy he presents his new theory, which he modifies once again just one week later. This modified theory is the first one that does not give preference to a specific coordinate system, but is still not quite correct from today's perspective. On the basis of this theory, Einstein calculates the precession of the perihelion of Mercury's orbit. It is not until 25 November that he publishes the correct field equation for gravitation. A submission for the Göttingen Academy by mathematician David Hilbert dated 20 November also contains the correct field equation. Einstein writes to his friend Zangger on 26 November: "The theory is of unequaled beauty. But just one of my colleagues has really understood it, and that one is adroitly trying to appropriate it."

1 (Einstein-Besso Manuscript
Berlin and Gorizia, 1913/14
On the search for evidence supporting his new understanding of gravitation, Einstein hits upon the problem of the motion of the perihelion of Mercury's orbit, interpreted even before as a possible clue to deviations from the Newtonian law of gravitation. With his friend Michele Besso, who now lives near Trieste, Einstein calculates Mercury's perihelion motion on the basis of the *Entwurf* theory but remains unable to explain the observations. Toward the end of 1915, however, the method developed with Besso becomes the magic wand that shows the agreement between his (in the meantime revised) theory's predictions and the astronomical data. Hilbert, impressed by this result, writes to Einstein on 19 November: "my heartfelt congratulations on conquering the problem of perihelion motion. If I could calculate as quickly as you, in my equations the electron would correspondingly have to capitulate, while the hydrogen atom would have to show his note of apology about why it does not radiate."

2 (Einstein to Arnold Sommerfeld
Berlin, 28 November 1915
In November 1915, Einstein describes to Sommerfeld the three reasons why he finally abandoned the *Entwurf* theory. He was unable to explain inertial forces in a rotating system as the effect of a gravitational field, he found the wrong value for Mercury's perihelion motion, and the attempt to derive the field equations from mathematical principles failed.

st in dem betrachteten Raume »Materie« vorhanden, so tritt deren
ietensor auf der rechten Seite von (2) bzw. (3) auf. Wir setzen

$$G_{im} = -\varkappa\left(T_{im} - \frac{1}{2}g_{im}T\right),\qquad\text{(2 a)}$$

$$\sum_{i'}g^{i'}T_{i'} = \sum_{\nu}T_{\nu}^{\nu} = T\qquad\text{(5)}$$

t ist; T ist der Skalar des Energietensors der »Materie«, die rechte
von (2 a) ein Tensor. Spezialisieren wir wieder das Koordinaten-

3

3 (Albert Einstein: The Field Equation
of Gravitation
*Sitzungsberichte der Preußischen
Akademie der Wissenschaften zu Berlin,
25 November 1915*
In this paper presented at the Prussian
Academy of Sciences on 25 November
1915, Einstein puts forward the correct
field equation of gravitation, which he
had found (but rejected) three years
earlier. Only now does he succeed in
showing that it represents a generalization
of Newton's law.

4 (David Hilbert: The Foundations of
Physics (First Announcement)
*Galley proof with handwritten
corrections by Hilbert,
stamped 6 December 1915*
Stimulated by Einstein's *Entwurf* theory,
Hilbert takes up its approach and develops
it further. He tries to link it with the
speculative theory of matter developed by
the physicist Gustav Mie. As the partially
preserved galley proofs show, Hilbert initially upholds the restriction developed in
the *Entwurf* theory, according to which only
time-space coordinates are admitted that
meet a certain condition following from
energy conservation. Other essential elements of the general theory of relativity
are also still missing.

5) Einstein to Hugo Andres Krüss
Berlin, 10 January 1918
Einstein names three possibilities for experimental testing of relativistic effects: the
perihelion motion of Mercury, the redshift
of spectral lines, and the deflection of light
in a gravitational field.
The general theory of relativity delivers
the correct value for Mercury's perihelion
motion, unexplained until now. Despite
this success, astronomers are not enga-
ging themselves in testing the other two
effects. In his letter to Councilor Krüss of
the Prussian Ministry of Education, Einstein
complains about this situation.

6 (David Hilbert: The Foundations of
Physics (First Announcement)
*Nachrichten von der Königlichen Gesell-
schaft der Wissenschaften zu Göttingen,
Mathematisch-Physikalische Klasse, 1915,
published 1916*
The published version of Hilbert's article is
dated 20 November 1915 – like his original
lecture. It has been thoroughly revised with
regard to the proofs version, however.
There are no more restrictions imposed
on the admissible coordinate systems.
On the other hand, the gravitational field
equation is explicitly written down.

4a

Es seien w_s (s = 1, 2, 3, 4) irgendwelche die Weltpunkte wesentlich eindeutig benennende Koordinaten, die sogenannten Weltparameter. Die das Geschehen in w_s charakterisierenden Größen seien:
  1) die zehn Gravitationspotentiale $g_{\mu\nu}$ ($\mu, \nu = 1, 2, 3, 4$) mit
symmetrischem Tensorcharakter gegenüber einer beliebigen Transformation der Weltparameter w_s;
  2) die vier elektrodynamischen Potentiale q_s mit Vektor-
charakter im selben Sinne.
  Das physikalische Geschehen ist nicht willkürlich, es gelten
vielmehr zunächst folgende zwei Axiome.

6 – DEZ. 1915

4b

Axiom III (Axiom von Raum und Zeit). *Die Raum-Zeit-*
koordinaten sind solche besonderen Weltparameter, für die der Energie-
satz (15) *gültig ist.*
  Nach diesem Axiom liefern in Wirklichkeit Raum und Zeit
eine solche besondere Benennung der Weltpunkte, daß der Energie-
satz gültig ist.

4c

Unter Verwendung der vorhin eingeführten Bezeichnungsweise
für die Variationsableitungen bezüglich der $g^{\mu\nu}$ erhalten die Gravi-
tationsgleichungen wegen (17) die Gestalt

$$(26)\qquad [\sqrt{g}\,K]_{\mu\nu} + \frac{\partial\sqrt{g}\,L}{\partial g^{\mu\nu}} = 0.$$

Bezeichnen wir ferner allgemein die Variationsableitungen von
$\sqrt{g}\,J$ bezüglich des elektrodynamischen Potentials q_h mit

$$[\sqrt{g}\,J]_h = \frac{\partial\sqrt{g}\,J}{\partial q_h} - \sum_k\frac{\partial}{\partial w_k}\frac{\partial\sqrt{g}\,J}{\partial q_{hk}},$$

6

Unter Verwendung der vorhin eingeführten Bezeichnungsweise
für die Variationsableitungen bezüglich der $g^{\mu\nu}$ erhalten die Gravi-
tationsgleichungen wegen (20) die Gestalt

$$(21)\qquad [\sqrt{g}\,K]_{\mu\nu} + \frac{\partial\sqrt{g}\,L}{\partial g^{\mu\nu}} = 0.$$

Das erste Glied linker Hand wird

$$[\sqrt{g}\,K]_{\mu\nu} = \sqrt{g}\,(K_{\mu\nu} - \tfrac{1}{2}K g_{\mu\nu}),$$

The Triumph of the Theory of Relativity

From the start, the general theory of relativity is confronted with a number of expectations which any new theory of gravitation should satisfy. In particular, it is plausible to expect an explanation of the slow rotation of Mercury's orbit, long connected to a possible deviation from Newton's gravitational law. Einstein's explanation for Mercury's perihelion motion in November 1915 thus represents the first, triumphant success of the general theory of relativity. Still, there were then other possible explanations for this astronomical phenomenon.

This makes further confirmation of the theory's predictions very important, something Einstein has been working on since 1911, long before the final formulation of general relativity. Especially the radical changes to the concepts of space and time already implied by his heuristics, in particular by the equivalence principle, are a strong reason to search for such confirmations. There are other possible theories of gravitation, closer to classical physics, formulated by contemporaries. Is light really deflected in a gravitational field, as is implied by the equivalence principle? Does a gravitational field actually cause redshift in light, as time passes more slowly in such a field thus changing the light's wavelength? Einstein's efforts to convince German astronomers of the important challenges these questions pose are, for the most part, fruitless. He does find early support from Erwin Freundlich, an assistant at the Potsdam Observatory, as well as from the astrophysicist Karl Schwarzschild. Einstein visits Freundlich in 1912 in Berlin and makes notes about the possibility of light deflection giving rise to the effect of gravitational lensing, but he believes the effect to be unobservable. Only twenty-four years later does Einstein publish an influential work about this at the suggestion of an amateur scientist.

Parallel to the skepticism and isolating attitude of his colleagues come soon the additional difficulties of wartime. A solar eclipse expedition, during which the deflection of light by the Sun should be detected by measuring the displacement of the apparent positions of stars, cannot take place as

"A new celebrity of world history," *Berliner Illustrirte Zeitung*, 1919

planned in 1914. Schwarzschild dies in 1916 as a consequence of the war.

Not until 1919 does an English expedition under Eddington's direction bring the breakthrough. It confirms the deflection of light, and not just qualitatively but quantitively, yielding Einstein's predicted value within the range of error – in contrast to the value that would be obtained by explaining such a deflection in the framework of the Newtonian particle theory of light. The confirmation by an English expedition of a theory of a Jewish German holding a Swiss passport so shortly after the end of World War I becomes an international media event. The Einstein myth is born.

○ PERSONALIA

Sir Arthur Eddington

Arthur Stanley Eddington (1882 – 1944), son of English Quakers, is one of the most eminent astrophysicists in the first half of the twentieth century. Eddington studies in Manchester and Cambridge, works as an assistant at the Royal Observatory in Greenwich, and in 1913 becomes professor and director of the University Observatory in Cambridge. After early works about the systematics of star movement, he delves into the inner composition of stars and the energy source that powers them. Eddington is leader of one of two 1919 expeditions for the Royal Society that are able to detect the deflection of light by the Sun's gravitational field as predicted by Einstein's theory. Thus, he paves the way for the acceptance of the general theory of relativity in England. In addition he knows how to make modern astrophysical research popular for general consumption.

○ FOUNDATIONS

Light Deflection in a Gravitational Field

If you assume that light is also subject to gravitation, its rays must bend when close to massive bodies. The occurrence of such an effect is already postulated as a consequence of Newton's theory of light particles, but after the triumph of the wave theory, this effect is no longer taken into consideration. Not until Einstein's deeper insight into the correlation of light energy and mass and his thoughts on a new theory of gravity does such light deflection again become a topic. He had already postulated the idea in 1907 with help from his equivalence principle, although the predicted value at that time is too low. In 1915, based on his general theory of relativity, Einstein predicts that a starlight ray bends by about 1.75 arcseconds when passing close to the Sun. This deflection is only visible during a total solar eclipse, when stars are visible near the Sun's edge.
The effect can also be measured by comparing two photographs of stars near the Sun. One photo shows the stars during a total solar eclipse, the second shows the same stars as seen under normal nighttime conditions.

Vossischen Zeitung. 1919 29. Mai

...aben. Für diese Privatdozenten ... nichts geschehen; man hat offen... auch gegenüber den sonstigen Be... geäußerten, Ansicht gezogen, es ...ivatdozenten zu werden, weshalb ...fel. Noch schwieriger ist die Lage ...ischen und an der Landwirtschaft... Stellung innerhalb der dortigen ...uf die hier nicht weiter ein... ...er Privatdozent das Glück, vom ...lten, dann erfährt er erst richtig, ...ist. Schreiber dieses hatte einen ... Vergütung von 150 Mark pro ... Dafür waren zu halten etwa ...r, deren jede etwa zwei Stunden ... Stunde Gelehrtenarbeit wurde

...nen Lohnstreik der Hochschullehrer ...resse der Oeffentlichkeit auf diese ...en. Seine Schlüsse mag jeder

Dr. Leopold Casper, der be... feiert am 31. Mai seinen 60. Ge... ner Berliner, ging nach erlangter Ausbildung in seinem Sonderfache ... sich 1883 in Berlin nieder, wo ...ielt er den Titel Professor. 1916 ...nannt. Casper hat sich besondere

Die Sonne bringt es an den Tag?

Eine Himmelsentscheidung in der Relativitätstheorie.

Von
Kurt Joël.

Sonnenfinsternisse sind sicherlich nichts Seltenes. Wiederholt sind in den letzten hundert Jahren wissenschaftliche Expeditionen ausgerüstet worden, um sie zu beobachten und die Ergebnisse dieser Beobachtung zu verarbeiten. Und doch sieht man der Verfinsterung unseres Zentralgestirns, die heute, am 29. Mai, eintritt und 8 Stunden 17 Minuten währt, mit besonderer Spannung entgegen. Nicht etwa wegen der langen Dauer dieser Finsternis, die mit der schmalen Zone ihrer Totalität das nördliche Brasilien und Mittelafrika durchschreitet und zu deren Erforschung von England aus zwei Unternehmungen — die eine mit dem Standort in Sobral (Brasilien), die andere nach der Insel Isla do Principe, etwa 180 Kilometer von der afrikanischen Küste — ausgerüstet worden sind. Nicht bloß die Astronomen, auch Physiker, Mathematiker, selbst Philosophen harren auf die endgültigen Ergebnisse dieser Himmelsbeobachtung, da sie mittelbar helfen sollen, eine der wichtigsten neueren physikalischen, ja erkenntnistheoretischen Fragen, die Einsteinsche Gravitationstheorie, zu beantworten.

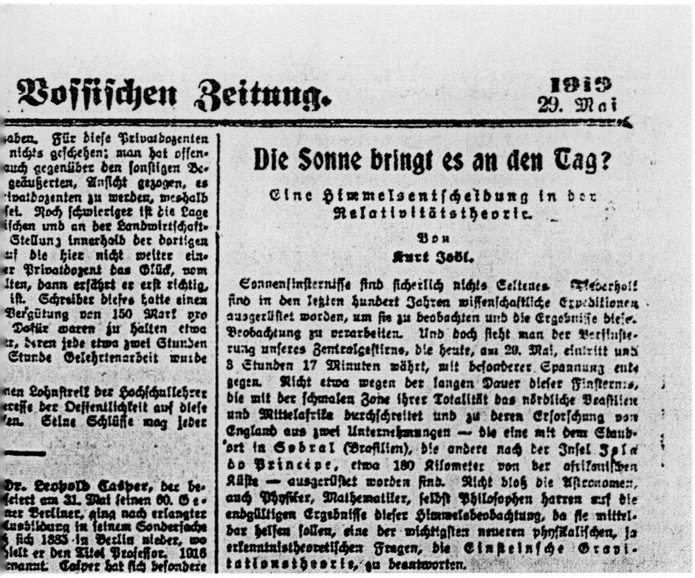

The triumph of the theory of relativity in the press, *Vossische Zeitung*, 1919

Zürich. 14. X. 13.

Hoch geehrter Herr Kollege!

Eine einfache theoretische Überlegung macht die Annahme plausibel, dass Lichtstrahlen in einem Gravitationsfelde eine Deviation erfahren.

Grav. Feld
Lichtstrahl

Am Sonnenrande müsste diese Ablenkung 0,84" betragen und wie 1/R abnehmen (R = ~~Sonnenradius~~ Entfernung vom Sonnen-Mittelpunkt).

0,84"
Stern
Sonne

Es wäre deshalb von grösstem Interesse, bis zu wie grosser Sonnennähe helle ~~grosse~~ Fixsterne bei Anwendung der stärksten Vergrösserungen bei Tage (ohne Sonnenfinsternis) gesehen werden können.

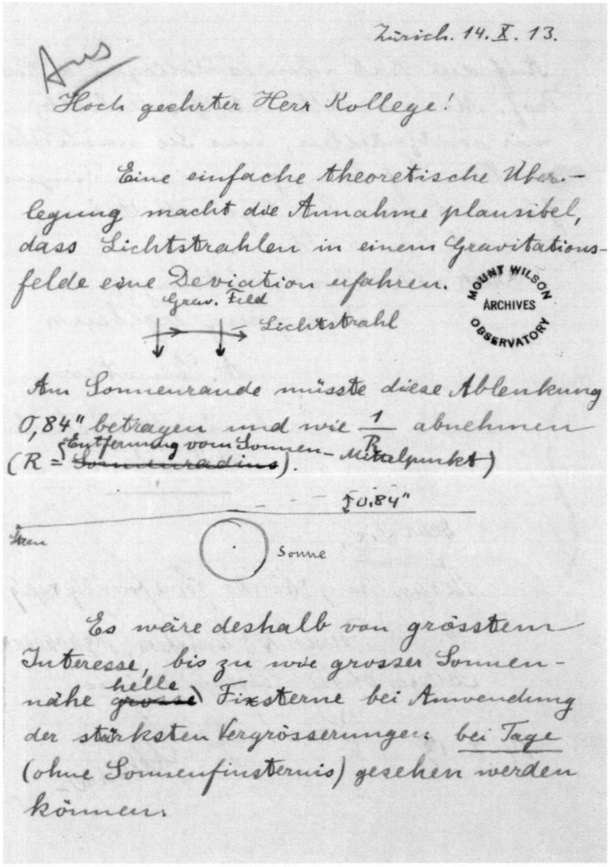

Albert Einstein asks the astronomer George E. Hale for help, 1913

The Struggle for Confirmation

Einstein predicts two previously unknown effects that can help test the general theory of relativity. The first effect is the deflection of light in a gravitational field. The second is a change in the color of light in a gravitational field, the so-called gravitational redshift. Both of these phenomena, however, are so small that at first only a few astronomers even dare attempt verification.

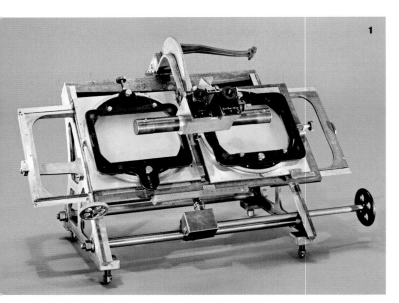

1 (Stereo Comparator by Carl Pulfrich
Prototype by Zeiss, 1899
The stereo comparator allows the comparison of two photographic plates. By looking at two photos taken at different times, the actual or apparent movement of astronomical objects can be ascertained. According to Einstein, the positions of stars during a solar eclipse should appear to shift since light rays are bent in the proximity of the Sun.

2 (Telegram from Hendrik A. Lorentz to Albert Einstein
The Hague, 22 September 1919
Writing from the neutral Netherlands Lorentz informs Einstein about the preliminary results of the English eclipse expedition. Although it is a year after the end of World War I, direct communication between England and Germany is still practically impossible.

Eddington has reported on 12 September that the bending of light passing close to the Sun has been proven, but the value is not yet very accurate. It can also be explained by Newton's theory of gravity, which predicts a value of 0.87 arcseconds (when one assumes that light has mass). It is not until 6 November that the final value is announced – it is in good agreement with Einstein's theory, which predicts twice as much, a deflection of 1.75 arcseconds.

4 (Donor List for the Einstein Donations

4

Auszug aus der Spender-Liste zur Einstein-Spende.

Badische Anilin- und Sodafabrik, Ludwigshafen a/Rh. . . . . M.	24 000.—
Farbwerke vorm. Meister, Lucius & Brüning, Höchst a/Main „	24 000.—
Hauptkasse d. Farbenfabriken vorm. Friedr. Bayer & Co., Leverkusen „	24 000.—
*) Carl Zeiss-Stiftung Jena . . . . „	20 000.—
Robert Bosch, Stuttgart . . . . „	20 000.—
F. Herrmann & Co., Luckenwalde . . . . „	20 000.—
N. N. . . . . „	20 000.—
Mendelssohn & Co., Berlin . . . . „	20 000.—
Otto Wolff, Berlin, Paulstr. 20 . . . . „	20 000.—
Berliner Handelsgesellschaft . . . . „	15 084.40
Wolf Netter & Jacobi, Berlin . . . . „	10 000.—
Großloge für Deutschland, Berlin W 62, Kleiststr. 12 . . . . „	10 000.—
Firma Leopold Cassella & Co. G. m. b. H., Frankfurt a/M. . . . „	10 000.—
Außenhandelsstelle für Eisen- u. Stahlerzeugnisse . . . . „	10 000.—
N. N. . . . . „	10 000.—
A. G. für Anilinfabrikation, Berlin SO . . . . „	9 000.—
Chemische Fabrik Griesheim-Elektron, Frankfurt a/M. . . . „	6 000.—
Vereinigte Spitzen- und Klöppel-Iudustrie . . . . . . „	6 000.—
**) Firma Siemens & Halske A. G., Siemensstadt . . . . . „	5 000.—
Verein deutscher Maschinenbau-Anstalten . . . . . . „	5 000.—
**) Sekretariat der Allgemeinen Elektrizitätsgesellschaft, Berlin „	5 000.—
Gebrüder Junghans A. G., Schramberg . . . . „	5 000.—
Bankdirektor Dr. Dannenbaum, hier . . . . „	5 000.—
Max M. Warburg, Hamburg . . . . „	5 000.—
Adolf Pitsch Berlin C 19, Hausvogteiplatz 6/7 . . . . „	5 000.—

*) Die Firma C. Zeiß hat Optik im Werte von Mk. 300.000 gestiftet und baut die geplante Anlage, einen Turmspektrograph, zu Gestehungskosten. Unter den gleichen Bedingungen liefert die Firma Schott & Genossen das ganze Glas.
**) Es ist von Seiten der großen Elektrizitätswerke ein Zusatzbeitrag zur Spende in Form der Lieferung der elektrischen Anlage zu besonderen Bedingungen in Aussicht genommen, die als Leistung einem Vielfachen des obigen Betrages entspricht.

3 (Model of the Einstein Tower
Erich Mendelsohn, 1920
Already during World War I the Berlin astronomer Freundlich plans the construction of a new tower telescope to measure gravitational redshift. He asks architect Erich Mendelsohn to design a building appropriate for the project. In the following years Mendelsohn produces hundreds of design sketches and three plaster models. This third and only surviving plaster model is finished sometime in the summer of 1920, the same time as construction starts on top of the *Telegraphenberg* in Potsdam. It shows the Einstein Tower as it is actually realized.

4 (Donor List for the Einstein Donations
Berlin, 1920
Since state funds for scientific projects in Germany after the end of World War I are scarce, in December 1919 Freundlich, with Einstein's approval, composes an "Appeal for Einstein Donations". The monies are to be used to build a tower telescope. After Freundlich manages to collect the majority of the necessary funds from countless private citizens, banks, and companies, the Prussian government contributes as well. Construction starts in the summer of 1920.

The Anti-Relativists

Experimental physicists such as Gehrcke and the Nobel Prize winner in physics Lenard view the growing importance of theorists and the impending revolution in physics at the beginning of the twentieth century as a challenge. They try to confront this trend with a form of physics more strongly oriented towards the direct observation of nature. Segments of the public also accuse the theory of relativity of over-mathematizing physics: according to them, what is not simple and clear cannot be true. Such resentments frequently go hand-in-hand with anti-Semitic and nationalist tendencies.

The theory of relativity enters the public discussion in a major way after the solar eclipse expedition of 1919 confirms the deflection of light in the gravitational field predicted by Einstein. Opponents of the theory of relativity, unable to prevail within the field of physics, now try a new argument: they allege that their longstanding scientific critique has been ignored because the relativists pursue an exagerated publicity campagne and mass suggestion. To prove this, Gehrcke collects more than 5,000 newspaper articles about Einstein or the theory of relativity.

The anti-relativists claim that the triumph of the theory of relativity in the public arena is well-suited to the intellectual climate of the Weimar Republic, with its parallel revolutions in art and politics. To them,

The nationalists take over the streets: the Kapp putsch in Berlin, 13–17 March 1920

the "conspiracy" of the relativity theorists seems to definitely need a response from genuine natural scientists. Einstein's opponents see themselves as victims, not aggressors. On 24 August 1920, a public lecture series criticizing the theory of relativity is launched at the *Philharmonie* in Berlin. It is organized by the anti-Semitic agitator Weyland. Einstein attends the event and hears Weyland refer to the theory of relativity as "mass suggestion" and "scientific Dadaism." Gehrcke, too, expresses his criticisms of the theory of relativity.

Einstein's colleagues Laue, Nernst, and Rubens then publish an advertisement speaking out against this way of conducting a scientific controversy. The Prussian Minister of Education demands in vain that the Academy of Sciences publicly stands up for its member Einstein. Einstein publishes his own response in the *Berliner Tageblatt* newspaper under the title *My Reply. On the Anti-Relativity Theory Company*. Shortly thereafter Lenard and Einstein have a dispute at a meeting of the Society of German Natural Scientists and Physicians in Bad Nauheim, illustrating once again the irreconcilability of their positions. In the spring of 1922 Einstein incurs the wrath of nationalist activists when he travels to France to revive scientific relations interrupted by the war. After the murder of Rathenau on 24 June by nationalist extremists, the situation becomes increasingly precarious for the committed democrat Einstein. He decides to withdraw from public life for a time, and even considers giving up scientific life in Berlin for good. The speaker at the centenary celebrations of the Society of German Natural Scientists and Physicians that same year is Laue, instead of the originally invited Einstein. Planck comments that this change of plans perhaps has the advantage "that those who still believe that the principle of relativity is basically a Jewish advertisement for Einstein will be proved wrong."

◗ PERSONALIA

Walther Rathenau

After studying physics, chemistry, philosophy, and mechanical engineering, the versatile industrialist, politician, and writer Walther Rathenau (1867–1922) becomes a leading figure at the electrotechnical company *AEG*. During World War he becomes an important organizer of the wartime economy. He supports moderate war aims, however, and hopes for a modernization of the German state and society. As a member of the German Democratic Party, he dedicates himself to the new democracy after the war, and as minister of reconstruction designs the policies for fulfilling the terms of the Treaty of Versailles. He becomes foreign minister in 1922. Rathenau has enemies in many political camps. Early on, as a German-Jewish patriot he calls upon Germany's Jewish population to assimilate, a position not shared by his friend Einstein. Rathenau ultimately falls victim to right-wing radical and anti-Semitic forces who see him as a traitor and murder him in June 1922.

Ernst Gehrcke

The experimental physicist Ernst Gehrcke (1878–1960) spends his professional life as a specialist in optics at the National Institute for Physics and Technical Standards in Berlin. His scholarly interests are broad, however, and range from physics, medicine, philosophy, and geology to acoustics and patent law. Gehrcke starts publishing against the theory of relativity in 1911. He believes that relativizing simultaneity would lead to a rela-

Ernst Gehrcke, around 1930

tivization of our entire world view, and that the experimental confirmations of the theory of relativity can be explained just as well with classical physics.

In the early 1920s Gehrcke is the key figure in a network of Einstein opponents who wish to proceed jointly against modern physics. Their public campaign begins with the lecture series against the theory of relativity in August 1920 in the Berlin *Philharmonie*.

Gehrcke never accepts modern physics, and writes satirical poems about Einstein even in old age.

Paul Weyland

In 1920 the press dubs Paul Weyland (1888–1972) the "Berlin Einstein Killer." He is completely unknown among scientists until August 1920, when he lectures against the theory of relativity in the Berlin *Philharmonie*. Now he

takes up arms against the theory of relativity and founds for this purpose his own organization, the Working Group of German Natural Scientists for the Preservation of Pure Science. Weyland is an anti-Semitic agitator who muddles along with temporary jobs and titles himself, among other things, an engineer, a chemist, and a writer. Max von Laue describes this shady character as a typical "hustler" of the postwar period. After an

adventurous life that takes him as far afield as Africa and is often financed by swindles, Weyland crosses Einstein's paths again in the 1950s in the United States: at the height of the McCarthy era Weyland denounces Einstein to the FBI as a Communist.

Nationalist attacks on Einstein in the headlines, early 1920s

Violent Words and Actions

1 (Gehrcke's Lecture in the Philharmonie
Lecture brochure with notes, 1920
On 24 August 1920, the opening event in a public lecture series against Einstein's theory of relativity takes place at the Berlin

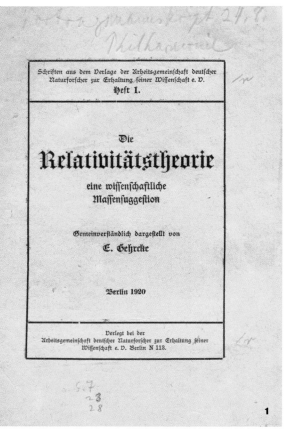

Philharmonie. The series is organized by the Working Group of German Natural Scientists for the Preservation of Pure Science, a short-lived association of dedicated opponents to Einstein. The first speakers are the engineer Weyland and the physicist Gehrcke. Ernst Gehrcke asserts in his lecture that the theory of relativity is nothing more than a "scientific mass suggestion." In his view, the success of theoretical physics is the product of public promotion, and he uses this argument to justify making public his own objections, based on classical physics, to the theory of relativity. Max von Laue, who like Einstein is in the audience, notes that these objections amount to "an old story rehashed." Gehrcke reads from a pamphlet that he has expanded with handwritten notes into a lecture. Like the typescript of Weyland's lecture the brochure has survived in Gehrcke's papers.

2 (Is Einstein a Plagiarist?
The Dearborn Independent, 30 April 1921
Two subjects are so important to the editors of *The Dearborn Independent* in 1921 that they are made into front-page items: *Is Einstein a Plagiarist?* and *Jew Admits Bolshevism!* Both headlines reflect the program of the weekly owned by the automobile magnate Henry Ford, a confirmed anti-Semite. By the early 1920s Ford is already using his weekly to rail against the "international Jewish conspiracy." In 1920 he publishes the notorious anti-Semitic pamphlet *The International Jew: The World's Foremost Problem.* The German edition, *Der internationale Jude*, is soon disseminated among anti-Semites in Germany. *The Dearborn Independent* has already published several articles opposing the theory of relativity. The accusation of plagiarism is often raised against Einstein in anti-Semitic circles. They claim that not the Jew Einstein, but a German actually developed the theory of relativity. As evidence, these critics cite alleged forerunners, who actually only dealt within the framework of classical physics with singular effects explained by the theory of relativity, such as the motion of the perihelion of Mercury (Gerber) or gravitational light deflection (Soldner).

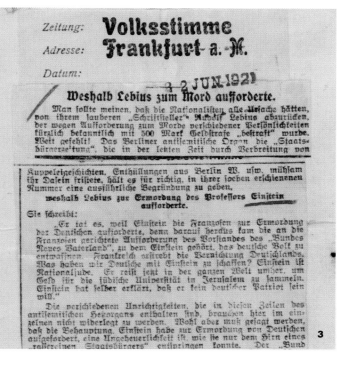

3

4

(Death Threats against Einstein in the Headlines

Volksstimme, Frankfurt am Main,
12 June 1921

In 1921 Rudolf Lebius (1868–1946) openly calls for Einstein's murder in the anti-emitic *Staatsbürgerzeitung*. The democratic press is outraged by the mild fine imposed on the journalist. The *Volksstimme* in Frankfurt comments: "It is shameful, but it must be said: in no other country in the world would such a degrading campaign of hatred against a great mind of Einstein's stature be possible such as has occurred in Germany." Einstein's life is threatened

again after the murder of Rathenau. In 1922 he tells Planck and Laue that several people have warned him that he is on the hit list of Rathenau's murderers. Einstein thereupon withdraws from public life, cancels his university courses and public lectures, and even considers leaving Germany. Democratic newspapers report with alarm on this "German cultural disgrace," while nationalist papers accuse Einstein of staging the threats himself as a form of self-promotion. Einstein decides to avoid this politically charged atmosphere by making a long journey abroad, to Japan.

4 (MP 18/1 Submachine Gun
T. Bergmann Arms Factory, 1918
On 24 June 1922 the foreign minister Walther Rathenau is gunned down in the street. The murder weapon is an MP 18/1 submachine gun, the first German weapon of its type, which already wreaked havoc among enemy ranks during World War I. With its range of 100–200 meters and an ability to fire up to 450 rounds a minute it is far easier to deploy in trench warfare than the unwieldy machine gun. Rathenau's murderers are former army men, members of the *Organisation Consul*, one of the paramilitary units that severely threatens the unstable democracy of the early Weimar years. Right-wing radicals murder more than 350 persons between 1919 and 1922, and many prominent democrats such as Einstein receive death threats.

Albert Einstein as a Public Figure

Einstein deliberately lends his famous name to political causes, but also to the public dissemination of scientific knowledge. Both interests are closely associated with his personal experience, from his early education to his politicization at the beginning of World War I.

He succeeds like no other scientist in conveying the results of his work to a broad public. He publishes not just popular works and newspaper articles on his theory, but also holds generally comprehensible lectures in venues accessible to many people, such as adult education institutions and planetariums. In 1925, for example, he lectures on *What a Worker Needs to Know About the Theory of Relativity*. For him, science is part of culture, and Einstein is a part of cultural life.

In Berlin Einstein mixes with prominent people from the fields of science, culture, politics, and business. The city of Berlin is planning to present him with a house for his fiftieth birthday, but, as it turns out, it is not the city's to give. Eventually Einstein himself purchases a plot of land not far from a lake near Berlin

Einstein visits the ruins of a French village destroyed during World War I, April 1922

and commissions the architect Wachsmann to build him a small wooden summer house. This gives him the opportunity to escape from hectic city life, concentrate on his work, and also take little excursions on his sailboat. At the same time, it becomes a legendary meeting place for intellectuals who meet with Einstein to discuss their concerns about the challenges of the times. He not only cultivates contacts with other scientists, but also becomes involved in politics. The Ministry of Foreign Affairs considers him a "first-rate cultural factor," and hopes that he will help to reestablish international ties. After World War I Einstein travels to Prague, Vienna, Leiden, Oslo, and Copenhagen. In 1921 he accompanies Weizmann on a fundraising tour through the United States. On the way home he visits England, where he breaks through the initially frosty atmosphere with a lecture on the international nature of science. At Rathenau's request Einstein accepts an invitation to Paris. In 1922 he is the first German scientist to be received in France after the war. Following Rathenau's murder and the death threats to himself, the journey to Japan in the autumn of 1922 provides a welcome escape from Germany. On his return journey at the beginning of 1923 Einstein stops in Palestine and Spain. His next major journey abroad in 1925 takes him to South America. On this trip he again inspires scientific life, supports the founding of the Hebrew University in Jerusalem by collecting donations, encourages Jewish communities, and strengthens the reputation of the fledgling German democracy. From within Germany Einstein also contributes to international understanding and promotes human rights, pacifism, and Zionism, as well as relations between Germany and the Soviet Union. His championing the right to conscientious objection plays a special role here. He considers it to be in accord with the highest ethical traditions of Judaism. Later on, in 1949, he also demands the state of Israel to uphold this right. The causes and prevention of war become the subject of a public exchange between Einstein and Freud in the summer of 1932. Despite his numerous political commitments, Einstein still accords great importance to the fates of individuals, as is evident in his tireless efforts on behalf of those who are persecuted, condemned, or forced to emigrate.

○ HISTORICAL CONTEXT

Mass Media

The term mass media is coined in the 1920s, when the press and radio begin reaching broad sectors of the population. The number of newspapers rises sharply to more than 4,000 daily and weekly papers with a total circulation of 25 million towards the end of the Weimar Republic. The lively press landscape reflects the political struggles in the country's public sphere, with

Einstein gives a speech at the Radio Exhibition, Berlin, 1930

many newspapers closely aligned to political parties or financially dependent. A few large publishing companies dominate the opinion market. Liberal papers such as the *Vossische Zeitung* or the *Berliner Tageblatt* do not reach a mass audience. Without the modern mass press Einstein would never have become so famous after the verification of the theory of relativity in 1919. Thousands of articles and other items appear within a brief period, from popular introductions to Einstein's theory to satires and even a *Waltz According to the Principles of the Theory of Relativity*. In 1930 Einstein opens the Radio and Phonograph Exhibition in Berlin. He expresses his conviction that radio will act as an instrument of social progress. This hope is later bitterly dashed during the period of National Socialist rule.

Avant-garde

The highly charged and tense atmosphere of the interwar period sees the emergence of new artistic movements with aspirations to purify and liberate society by radically overcoming conventions. They are perceived at the same time as signs of the disappearance of cultural and aesthetic familiarity. Expressionism, New Objectivity, and atonal music influence modes of expression which the insecure middle classes perceive purely as phenomena of a cultural crisis. Architects design smooth cubes and coolly elegant curtains of glass, while aspiring to a new ideal of socially responsible construction and a radical recasting of urbanity. Two monuments of avant-garde design are associated with Albert Einstein, who is actually rather conservative in such matters: his simple summer house in Caputh, designed by the *Bauhaus* architect Konrad Wachsmann, and Erich Mendelsohn's expressionistic Einstein Tower in Potsdam.

Einstein as an Ambassador for Germany

The behavior of German scholars and scientists during World War I leads to an international boycott which continues into the second half of the 1920s. The greatest bitterness exists in France, and scholars from the neutral countries try to mediate. Einstein's fame and the sensation surrounding his theory of relativity provide the German Ministry of Foreign Affairs with a welcome opportunity to accord him the role of a rather reluctant cultural ambassador. His internationalist stance during the war is also of importance here. In the eyes of the diplomats, he represents the best and most productive aspects of German intellectual activity. Einstein's visit to Paris in 1922 is a success, but most of his German colleagues now distrust him politically all the more. What they fail to recognize is that Einstein also criticizes the politics of the *Entente* powers, and expects fairness from all sides in the conflict.

The Radicalization of Politics

The early years of the Weimar Republic are overshadowed by political violence and economic crisis. The militancy of the conflicts culminates in the murders of prominent politicians such as Karl Liebknecht, Rosa Luxemburg, Matthias Erzberger, and Walther Rathenau. The extreme right makes a number of attempts to seize power by coup d'état.

Einstein with guests at a reception in his honor, Argentina, 1925

In France, in the meantime, people have fears about the material basis from which Germany is supposed to meet the enormous reparations demands. For this reason, the Ruhr is occupied in 1923. The political situation in Germany escalates. The population is called upon to engage in civil resistance, and the *Reichsmark* plummets in value. The new democracy only enters calmer waters in 1924. The world economic crisis of 1929 brings an abrupt end to five short years of fragile stability. Brutal conflicts between various political groupings accompany the Weimar Republic until its demise.

Albert Einstein as a Public Figure

International Understanding

1 (Einstein Honored in London.
The Scientific International
National-Zeitung, 15 June 1921
On his return journey from America Einstein
visits England. He lectures in Manchester
and at King's College London. The German
professor again receives a frosty reception

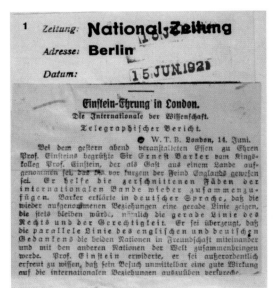

2 (Einstein's Sailboat
Model by M. Keyser, 2005
For his fiftieth birthday on 14 March 1929,
friends give Einstein the "Tümmler" (Dolphin), a 7-meter-long and 2.35-meter-wide
yawl designed by the ship construction
engineer Adolf Harm. The "Tümmler" is
moored on the Templiner Lake, only a few
minutes from Einstein's summer house in
Caputh, near Potsdam. Einstein enjoys
inviting friends to Caputh, and often takes
them out on the water: sailing with Einstein
is both a private pleasure and a social
event about which the daily papers publish
photo reports. The "Tümmler" is confiscated and sold off by the *Gestapo* in 1933.
In the late 1930s it disappears from the
records. Einstein tries in vain to locate
his "fat sailboat" after the war.

there, but Einstein succeeds in convincing
his audience of the international nature of
science, not least because of his emphasis
on how much he owes to Newton. The official receptions for the famous German are
important political signals in the postwar
period. Einstein stays at the home of the
former minister of war, Lord Haldane,
where he meets leading figures including
the archbishop of Canterbury, George
Bernard Shaw, and Arthur Eddington.
Even the nationalist German newspapers
acknowledge the reestablishment of
scientific and political relations promoted
by Einstein's activities.

3 (Einstein's Declaration
on International Understanding
1929
In April 1929 Einstein writes a brief dedication on international understanding for
No More War, a British pacifist newspaper:
"The peoples themselves must muster the
initiative to avoid being led to the slaugh-
terhouse again. To expect such protection
from their governments is folly."

4 (Einstein Honored
Arbeiter-Illustrirte-Zeitung, no. 12, 1929
The paper honors Einstein's scientific work
and political commitment on the occasion
of his fiftieth birthday. Einstein is "a sincere friend of the suffering, oppressed, and

efforts on behalf of the organization *Rote
Hungerhilfe* (Red Hunger Aid): "In 1919,
as we know, the *Entente* tried to bring
down the Soviet Union with a hunger
blockade. And we shall never forget that
the at times energetic protests against

5 (Einstein Petitions for the Pardon
of Carl von Ossietzky
Pasadena, 8 February 1932
Carl von Ossietzky (1889–1938) is editor-
in-chief of the *Weltbühne* and one of the
most influential critical reporters on the

Albert Einstein auf einer Reise in Südamerika, wo man den großen Gelehrten durch vielerlei Auszeichnungen ehrte

In Japan ehrte man die Verdienste des Wissen-schaftlers Einstein durch die Aufstellung dieser Büste in Kobe

Einstein in der Akademie der Wissenschaften in Buenos-Aires

Professor Einstein hält in der Pariser Sorbonne einen Vortrag über seine allgemeine Relativitätstheorie, der die Newtonschen Forschungsergebnisse über das Gravitationsgesetz angreift

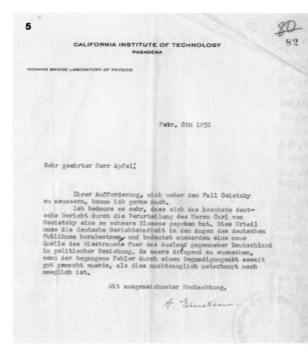

persecuted," who has always promoted
peace among peoples. Besides Einstein's
rejection of German militarism during
World War I, the article emphasizes his

this blockade by German, French, and
English pacifist organizations can be
attributed in large part to the inspiration
of Albert Einstein."

alarming erosion of the Weimar Constitution. On 12 March 1929, the article *Windy
Goings-On in German Aviation*, a report on
secret rearmament activities, appears in
the *Weltbühne*. As the responsible editor
and publisher, Ossietzky is charged with
treason together with the author, Walter
Kreiser (1898–1954). The trial causes a
sensation. Many prominent individuals,
including Thomas Mann and Albert Einstein, speak out on Ossietzky's behalf at
the request of his lawyer, Alfred Apfel.
Ossietzky is sentenced to eighteen months
in prison. He is released as part of an
amnesty in December 1932. In 1992 an
attempt to appeal against the verdict is
rejected by the Federal German Supreme
Court of Justice.

Wave and Particle as a Persistent Contradiction

Although the idea of light as particle allows Einstein to explain puzzling experimental results like the photoelectric effect, his light quantum hypothesis encounters widespread criticism. His arguments about the shortcomings of the classical wave image are based especially on the analysis of the statistical properties of radiation. Direct experimental evidence for light's particle properties is at first unavailable. Therefore most of Einstein's colleagues find it a bit quick to question the established idea of light as wave. Planck identifies the significance of the quantum of action he introduced in 1900 in a new understanding of the interaction between light and matter. He is convinced that the classical wave theory of light will remain valid for empty space.

Wave or particle?

Einstein himself knows that the properties of radiation cannot be fully understood as a particle phenomenon, and he searches long in vain for a theory of light that can be understood as a fusion of wave and particle theories. In 1916, he develops a model of the interactions between atoms and radiation, providing an additional clue about the particle properties of light quanta. Einstein shows that they must possess a direction and an impulse, similar to particles of matter. He further introduces the idea of a stimulated emission of light quanta, later prompting the invention of lasers.

Although Einstein receives the 1921 Nobel Prize for Physics in 1922 for his law of the photoelectric effect, the light quantum hypothesis that is the basis of the law still meets with widespread rejection. Only Compton's experiment brings the breakthrough in 1923. The experiment shows that x-rays collide with electrons in conformity with the classical model of particle collision and in blatant contradiction to the idea of light as a wave. His results are the first that convince a majority of physicists of the justification of the light quantum hypothesis.

Also in another way, the year 1923 is a turning point for the understanding of the role the intuitive ideas of waves and particles will have in the emerging field of modern physics. In his doctoral dissertation, de Broglie shows that it makes sense to also assign wave properties to particles. With the help of his idea of matter waves, he can explain significant traits of Bohr's atomic model. Einstein is elated by the work.

But to what extent, if at all, do the familiar ideas of our sensory sphere translate to the microworld of atoms? In 1924, with a new derivation of Planck's radiation law, Bose clearly shows that light quanta are not classical particles, independent and distinguishable from each other. Einstein transfers this insight – thermal radiation as a gas of light quanta – to a material gas. This provides a basis for understanding thermal behavior of bodies at temperatures near absolute zero. Beyond this, the notion of indistinguishable particles leads to the prediction of a previously unknown material state, the so-called "Bose-Einstein condensate." Decades then pass before the first such condensate is produced.

○ PERSONALIA

Louis de Broglie

The physicist Louis-Victor de Broglie (1892–1987) is the descendent of a famous French noble family. He initially plans to follow the family tradition of a diplomatic career and begins studying history. Influenced by his older brother Maurice, who has become a physicist, he also turns to this field. His research concerning the photoelectric effect of x-ray radiation performed in his brother's private laboratory leads him to the fundamental recognition that matter particles have wave properties (wave-particle duality).

With this thesis he gains his doctorate at the Sorbonne in Paris in 1924 and the Nobel Prize for Physics in 1929.

De Broglie is appointed professor of physics at the Henri Poincaré Institute in Paris in 1928 and teaches from 1932 to 1962 at the Sorbonne. He continues to conduct research into problems of quantum mechanics, is advisor to the French atomic energy commission after World War II, and writes popular books about modern physics.

○ FOUNDATIONS

The Compton Effect

Many physicists reject Einstein's idea of light quanta for quite some time. Only the discovery of the Compton effect in 1923 delivers an impressive confirmation of the light particle hypothesis. The American physicist Arthur H. Compton scatters light off electrons and measures the subsequent direction and energy of its scattered quanta. He proves that the classical laws of mechanical collision are also valid for light. The light quanta act like spheres shot at other spheres. Einstein is so pleased by Compton's experiments that he writes an article about them for the newspaper *Berliner Tageblatt*. In 1926, the name "photon" is adopted for light quanta. Compton receives the Nobel Prize for Physics in 1927 for the effect he discovered.

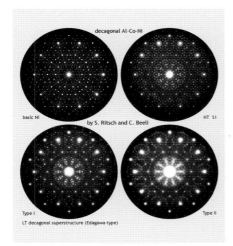

Arthur H. Compton, around 1940

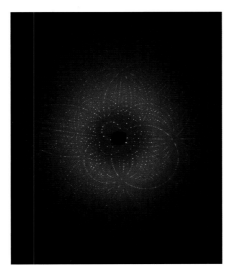

Above left:
Images of electron scattering patterns

Below left:
Image of x-ray scattering

The Delayed Success of the Light Quantum Hypotheses

1 (Proposal for Einstein's Membership in the Prussian Academy of Sciences
Berlin, 12 June 1913
The proposal is submitted by Planck, Nernst, Rubens, and Warburg. Even in this enthusiastic recommendation of Einstein, it is striking to see how little recognition his light quantum hypothesis has yet gained. The proposal's authors feel compelled to apologize that in this case Einstein "may have overshot the mark […] in his speculations." In contrast, at about the same time Einstein writes about Planck's attempt to deduce a formula for radiation without assuming the existence of light quanta: "Planck's new theory contains so many hypotheses it appears almost worthless to me."

2 (Wave or Particle – an Experiment
Einstein to Arnold Sommerfeld, Berlin, 9 October 1921
Einstein deals not only with theoretical questions concerning the wave or the quantum nature of light. For many years he also searches for experiments that could help clarify the matter. One of these attempts is documented in this letter. Einstein hopes that light emitted by fast moving atoms has unusual refractive properties that can help determine its nature as wave or particle. At Einstein's suggestion, Hans Geiger and Walther Bothe conduct the experiment at the *Physikalisch-Technische Reichsanstalt* (National Institute for Physics and Technical Standards) in Berlin. In his letter to Sommerfeld Einstein indicates his uncertainty about which result to expect. Several months later Ehrenfest shows that Einstein has made an error – the wave theory does not imply the effect Einstein predicted. Thus the experiment has no decisive value and Einstein has to retract his previous announcement.

3 (A New Statistics
Einstein to Erwin Schrödinger, Berlin, 28 February 1925
In his reply to a letter from Schrödinger, Einstein remarks that, contrary to Schrödinger's opinion, he has in no way made an error in the derivation of his quantum statistics for gases. Rather, he has introduced a very surprising hypothesis without explicitly stating it: statistics in quantum theory is categorically different from that in everyday life. Einstein uses the diagram shown here to explain the difference. The question is: How many equally probable possibilities are there to distribute two particles in two cells? According to classical statistics there are four (diagram right), but according to Bose-Einstein statistics there are only three (diagram left). For in the case that one particle is in each cell, classical statistics has two possibilities: particle I is in the first cell and particle II is in the second, or vice-versa.
In contrast, Bose-Einstein statistics does not recognize this difference. There is only one possibility. Today this property with far-reaching consequences is known as the indistinguishability of elementary particles.

was aber bei mir im vorliegenden Falle nicht zutrifft. Ich würde also diese Absage mit einem Gefühl ungeteilten Bedauerns absenden, wenn sie nicht persönlich an Sie gerichtet werden müsste, aber auch so bleibt mir nichts anderes übrig. –
Ich bin sehr neugierig auf den Ausgang unseres Kanalstrahlen-Experiments. Meine Erwartung ist total unsicher. Dass eine Art Kugelwelle abgeht, das bin ich überzeugt wegen der Interferenz-Fähigkeit für grosse Öffnungswinkel. Aber da die Emissionszeit klein ist gegen die Kohärenzzeit für Interferenz mit Gangunterschieden, bin ich nicht überzeugt, dass dasjenige, was ausgesendet wird, unmittelbar oszillatorischen Charakter hat. Der Mechanismus könnte indirekter und ein total anderer sein, als wir vorauszusetzen gewohnt
1977-28/A,78(20

2

sind, zumal eben der Widerstreit zwischen Quanten- und Undulationstheorie mit unverminderter Heftigkeit fortbesteht. Jedenfalls kann auf Grund des bisher Bekannten der Ausgang des Experimentes nicht vorausgesagt werden. Auch die Existenz des Dopplereffektes, wie ihn Stark nachgewiesen hat, erlaubt keine Voraussicht, was ich aber in diesem kurzen Brief nicht begründen kann.
Es grüsst Sie herzlich
Ihr
A. Einstein.

2

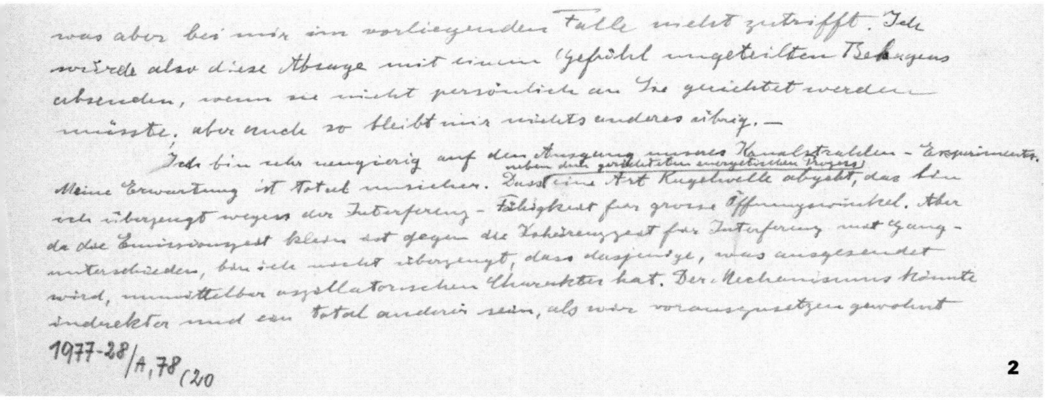

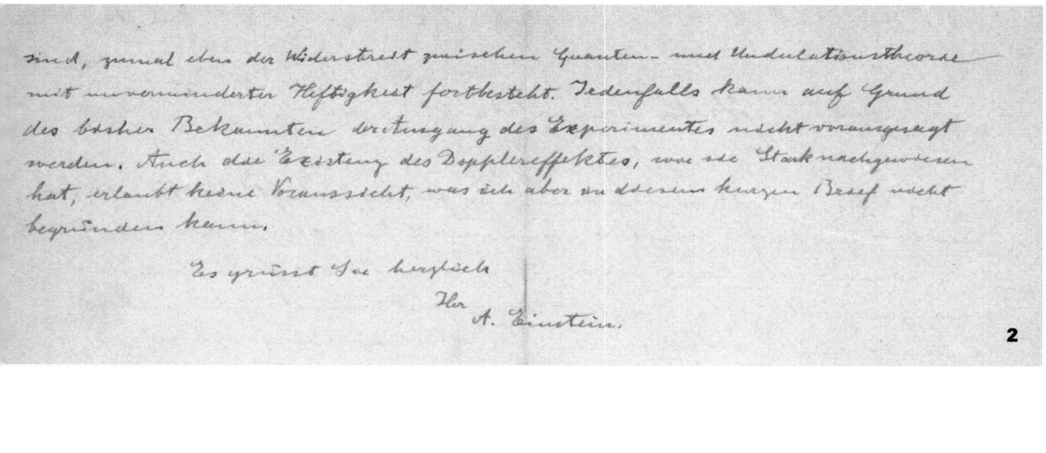

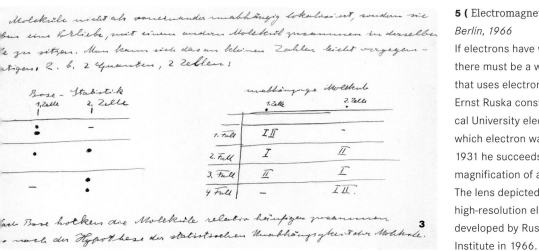

3

5 (Electromagnetic Lens (Cut)
Berlin, 1966
If electrons have wave properties then there must be a way to build a microscope that uses electrons. The young engineer Ernst Ruska constructs at Berlin's Technical University electromagnetic lenses with which electron waves can be bundled. In 1931 he succeeds in displaying a 16-fold magnification of a wire lattice.
The lens depicted here comes from a high-resolution electron microscope developed by Ruska at the Fritz Haber Institute in 1966.

5

4

4 (Einstein's Nobel Prize Diploma
Stockholm, 1922
On 9 November 1922 Einstein is on his way to a lecture tour through Japan when it is announced in Stockholm that he is to receive the Nobel Prize for Physics for 1921. The award in itself is not so surprising, but some people find the reason baffling. Einstein is honored not for his theory of relativity, but "for the discovery of the law of the photoelectric effect."

Towards a New Cosmology

Einstein's general theory of relativity is, today, the foundation of cosmology. It explains the expansion of the universe and allows the description of strange phenomena like black holes, gravitational lensing, and gravitational waves, none of which are known at the time of its formulation. At that point, it is not even sure if there are galaxies beyond our own Milky Way. For the most part, Einstein's famous *Cosmological Considerations* of 1917 concern the internal consistency of applying the theory of relativity to the cosmos as a whole, and the philosophical demands on such an application. Einstein expects that the theory will describe a static universe where, in accordance with Mach's critique of classical mechanics, a body's in-

ertia is not a property of absolute space but a result of the interaction with other bodies. Starting in 1916 he initially tries unsuccessfully to reach this goal by choosing suitable boundary conditions for a solution to the field equations of the general theory of relativity. In 1917 he takes a bold step – in order to make possible a static, closed universe with a uniform distribution of matter that meets his philosophical requirements, Einstein changes his theory once again by adding the "cosmological constant," as it is later called. In 1918 Einstein calls the idea of a complete determination of the properties of space through matter as "Mach's principle." He hopes to realize this idea through the modified theory. But in exchanges with the Dutch astronomer de Sitter it soon becomes apparent that even Einstein's altered theory is not up to the task. In 1922 the Russian mathematician Alexander Friedmann presents a non-static solution for the field equations of general relativity that Einstein initially believes flawed. In 1927, independent of Friedmann, the Belgian astrophysicist Lemaitre publishes dynamic solutions to the equations describing an expanding universe. Gradually, the cosmological question moves from a philosophical problem to a question that only astronomical observation can answer.

In the meantime Einstein's considerations about a unified field theory also provide a reason to doubt the primary role of matter in respect

Einstein with the observatory's director Walter S. Adams (center) and William W. Cambell (right) at the Mount Wilson Observatory, 1931

to space. Would space really disappear if all the matter in it were completely removed? Isn't matter rather a property of space itself, which due to its physical effects is equivalent to the all-pervading aether of classical physics, even if it cannot be assigned any mechanical properties any more?

Toward the end of the 1920s, new advances in astronomy are shifting the discussion into a different direction. New determinations of the distances of galaxies, and Hubble's discovery that they are moving away from us, point to an expanding universe. Thus the debates about static solutions to the field equations and about "Mach's principle" lose importance. In 1932 Einstein finally abandons the cosmological constant. Instead, he now contributes with de Sitter to the development of dynamic solutions to the field equations. The development and structure of the universe become the most important subject of the general theory of relativity.

◯ PERSONALIA

Willem de Sitter

The Dutch astronomer Willem de Sitter (1872–1934) is one of the founders of modern cosmology. He studies in Groningen, works from 1897 to 1899 at the Cape Town Observatory, and then returns to Groningen where he gains his doctorate in 1901. From 1908 on, he works is still a commonly used model, the "Einstein-de Sitter Universe." During World War I, de Sitter's works are an important source of information about the general theory of relativity in English-speaking countries. He provides the decisive stimulus for Eddington's eclipse expedition in 1919.

Alexander Friedmann

Alexander Friedmann (1888–1925) comes from a St. Petersburg family of artists and studies mathematics at the university there. After graduating in 1910, he works at different research institutes, including the Aerological Observatory in Pavlovsk. For some months he also stu-

◯ FOUNDATIONS

The Cosmological Constant

In 1917 Einstein introduces an additional term into the original field equations in the general theory of relativity and obtains to the field equation with cosmological constant. This additional term represents – in the case of a positive cosmological constant – a tension

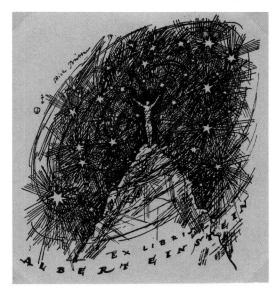

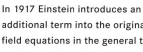

Erich Büttner:
ex libris for Albert
Einstein, 1917

in Leiden and becomes the director of the observatory there in 1919. Among other things, he reviews the values of astronomical constants and discusses the influence of tidal frictions on the orbits of the Earth and the Moon. De Sitter is among the first astronomers to take up the general theory of relativity. A groundbreaking debate with Einstein on the subject heralds the founding of relativistic cosmology. Together in 1932 they publish what

From left to
right standing:
Albert Einstein,
Paul Ehrenfest,
Willem de Sitter;
from left to
right sitting:
Arthur Eddington,
Hendrik Antoon
Lorentz.
Leiden, 1923

dies meteorology in Leipzig. In 1918 he is appointed professor at the University of Perm. Finally, in 1920, he receives an appointment at the Geophysical Observatory of St. Petersburg and becomes its director shortly before his death. In post-revolutionary Russia Friedmann develops interest in Einstein's relativity theory. He publishes two seminal works in 1922 and 1924 that point out time-dependent solutions to Einstein's field equations of gravitation, which Einstein initially greets with skepticism. These solutions correspond to universes where geometry and the distribution of matter change with time. The resulting model of a non-stationary, expanding cosmos is confirmed by Hubble's observations of galaxies.

pulling the universe apart, which on a large scale could outweigh gravitation. Einstein uses this term in what turns out to be an unsuccessful search for a static model of the universe with a finite mass density. He discards the cosmological constant after the discovery that the universe is expanding. Today, however, it plays a key role in explaining the accelerated expansion of the universe, as deduced from the observations of far-away galaxies. In this context one also speaks of "dark energy."

From Philosophy to Astronomy

1 (Albert Einstein: Cosmological Considerations in the General Theory of Relativity
Sitzungsberichte der Königlich Preußischen Akademie der Wissenschaften, 1917

defined at infinity in an infinite universe, meaning uninfluenced by physical masses. As Einstein finds the introduction of such boundary conditions unsatisfactory, he postulates a closed, finite universe. In 1917

2 (Mount Wilson Observatory
Pasadena, California
In 1904 George Ellery Hale initiates the construction of the observatory on top of Mount Wilson near Pasadena with funds

EINSTEIN: Kosmologische Betrachtungen zur allgemeinen Relativitätstheorie 151

müßten wir wohl schließen, daß die Relativitätstheorie die Hypothese von einer räumlichen Geschlossenheit der Welt nicht zulasse.

Das Gleichungssystem (14) erlaubt jedoch eine naheliegende, mit dem Relativitätspostulat vereinbare Erweiterung, welche der durch Gleichung (2) gegebenen Erweiterung der Poissonschen Gleichung vollkommen analog ist. Wir können nämlich auf der linken Seite der Feldgleichung (13) den mit einer vorläufig unbekannten universellen Konstante — λ multiplizierten Fundamentaltensor $g_{\mu\nu}$ hinzufügen, ohne daß dadurch die allgemeine Kovarianz zerstört wird; wir setzen an die Stelle der Feldgleichung (13)

$$G_{\mu\nu} - \lambda g_{\mu\nu} = -\varkappa \left(T_{\mu\nu} - \frac{1}{2} g_{\mu\nu} T \right). \qquad (13\,a)$$

Auch diese Feldgleichung ist bei genügend kleinem λ mit den am Sonnensystem erlangten Erfahrungstatsachen jedenfalls vereinbar. Sie befriedigt auch Erhaltungssätze des Impulses und der Energie, denn man gelangt zu (13a) an Stelle von (13), wenn man statt des Skalars des RIEMANNschen Tensors diesen Skalar, vermehrt um eine universelle Konstante, in das HAMILTONsche Prinzip einführt, welches Prinzip ja die Giltigkeit von Erhaltungssätzen gewährleistet. Daß die Feldgleichung (13a) mit unseren Ansätzen über Feld und Materie vereinbar ist, wird im folgenden gezeigt.

One of the fundamental ideas of Einstein's theory of relativity is what he later calls Mach's principle: the inertia of heavy bodies is not an effect of absolute space, as Newton postulated, but an effect of all the other masses in the universe, just like gravitation. Therefore, in an empty space there should also be no inertial forces. The general theory of relativity as Einstein finalizes it in 1915 only partially fulfills this requirement. Although inertia is determined by the metric field as opposed to absolute space, the metric field itself does not disappear in an empty universe. Instead, like all fields, it can be determined by boundary conditions that are

he publishes an article introducing a new field equation for his general theory of relativity and shows that a closed universe of constant size satisfies this field equation. The difference in the new field equation is the introduction of the cosmological term. This starts a debate with de Sitter who proves that this modification also allows another solution unwanted by Einstein – an empty universe that still possesses a well-defined structure of spacetime. Although Einstein eventually discards the cosmological constant because of this, it still plays a key role in science today. Einstein's essay marks the birth of modern cosmology.

donated by the Carnegie Foundation. It is the first modern observatory built especially to study the Sun and becomes a prototype for other solar observatories. The reflecting telescopes (the first with a 1.5-meter mirror diameter, later in the 1920s another with 2.5-meter mirror diameter) are the largest in the world and make pioneering discoveries possible. In 1929, after considerable preparatory work, Edwin P. Hubble and his colleagues prove that the spectrum of distant galaxies is shifted into the red end and that this redshift is proportional to the distance of the galaxies. This is the decisive argument for the expansion of the universe.

3 (Einstein's Manuscript for an Addendum to "On the Special and General Theory of Relativity (A Popular Account)"
Princeton, 1946
In this addendum for the fourteenth English printing of his popular scientific book about the theory of relativity Einstein writes about the changes in cosmology since the book's first printing in 1917. In particular he points out that an assumption he made at that time, namely that the universe as a whole is essentially static, has meanwhile been proven false. He calls attention to the theoretical work of Alexander Friedmann and the astronomical observations of Edwin P. Hubble that show there are good theoretical and experimental reasons to believe that the universe as a whole is expanding. For this reason Einstein considers the cosmological constant he introduced at the time to be no longer justified.

4

3

4 (Black Holes
European Southern Observatory, 2003
Black holes are not invisible – the matter that falls into them emits high-energy radiation, for example x-rays.
This photograph shows a small section of the southern night sky taken at the European Southern Observatory in La Silla, Chile. The x-ray radiation from this region was "photographed" by the satellite Chandra in the year 2000 with an exposure time of 11.5 days. The comparison with this photo makes it possible to identify the x-ray sources as giant black holes at the centers of far-off galaxies.

The Paradoxes of Quantum Mechanics

After initial successes, quantum theory is in crisis in the early 1920s. For one thing, it is little more than an unclear mixture of fragments of classical physics and rules for their modification as applied to problems of atomic physics. For another, it fails in the attempt to understand the properties of complex atoms. The need for new foundations is becoming ever more apparent. The path leading there can be found in the analysis of the interaction of light and matter, in which Einstein's ideas from 1916 about this interaction play an important role. The starting point for a new conceptual beginning is found especially in the reinterpretation of formulas describing the scattering (dispersion) of light on atoms. Based on this, Heisenberg introduces a new mechanics in 1925 that abstains from assigning position and velocity to the elec-

Niels Bohr and Albert Einstein, end of the 1920s

trons in the atom. Together with Born and Jordan he shows that this theory, instead of using numbers, involves infinitely large tables of numbers, called matrices.

Following Einstein's thoughts about the dualism of waves and particles, Schrödinger takes a different route to the completion of quantum theory. In 1926 he expands de Broglie's idea of matter waves by introducing a wave equation for electrons. Its solutions are states of the Bohr-Sommerfeld atom. Soon afterwards, Schrödinger finds the proof that his formalism is essentially equivalent to that of Heisenberg.

Both of their approaches suffer from problems of physical interpretation. Schrödinger's wave mechanics cannot explain the particle characteristics of electrons as their waves diverge. In reaction to this, Born develops the statistical interpretation of quantum theory in which the waves only describe the probability of an electron being found at a particular location. On the other side, Heisenberg's theory is so abstract that it proves difficult to use for concrete problems. In 1927 he explains the theory's novel mathematical properties with novel physical properties of particles: he formulates the uncertainty principle, according to which certain pairs of properties (which Bohr later calls "complementary"), such as position and velocity, are not simultaneously determinable to an arbitrary degree of accuracy.

Einstein is convinced that a statistical theory can only be an incomplete description of physical reality. The question of completeness becomes the central point in his discussions with the defenders of this new theory, especially Bohr. In 1927 Bohr interprets the uncertainty principle as expression of the limited applicability of our concepts to the microworld. Einstein reacts by conceiving thought experiments aimed at proving that it actually is possible to measure complementary qualities simultaneously. Bohr rebuts his arguments by pointing to the fact that measuring instruments are also subject to the uncertainty principle. In 1935 Einstein uses the existence of "entangled states" to support his conviction that quantum theory is incomplete. In such states the properties of different particles depend on one another. Einstein argues that because of this, measurements of one particle's location or velocity should allow one to infer the values for the other particle without actually disturbing it. Thus it must have both a definite velocity and a definite position. However, that would mean that quantum mechanics is incomplete since, according to its rules, only one of the two values can be exactly defined ("Einstein-Podolsky-Rosen Paradox"). The theoretical work of John S. Bell creates the possibility to empirically test this assertion in 1964. The experiment turns out in quantum mechanics' favor.

○ PERSONALIA

Max Born

Max Born (1882–1970) is born in Breslau (now Wrocław, Poland), the son of an anatomy professor. He studies in Breslau, Heidelberg, Zurich, and Göttingen, where he gains his doctorate in 1906. After further stations in Cambridge, Breslau, and Göttingen, he receives a professorship at Berlin University in 1914 where he becomes friends with Einstein. Born is appointed full profes-

Niels Bohr

Niels Bohr (1885–1962) comes from a Danish patrician family. He studies physics in Copenhagen and gains his doctorate in 1911. During semesters abroad in Cambridge and Manchester, Bohr receives decisive inspiration from Rutherford, leading to his theoretical quantum model of the atom in 1913 for which he receives the Nobel Prize for Physics in 1922.

Erwin Schrödinger

Born in Vienna, Erwin Schrödinger (1887–1961) studies physics there and receives his doctorate in 1910. He is appointed professor at the University of Zurich in 1921 and in 1927 becomes Planck's successor in Berlin.

In winter 1925/26 Schrödinger follows up on de Broglie's idea of matter waves to develop his wave mechanics for describing atomic

Werner Heisenberg

Werner Heisenberg (1901–1976) is born in Würzburg, studies physics in Munich and gains his doctorate with Sommerfeld in 1923. In 1924 he joins Bohr in Copenhagen. Working together with Born in Göttingen in 1925 he plays a central role in the foundation of quantum mechanics. He develops the so-called matrix mechanics and, in 1927 formulates the uncertainty principle

From left to right: Max Born, around 1925; Erwin Schrödinger, 1926; Werner Heisenberg, around 1927

or for theoretical physics in Göttingen in 1920. His role is decisive in Göttingen becoming one of the centers for fundamental work on the creation and application of quantum mechanics. In 1926 he formulates the statistical interpretation of quantum mechanics, but does not receive the Nobel Prize for this until 1954.

Born is forced to emigrate in 1933 due to his Jewish descent. He receives a professorship in 1936 in Edinburgh and returns to Germany after his retirement in 1953. In his final years he is an active advocate of peace and disarmament.

In 1916 Bohr is appointed professor in Copenhagen. A chair for theoretical physics is created especially for him in 1920 and, one year later an entire institute is built, soon becoming the "Mecca" of atomic research. In discussions with countless students, he develops the foundations for quantum mechanics. In 1927 he formulates the principle of complementarity, leading to a long-lasting and fundamental dispute with Einstein.

In 1939 Bohr works on the theoretical foundations of nuclear fission. Starting in 1933 he helps many scientists who have emigrated from Germany, but is himself forced to emigrate in 1943. After the war, he returns to his institute.

processes – in contrast to the Heisenberg-Born matrix mechanics. The "Schrödinger equation" is soon accepted as the preferred means of representing quantum mechanics and he receives the Nobel Prize in 1933 for this work. Although Schrödinger is less successful with his realistic interpretation of the wave function, he is nevertheless unable to accept Born's statistical interpretation. Schrödinger leaves Germany in 1933. In 1939, he is appointed director of the newly founded Institute for Advanced Studies in Dublin. He returns to his native Austria in 1956.

for its physical interpretation. He receives the 1932 Nobel Prize in Physics for these two contributions.

In 1927 Heisenberg is called to Leipzig. He works out a model of the atomic nucleus in 1932. Under National Socialism he starts working on the German nuclear energy program in 1939 and, in 1942 is appointed director of the Kaiser Wilhelm Institute for Physics in Berlin. After World War II he establishes the Max Planck Institute for Physics in Göttingen which is relocated to Munich toward the end of the 1950s. In 1958, he attempts a unified description of the physical world with his "world formula."

Thought Experiments

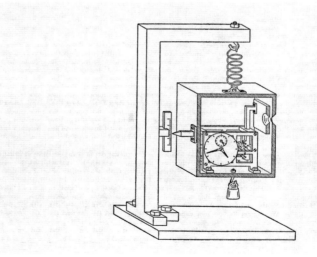

1 (Einstein on the 200th Anniversary
of Newton's Death
Manuscript, March 1927
Einstein writes a brief statement published
in *Nature* on the occasion of the 200th anniversary of Newton's death. The manuscript contains an appendix which Einstein
evidently did not find easy to write, as is
shown by the amount of crossed-out text
and the existence of a second version.
The appendix is an emphatic confession
of faith in the determinism of classical physics. Einstein invokes the ghost of Newton
in almost religious tones and expresses his
hope that the indeterminism of quantum
mechanics will be overcome.

2 (The Einstein-Bohr Debate
At the Solvay Conferences in Brussels in
1927 and 1930, Einstein and Bohr hold
numerous discussions about the interpretation of quantum mechanics that deeply
impress the attendant physicists. Heisenberg and Bohr justify the formalism of
quantum mechanics with the impossibility
of simultaneous exact measurements of
certain physical observables that are
"complementary" to each other, like position and velocity. Einstein attempts to
rebut this argument with various thought
experiments in which complementary observables can be measured to an arbitrary
level of accuracy. Bohr counters that measuring devices also show the uncertainty
of complementary observables, thus one
cannot use the same measuring device to

measure two complementary observables.
This makes the measuring situations used
in Einstein's thought experiments impossible in practice. In 1949 Bohr publishes a
written account of these talks with lovingly
drawn diagrams to illustrate Einstein's
thought experiments.

3 (Quantum Mechanics and Reality
Einstein's Essay for Dialectica, 1948
In this short essay, Einstein summarizes
what he sees as the essentials of the Einstein-Podolsky-Rosen argument. He starts
from a "entangled state" of two particles.
In such a state, according to quantum mechanics the individual particle has no definite state. However, as soon as the state of

4 (Einstein-Podolsky-Rosen Experiment
*Institute for Experimental Physics of the
University of Vienna (Zeilinger Group), 2005*
A modern realization of the Einstein-
Podolsky-Rosen thought experiment with
entangled pairs of photons created in an
anisotropic crystal from ultraviolet light.
The decisive property of the photons in this
experiment is their polarization angle. It is

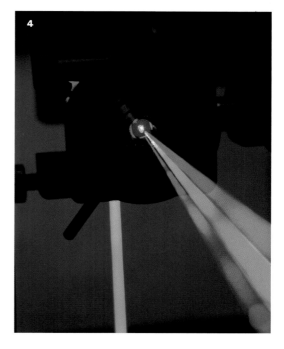

one of the particles is measured the state
of the other is also known. The decisive
factor is that one can take various complementary measurements of the first particle. The state of the second particle varies
depending on what measurement is taken.
With this situation Einstein creates a dilemma for the assumption that quantum mechanics provides a complete description of
natural reality. Either: the various quantum
mechanical states of the second particle
describe no objective difference in the
facts, in which case they cannot be a complete description of natural reality. Or: in
the case that they do deliver a complete
description, then the objective state
changes with the quantum mechanical
state of the second particle, which depends on which measurement is carried
out on the first particle. But this second assumption means there must be an immediate action at a distance occurring between
the two particles. According to Einstein,
this stands in contradiction to our fundamental knowledge about physical space.

measured by two adjustable rotary polarization filters, behind which are detectors.
When the filter is aligned in the same direction as the photon's polarization, then ide-
ally the photon will pass through the filter
and be registered by the detector. When
the filter is aligned at a right angle to the
photon's angle of polarization, the photon
is absorbed. The entanglement of the
photon pairs is such that the polarization
angle of one is always at a right angle to
the other, although according to quantum
mechanics neither one of the photons has
a particular polarization angle. This is observable when the two filters are adjusted.
When the filters are set at right angles to
each other, it is always the case that if one
of the detectors registers a photon the
other does as well. If one of the filters is
now turned 45 degrees, only half the time
do photons show up simultaneously on
both detectors. If the second filter is also
now turned 45 degrees, so that the two
filters are again perpendicular to each other (although at different angles than in the

first case) the photons again hit both detectors simultaneously. But how does a
photon "know" whether or not the other
photon will pass through a given filter
setting? Einstein argues that either the
two photons must communicate with
one another through a "spooky action at
a distance" when they pass through the
filters, or it is already predetermined
from the outset whether the photons
pass through the filters (or not) for every
possible setting of the filters. However,
information like this is not included in
quantum mechanical states.

The Search for the Unity of Nature

The search for the unity of nature is the central theme of Einstein's scientific biography. In the last three decades of his life, he tries to reach this goal by creating a unified theory of gravitation and electromagnetism. Time and time again he devises new approaches to such a theory only to fail, time and time again. However, the methods and insights he develops do prove productive, if, for the time being, only in the field of mathematics.

What drives Einstein? And why is he now unable to match any of his earlier scientific breakthroughs? The world of classical physics is based on two fundamentally different building blocks, particles and fields. The triumph of field theory suggests attempts to transcend the dualism and to reduce particles to properties of fields. Since 1900, the quantum problem represents an additional challenge. In 1909, Einstein hopes that the particle properties of light he discovered in 1905 can be explained by a modification of

Maxwell's electromagnetic field theory. Gustav Mie suggests a similar modification in 1912 in order to represent elementary particles as properties of the electromagnetic field. Hilbert joins the program and tries to combine it with Einstein's *Entwurf* theory of gravitation of 1913. Although this attempt fails it initiates other efforts to develop a unified theory of electromagnetic and gravitational fields, the two basic forces of nature known at the time. However, in contrast to Einstein's revolution of 1905, there are no known

At the sea of possibilities

borderline problems in which both forces play a role. This makes the endeavor of their unification appear not very promising. For the same reason, mathematical considerations are increasingly winning importance for the formulation of new proposals. In 1918 the mathematician Weyl proposes a new geometry that would combine electrodynamics and the general theory of relativity. The mathematician Kaluza also attempts such a combination in 1919 by introducing a fifth dimension. In both cases, Einstein is impressed with the mathematical elegance of these attempts but initially judges their physical relevancy with skepticism. At the same time, he takes his first own steps in this direction. These are shaped by his experience with the general theory of relativity on the one hand, and his struggle with the quantum problem on the other. In hindsight, it increasingly appears to him that his pursuit of mathematical simplicity was the true secret of success for his formulation of the general theory of relativity in 1915. This provides him with the confidence to continually try new mathematical approaches in order to discard them if they do not lead him to the expected results. From 1925 on, Einstein sees his program as an alternative to quantum mechanics, which he does not believe is able to deliver a satisfactory solution to the quantum problem, due especially to its statistical character. However, Einstein is increasingly alone in his efforts. The majority of physicists see quantum mechanics as the avenue to the future.

○ PERSONALIA

Hermann Weyl

Born near Hamburg as the son of a bank director, Hermann Weyl (1885–1955) studies mathematics in Göttingen, where he gains his doctorate and the habilitation as a lecturer. He is regarded as the most distinguished student of David Hilbert. In 1913 Weyl is appointed professor at the Swiss Federal Institute of Technology in Zurich. In 1930 he becomes Hilbert's successor in Göttingen. In 1933 he emigrates from Germany and receives a position at the Institute for Advanced Study in Princeton.
Weyl's wide-ranging mathematical works can be separated into three

Valentine Bargmann

Born in Berlin, Valentine Bargmann (1908–1989) is the son of Russian-Jewish parents. He studies physics in Berlin and after emigrating gains his doctorate in 1936 at the University of Zurich. He then goes to the United States. From 1937 to 1946 he works at the Institute for Advanced Study in Princeton, afterwards as professor for mathematical physics at Princeton University. In 1988 he receives the Max Planck Medal from the German Physical Society.
Bargmann is Einstein's assistant from 1937 to 1938 and works with him on problems of the unified

Peter Bergmann

Peter Gabriel Bergmann (1915–2002), born in Berlin, emigrates to Prague in 1933 where he studies physics with Philipp Frank. He gains his doctorate there in 1936. On Frank's recommendation he goes to the Institute for Advanced Study in Princeton where he works as an assistant to Einstein from 1936 to 1941. After several other stops, he receives a professorship for theoretical physics at Syracuse University in the state of New York and teaches there from 1947 to 1982. He also holds several visiting professorships, for example at the

University of Marburg in 1955.
In the framework of their research on a unified field theory, Bergmann and Einstein postulate the reality of a fifth spacetime dimension. In 1942 Bergmann publishes his first textbook on the general theory of relativity, for which Einstein writes the foreword. In Syracuse he founds the first research center for the general theory of relativity in the United States. With his collaborators there, he researches in particular the quantization of the gravitational field, a problem that remains unsolved to this day.

main categories: problems of pure mathematics in the spirit of Hilbert's school, number theory, and mathematical physics. In 1918 he publishes *Space – Time – Matter*, one of the first books about the general theory of relativity. His book *The Theory of Groups and Quantum Mechanics* from 1928 is probably his most influential contribution to modern physics. He is also concerned with philosophical problems of mathematics and mathematical physics.

field theory. During World War II he works with John von Neumann, on shock waves, and after the war with Eugene Wigner on relativistic wave equations.

Above:
Albert Einstein, Valentine Bargmann (right), and Peter Bergmann in a seminar room at the Institute for Advanced Study, Princeton, 1940

Right:
Albert Einstein, Valentine Bargmann (right), and Peter Bergmann, Princeton, 1940

From Physics to Mathematics

1 (On the Current State of Field Theory
Einstein's contribution to the festschrift for Aurel Stodola, Zurich, 1929
In this article, Einstein expounds the motivation and goals of a research program that will occupy him until the end of his life. He establishes a historic arc from Faraday's introduction of the field concept to the current state of research, and at the same time reveals his personal goals for further advancement to a unified field theory. Of central importance to Einstein is the logical unification of all of the laws of physics. This is bound to a fundamental requirement: "We don't just want to now *how* nature is [...], but if possible, although the goal may appear perhaps utopian and presumptuous, we also wish to know why nature *is as it is and not otherwise.*" He argues that the structure of his general theory of relativity arises directly and completely from the demand that it describe inertial and gravitational forces in a unified way. Einstein similarly hopes that the requirement of an even more comprehensive unification will lead him unambiguously to the form of the physical laws of what he calls unified field theory. Even if Einstein's attempts to achieve a unified theory without including quantum laws are only of historical importance today, his motivations remain very much alive in research areas like string theory.

Über den gegenwärtigen Stand der Feld-Theorie.

Die Feldtheorie, welche nach meiner Ansicht die tiefste Konzeption der theoretischen Physik seit deren Grundlegung durch Newton ist, ist aus Faradays Kopf entsprungen. Wie einfach scheint diese Idee a posteriori und wie sublim ist sie doch! Statt zu denken: „Ein elektrisches Teilchen e_1 wirkt auf ein zweites e_2 durch den Raum hindurch und übt auf letzteres eine bewegende Kraft aus", denkt Faraday: „Ein elektrisches Teilchen bewirkt durch seine blosse Existenz eine Modifikation des Zustandes des Raumes seiner unmittelbaren Umgebung und zeitliche Änderung (elektrisches Feld). Die räumliche Verteilung eines solchen Feldes ist beherrscht durch Gesetze, die dem Raum anhaften. Vermöge dieser Gesetze erstreckt sich das dem Teilchen e_1 entstammende Feld bis zu dem Teilchen e_2 und wirkt dort auf dieses." Diesem Gedanken entsprang bald darauf Maxwells wunderbare Gesetze des elektromagnetischen Feldes. Hertz zeigte entgültig, dass dieser Feld-Theorie gegenüber Newtons Fernwirkungs-theorie der Vorrang gebührte, und kurz darauf H. A. Lorentz, dass das Feld überall, auch im Innern der Materie seinen Sitz im leeren Raume habe, ja dass die elementaren Bausteine der Materie – wenigstens in elektromagnetischer Hinsicht – nichts anderes seien als Quellpunkte des elektrischen Feldes. Dies war der Stand der Auffassung um die Jahrhundert-Wende.

Bevor ich nun die Entwicklung der Feldtheorie weiter ins Auge fasse, möchte ich eine kurze Bemerkung über Ziel und Tendenz der theoretischen Forschung überhaupt einschalten. Die Theorie hat zwei Sehnsüchte:
1) möglichst alle Erscheinungen und deren Zusammenhänge zu umfassen (Vollständigkeit).
2) Dies zu erreichen unter Zugrundelegung möglichst weniger von einander logisch unabhängiger Begriffe und willkürlich gesetzter Relationen zwischen diesen (Grundgesetze bezw. Axiome). Ich will dies Ziel das der logischen Einheitlichkeit

2 (Einstein's Belief in the Creative Power of Mathematics

Sitzungsberichte der Preußischen Akademie der Wissenschaften, Berlin, 1932

This work represents a conclusion of sorts to Einstein's time in Berlin – it is the last article that the academy member publishes in the house organ of his employer. The co-author is Walther Mayer, with whom Einstein has been working together on unified field theory since 1930. The fact that Mayer is not a physicist but a mathematician is a trend that continues until the end of Einstein's life: "The actual creative principle rests [...] in mathematics," he declares during a lecture at Oxford University one year later. The work at hand delivers quite an abstract analysis of the mathematical foundations of the spinors introduced by Paul A. M. Dirac in order to describe the electron. Einstein's approach leads only incidentally to possible physical applications – it is no longer physical, but mathematical intuition that is supposed to lead him to new insights.

522

Semi-Vektoren und Spinoren.

Von A. Einstein und W. Mayer.

(Vorgelegt am 10. November 1932 [s. oben S. 473].)

Bei der großen Bedeutung, welche der von Pauli und Dirac eingeführte Spinor-Begriff in der Molekularphysik erlangt hat, kann doch nicht behauptet werden, daß die bisherige mathematische Analyse dieses Begriffes allen berechtigten Ansprüchen genüge. Dem ist es zuzuschreiben, daß P. Ehrenfest bei dem einen von uns mit großer Energie darauf gedrungen hat, wir sollten uns bemühen, diese Lücke auszufüllen. Unsere Bemühungen haben zu einer Ableitung geführt, welche nach unserer Meinung allen Ansprüchen an Klarheit und Natürlichkeit entspricht und undurchsichtige Kunstgriffe völlig vermeidet. Dabei hat sich — wie im folgenden gezeigt wird — die Einführung neuartiger Größen, der »Semi-Vektoren«, als notwendig erwiesen, welche die Spinoren in sich begreifen, aber einen wesentlich durchsichtigeren Transformationscharakter besitzen als die Spinoren. In der vorliegenden Arbeit haben wir uns absichtlich auf die Darlegung der rein formalen Zusammenhänge beschränkt, um das Mathematisch-Formale in seiner Reinheit klar hervortreten zu lassen.

Das Wesentliche des in dieser Arbeit ausgeführten Gedankenganges läßt sich folgendermaßen skizzieren. Jede reelle Lorentz-Transformation $\mathfrak{D}$ läßt sich eindeutig in zwei spezielle Lorentz-Transformationen $\mathfrak{B}$ und $\mathfrak{C}$ zerlegen, deren Transformationskoeffizienten $b^i{}_k$ und $c^i{}_k$ zueinander konjugiert komplex sind, wobei die Transformationen $\mathfrak{B}$ und $\mathfrak{C}$ Gruppen $(\mathfrak{B})$ und $(\mathfrak{C})$ bilden, die zur Gruppe $(\mathfrak{D})$ der Lorentz-Transformationen isomorph sind. Semi-Vektoren sind Größen mit vier komplexen Komponenten, welche bei Vornahme einer Lorentz-Transformation eine $\mathfrak{B}$- bzw. $\mathfrak{C}$-Transformation erleiden. Es gibt spezielle Semi-Vektoren, welche durch gewisse Symmetriebedingungen charakterisiert sind und nur zwei (statt vier) voneinander unabhängige (komplexe) Komponenten haben. Dieser Umstand gibt Anlaß zur Einführung von Größen mit nur zwei (komplexen) Komponenten, nämlich den Diracschen Spinoren.

§ 1. Drehung und Lorentz-Transformation.

Wir denken uns den Raum R_4 der speziellen Relativitätstheorie auf kartesische (nicht notwendig rechtwinklige) Koordinaten bezogen. Der Maßtensor (g_{ik}) hat bestimmte konstante Komponenten, welche gegenüber den nachher

2

Einstein and Judaism

Einstein's Jewish heritage and the fate of the Jewish people becomes an increasingly important source for his personal identity over the course of his life. As a child he is exposed to anti-Semitic stereotypes that first make him aware of the question of lineage. As an adolescent he turns against his parent's assimilation and, for a time, obeys the religious commandments. His encounters with the worldview of the natural sciences, however, mean the end of traditional religiosity for him. At the same time – with his characteristic self-irony – "God the Lord" remains a life-long conversational partner in his quest for objective natural knowledge: "God doesn't throw dice."

When Einstein comes to Berlin in 1914 he first becomes conscious of the Jewish people as a community united by fate. This leads him to turn against what he perceives as the undignified assimilation and "pussy-footing" of German Jews. He is especially committed to the "Eastern Jews," the Jews recently immigrated from Eastern Europe. His efforts for the Zionist movement, the creation of a Jewish state, and the founding of a Jewish university in Jerusalem also bear the hope of it opening up new perspectives for their future. He warns early of the ghastly injustice threatening the Eastern Jews in Germany. Einstein himself is the victim of anti-Semitic attacks and even receives death threats in Germany in the 1920s. Close colleagues of his foster anti-Semitic prejudices themselves.

In 1921 Einstein cancels his participation in the Solvay Conference and, together with the president of the World Zionist Organization Chaim Weizmann, he travels instead to the United States to help raise funds for the construction of the Hebrew University in Jerusalem. Once there, Einstein has the impression of meeting for the first time not only single

Einstein speaks at the Jewish Students Conference in Germany, 1924

Jews but the Jewish people. In 1923 he travels to Palestine, gives a lecture at the future site of the university, and then travels the country. Although he is deeply moved, he cannot bring himself to give in to his hosts' urgings and relocate to Palestine. Regardless, he remains closely connected to the idea of the founding of a Jewish state and is an advocate for the peaceful coexistence of Jews and Arabs.

After his emigration to the United States he devotedly commits himself to Jewish refugees. The Depression and the difficult situation on the academic job market make it problematic for displaced scientists to build a secure existence. Even help organizations have to be careful, as they encounter prejudices and the fear of increased competition for university jobs. Einstein writes affidavits for countless Jewish refugees – even many he does not personally know – acting as guarantor by using his own financial means to the greatest extent possible. He also supports efforts against the discrimination of Jews in the United States, for example by supporting the founding of Jewish organizations including Brandeis University. Deeply distressed by the Holocaust, Einstein calls for Germany's permanent disempowerment after the war. Following Weizmann's death in November 1952 Einstein receives the offer to become Israel's next president, which he declines. He speaks, however, in this context of his relationship with the Jewish people as his "strongest human bond."

○ HISTORICAL CONTEXT

The Hebrew University

At the beginning of the twentieth century an idea develops within the Zionist movement for the founding of a university in Jerusalem. Chaim Weizmann and Martin Buber are important advocates of the idea. Russian Jews buy a plot of land on Mount Scopus, upon which the cornerstone for the Hebrew University is laid in 1918. In 1925 the university opens its doors. At the start, it consists of three institutes – for

Zionism

The Dreyfus affair in France moves the journalist Theodor Herzl to develop a plan for the organized immigration of Jews from all over the world into their own national territory. At the first Zionist World Congress in Basel in 1897, he founds the World Zionist Organization. In 1905 the organization rejects the Uganda scheme, ultimately deciding on Palestine as the area for a future state. Since 1908, with help

○ PERSONALIA

Chaim Weizmann

Born in a village near Pinsk in Russia, Chaim Weizmann (1874–1952) is a chemist, leading Zionist, and first president of Israel. He studies in Darmstadt and in Freiburg, Switzerland. In 1904 he is appointed senior lecturer at the University of Manchester. During World War I he works on armament research for the British army and, at the same time, tries to win over the British public to the idea of a Jewish state.

Weizmann works on the foundation of the Hebrew University in Jerusalem. From 1920 to 1931 he is president of the World Zionist Organization. In 1947 he holds a speech to the United Nations that gains much attention, leading to the recognition of Israel by the United States. He is Israel's first president from 1948 until his death in November 1952.

chemistry, for microbiology, and for Jewish studies – and has thirty-three faculty members and 141 students, the first of whom graduate in 1931. The war for independence forces the university to relocate its facilities to another part of the city. After the six-day war the university returns to its original location where the main building stands again since 1981.

of the Palestine Office, a systematic colonization of the territory is organized. In 1929 the Jewish Agency is founded, representing the Jewish people before the British mandate authorities. During the 1940s the conflict with Great Britain escalates and, in November 1947, leads to the end of the mandate. The duties of the Jewish Agency are taken over by the agencies of the state of Israel after its foundation on 14 May 1948.

In 1917 he lands a great success with the declaration by Balfour, British foreign secretary, in support of the formation of a national homeland for the Jewish people in Palestine. At the beginning of the 1920s

Einstein and his secretary, Helen Dukas, during a concert in the synagogue on Oranienburger Straße, Berlin, 1930

Albert Einstein and Chaim Weizmann (first from left) in New York, 1921

Dispatches in the Daily Press

1 Die Zuwanderung aus dem Osten.

Von [Nachdruck verboten.]

Universitätsprofessor Dr. Albert Einstein.

Wir geben den nachstehenden Ausführungen des hervorragenden Gelehrten gern Raum und möchten dazu bemerken, daß wir seine Ansicht über eine Ausweisung der schon eingewanderten armen Ostjuden teilen. Ob für die Zukunft eine Beschränkung der Einwanderung möglich sein wird, ist eine andere Frage — eine solche gesetzliche Vorkehrung müßte aber in jedem Falle immer allgemein gehalten werden und sich nicht nur gegen bestimmte Religionsgenossenschaften und Kreise richten. Die Redaktion.

In der deutschen Oeffentlichkeit mehren sich die Stimmen, die gesetzliche Maßnahmen gegen die Ostjuden verlangen. Man behauptet, 70 000 Russen, d. h. Ostjuden, leben allein in Berlin; diese Ostjuden seien Schieber, Schleichhändler, Bolschewisten oder arbeitsscheue Elemente. All diese Argumente zielen darauf hin, schärfste Maßnahmen, d. h. Zusammenpferchung in Konzentrationslagern oder Auswanderung aller Zugewanderten zu erzwingen.

Eine so viele Existenzen vernichtende Maßregel darf nicht auf Grund schlagwortmäßiger Behauptungen erfolgen, um so weniger, als eine sachliche Nachprüfung ergibt, daß es sich um demagogische Agitation handelt, die weder dem wahren Sachverhalt gerecht wird, noch auch geeignet ist, tatsächlich vorhandene Schäden wirksam zu bekämpfen. Gerade bei der Agitation gegen Ostjuden liegt von vornherein die Vermutung sehr nahe, daß das ruhige Urteil durch die starken antisemitischen Instinkte getrübt und gleichzeitig auch die Methode gewählt wird, durch Einwirkung auf die Stimmung des Volkes die Aufmerksamkeit von den wirklichen Problemen und von den wahren Ursachen der allgemeinen Notlage abzulenken.

Eine Nachprüfung, die, soviel bekannt geworden, von den amtlichen Stellen noch niemals versucht worden ist, würde zweifellos die Hinfälligkeit der vorgebrachten Beschuldigungen erweisen. Wohl mag es richtig sein, daß in Berlin 70 000 Russen wohnen; von ihnen bilden jedoch nach den Angaben sachverständiger Beurteiler die Juden nur einen geringen Bruchteil, die überwiegende Mehrheit ist deutscher Abstammung. Seit dem Friedensschluß sind nach maßgebender Schätzung nicht mehr als 15 000 Juden aus dem Osten zugewandert. Diese sind fast ausnahmslos durch die furchtbaren Zustände in Polen zur Flucht gezwungen worden und wollen hier eine Zufluchtsstätte finden, bis ihnen die Möglichkeit zur Weiterwanderung gegeben wird. Hoffentlich werden viele von ihnen in dem neu entstehenden jüdischen Palästina als freie Söhne des jüdischen Volkes eine wahre Heimat finden.

Bolschewistische Agenten dürfte es wohl in Deutschland geben, diese haben aber zweifellos ausländische Pässe, verfügen über reiche Geldmittel und können durch irgendwelche organisatorische Maßnahmen nicht erfaßt werden. Die großen Schieber unter den Ostjuden haben sicher auch längst Vorkehrungen getroffen, die sie behördlichen Zugriffen entziehen. Getroffen würden ausschließlich jene Armen und Unglücklichen, die in den letzten Monaten unter unmenschlichen Entbehrungen den Weg nach Deutschland gefunden haben und hier Arbeit suchen. Nur diese, der deutschen Volkswirtschaft sicher unschädlichen Elemente werden die Konzentrationslager füllen und dort körperlich und seelisch verkommen. Dann wird man sich über die selbstgezüchteten „parasitären Existenzen" beklagen, die es verlernen, in einem normaler Wirtschaftsleben ihren Platz auszufüllen. Die verkehrte Politik, Tausende zwangsweise während des Krieges nach Deutschland gebrachte ostjüdische Arbeiter plötzlich zu entlassen und dadurch brotlos zu machen und sie systematisch von Arbeitsgelegenheiten fernzuhalten, mußte tatsächlich Menschen, die nicht mit ihrer Familie verhungern wollten, zum Schleichhandel zwingen. Auch dem deutschen Wirtschaftsleben wird sicherlich damit am besten gedient, daß die Oeffentlichkeit die Bemühungen jener Kreise unterstützt, welche es sich (wie zum Beispiel das schon öfters genannte „jüdische Arbeitsamt") zum Ziel gesetzt haben, die ostjüdischen Einwanderer produktiver Arbeit zuzuführen. Jede „Ausweisungsverfügung", die jetzt so stürmisch verlangt wird, hätte nur die Wirkung, daß die schlechten und schädlichen Elemente im Lande bleiben, dagegen die arbeitswilligen ins bitterste Elend und zur Verzweiflung getrieben werden.

Das Gewissen der Oeffentlichkeit ist gegen die Mahnungen zur Menschlichkeit so abgestumpft, daß sie das entsetzliche Unrecht, auf das hier hingearbeitet wird, nicht mehr unmittelbar empfindet. Ich unterlasse es, darauf näher einzugehen. Aber befremdend ist es, daß selbst führende Politiker nicht überlegen, wie schwer durch die von ihnen gewünschte Behandlung der Ostjuden die politische und wirtschaftliche Stellung Deutschlands beeinträchtigt wird. Hat man schon vergessen, wie die Deportierung belgischer Arbeiter den moralischen Kredit Deutschlands untergraben hat? Und heute ist Deutschlands Lage unvergleichlich kritischer. Trotz aller Bemühungen ist es überaus schwierig, die unterbrochenen internationalen Beziehungen wiederherzustellen; wenige geistige Menschen in allen Völkern machen erste Versuche; die Hoffnung auf neue wirtschaftliche Verbindungen (z. B. die materielle Hilfe Amerikas) ist heute noch sehr schwach. Die Austreibung der Ostjuden, welche namenloses Elend zur Folge hätte, würde aller Welt als ein neuer Beweis „deutscher Barbarei" erscheinen und einen Anlaß bieten, im Namen der Menschlichkeit den Wiederaufbau Deutschlands zu erschweren.

Die Gesundung Deutschlands kann wahrlich nicht durch Anwendung von Gewalt gegen einen kleinen wehrlosen Bruchteil der Bevölkerung herbeigeführt werden.

1 (Albert Einstein: Immigration from the East
Berliner Tageblatt, 30 December 1919
Many Eastern-European Jews flee from crushing poverty and persecution after World War I, especially from Poland and Russia. The considerable increase of their community in Berlin alarms anti-Semitic zealots who begin loudly demanding their internment and deportation. Shortly after Einstein becomes famous he uses his popularity to speak out on behalf of the immigrants. He confronts the common stereotypes and dreadful prejudices and warns of a loss of prestige for Germany internationally. He also fears that completely harmless and innocent people "will fill the concentration camps facing physical and mental wrack and ruin there." Quite in line with the purposes of the Zionist idea, with which he increasingly sympathizes, he hopes that the Jewish immigrants "will be able to find a homeland in the newly emerging Jewish Palestine as free sons of a Jewish people."

2

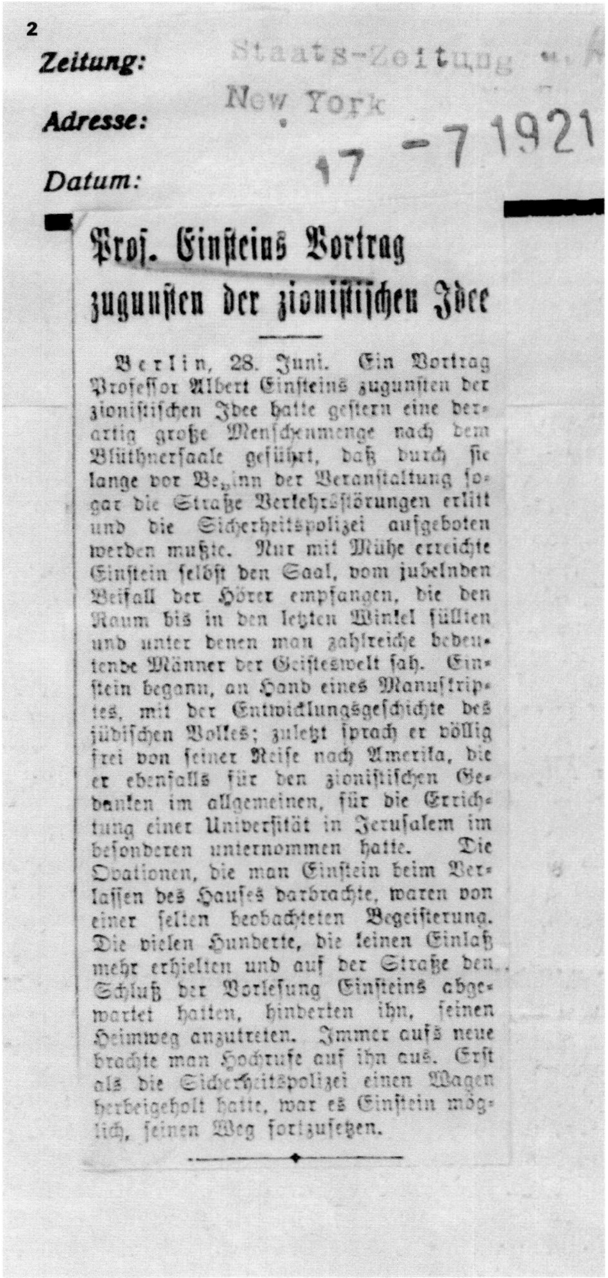

Zeitung: Staats-Zeitung

Adresse: New York

Datum: 17 – 7 1921

Proj. Einſteins Vortrag zugunſten der zioniſtiſchen Idee

Berlin, 28. Juni. Ein Vortrag Profeſſor Albert Einſteins zugunſten der zioniſtiſchen Idee hatte geſtern eine derartig große Menſchenmenge nach dem Blüthnerſaale geführt, daß durch ſie lange vor Beginn der Veranſtaltung ſogar die Straße Verkehrsſtörungen erlitt und die Sicherheitspolizei aufgeboten werden mußte. Nur mit Mühe erreichte Einſtein ſelbſt den Saal, vom jubelnden Beifall der Hörer empfangen, die den Raum bis in den letzten Winkel füllten und unter denen man zahlreiche bedeutende Männer der Geiſteswelt ſah. Einſtein begann, an Hand eines Manuſkriptes, mit der Entwicklungsgeſchichte des jüdiſchen Volkes; zuletzt ſprach er völlig frei von ſeiner Reiſe nach Amerika, die er ebenfalls für den zioniſtiſchen Gedanken im allgemeinen, für die Errichtung einer Univerſität in Jeruſalem im beſonderen unternommen hatte. Die Ovationen, die man Einſtein beim Verlaſſen des Hauſes darbrachte, waren von einer ſelten beobachteten Begeiſterung. Die vielen Hunderte, die keinen Einlaß mehr erhielten und auf der Straße den Schluß der Vorleſung Einſteins abgewartet hatten, hinderten ihn, ſeinen Heimweg anzutreten. Immer aufs neue brachte man Hochrufe auf ihn aus. Erſt als die Sicherheitspolizei einen Wagen herbeigeholt hatte, war es Einſtein möglich, ſeinen Weg fortzuſetzen.

2 (Einstein's Engagement for Zionism
New Yorker Staats-Zeitung, 17 July 1921
"Berlin, 28 June. A lecture Professor Einstein gave yesterday in support of the Zionist idea attracted such large crowds of people to the *Blüthnersaal*, that the traffic was seriously affected long before the start of the event and security police had to be called to the scene. It was only with great pains that Einstein was able to enter the hall to thunderous rounds of applause from the jubilant audience, who filled every last bit of space. Among them many distinguished gentlemen from the intellectual world were to be seen. Using a manuscript Einstein began with the history of the Jewish people. Later, speaking freely, he talked of his journey to America, which he had undertaken to gather support for Zionist cause in general and more specifically for the foundation of a university in Jerusalem.

The ovations that he received upon leaving the hall were of an enthusiasm seldom seen. The many hundreds who were unable to gain entry, and had awaited the end of Einstein's lecture on the street, prevented him from making his way home. Cheers for him repeatedly arose anew from the crowd. It was not until the security police were able to fetch a wagon that Einstein was able to continue his journey."

Emigration

As early as the beginning of the 1920s, in the context of public attacks and death threats, Einstein asks himself if it would not be better for him to withdraw from his academic positions and emigrate from Germany. He finds it a relief to be able to turn his back on Germany for a time in 1922 and 1923, with a longer stay in the Netherlands and an extended trip through Asia. However he cannot make the decision to definitively leave Germany. There is too much connecting him, both socially and scientifically to his surroundings in Berlin. Einstein is at the California Institute of Technology when the National Socialists seize power. In reaction to the violence against Jewish fellow citizens and the persecution of opponents of the new regime, Einstein delivers a statement, very moderate in tone, that is circulated in the American media. Not only those in power in Germany vilify his statements abroad, members of the Prussian Academy

Jewish citizens stand in line in front of the Palestine & Orient Lloyd in the Meinekestraße in Berlin, 1939

of Sciences also describe it as an "atrocious smear campaign," prompting discussions of Einstein's expulsion from the Academy. Einstein forestalls this step with his resignation on 28 March 1933. Despite this, one of the four Permanent Secretaries of the Academy of Sciences, Ernst Heymann, decides of his own accord to release an official declaration. Speaking for the entire institution, he condemns Einstein's statements and explicitly states that his resignation is not mourned. Only Laue takes action to prevent the subsequent endorsement of this declaration, whereas Planck, together with the majority of the members of the Academy, believes that Einstein has become untenable for the Academy due to his political statements. At the beginning of April 1933 Einstein is in Belgium when he finds out that the National Socialists have confiscated his bank accounts. At the end of May 1933 the Nazi paramilitary SA ransacks his apartment, but the valuable correspondence and scientific papers can be saved with the help of the French ambassador. Einstein travels to England for the time being where, together with colleagues like Ernest Rutherford and James Jeans, he gets involved with a project to help refugees. In mid-October he arrives in the United States, where he becomes Fellow of the Institute for Advanced Study in Princeton.

Einstein remains distant toward Germany for the rest of his life. Even among his former colleagues, he sees too few exceptions regarding the attitude toward National Socialism to justify maintaining official contact with German scientists again. Einstein rejects invitations of membership in scientific associations or academies – it is not just the time of National Socialism itself, but also the behavior of most of his colleagues after the war that plays a decisive role in his decision. Even old friends like Laue have great difficulties putting themselves into the shoes of the persecuted and displaced. He complains of Einstein's "resentment" and remains very cautious about mentioning the injustices out of respect for the regime's sympathizers and accomplices who are still living. Laue envisages the publication of a documentation of the infamous conduct of the Academy, but not until all "who were involved in this affair" have died, as he writes, in 1947, to one of the Permanent Secretaries in charge at the time of the "affair." As time passes Einstein's negative attitude toward Germany tempers. In 1949 he allows an elementary school in Caputh to be named after him. In the 1950s he also reacts positively to the request of Berlin students to name their school in his honor – an encouraging signal to the younger generation.

○ HISTORICAL CONTEXT

The National Socialist Seizure of Power

The surrender of the first German democracy is an insidious process. Even before 1933 the President of the Reich governs through emergency decrees beyond the control of parliament. The conservative powers believe that Hitler's appointment as Chancellor of the Reich will contain the National Socialist movement, underestimating the effective combination of propaganda and intimidation. The *Reichstag* fire is used as a pretense for the merci-

less persecution of political opponents, the NSDAP chalks up large election victories, the constitution is undermined bit by bit. The *Ermächtigungsgesetz* (Act of Enablement) ends the parliamentary system in Germany. Starting in April 1933 the isolation and persecution of Jews become systematic. The so-called Law for the Restoration of the Permanent Civil Service represents an additional escalation. Behind the legal façade is an arbitrary cynicism that costs thousands of judges, doctors, and scientists their jobs. The German educated elite puts up almost no resistance.

German Scholars under the Nazi Regime

In Germany there is little or only very restrained protest against the ostracism, persecution, and expulsion of Jewish and politically disfavored colleagues from the academic world. Max Planck complains to Hitler about the losses for German science, but at the same time seeks a tolerable arrangement with the new rulers. National Socialism is also often seen as a chance for science and scientific endeavors. Ideologically motivated scientists like Le-

nard and Stark are able to dominate the scene for only a short time. Many scientists face new tasks in military research, in eugenics, in the research on Eastern Europe and the so-called *Lebensraum* (the territorial space which the National Socialists hold to be indispensable for Germany's survival), as well as in other projects that aim to provide scientific support to the National Socialist claims for domination. While student numbers initially drop at universities, and certain areas (like theoretical physics) are devastated by persecution and ignorance, research budgets increase substantially when the preparations for war begin. The National Socialist movement also has the universities under its control through student bodies radicalized well before 1933. There are very few niches where skeptical spirits are tolerated, but only as long as they do not attract attention. During the war, scientific and intellectual resources are mobilized for the desired *Endsieg* (ultimate victory).

Above left:
The refugee Albert Einstein on arrival in New York, October 1933

Below:
Max Planck inaugurates the Kaiser Wilhelm Institute for Metal Research in Stuttgart, 1935

○ PERSONALIA

Max von Laue

The physicist Max von Laue (1879–1960) publishes the first monograph about relativity theory

in 1911. His work on a relativistic theory of extended solid and fluid bodies contributes substantially to the development of the insights contained in Einstein's famous formula about the equivalence of mass and energy. He remains skeptical of Einstein's general theory of relativity like many of his colleagues. In 1914 he receives the Nobel Prize for his discovery of the interference of x-rays in crystals. Under the Nazi regime he uses his scientific authority to stand up for beset colleagues. Einstein keeps in touch with Laue even after the National Socialist takeover, respecting him as one of the few colleagues who shows moral courage and has "remained upstanding."

Above right:
Max von Laue, around 1940

Moral Challenge

1 (Einstein's Resignation from the
Prussian Academy of Sciences
28 March 1933
"The prevailing conditions in Germany
prompt me to hereby resign my position at
the Prussian Academy of Sciences. For
nineteen years the Academy has provided
me with the opportunity to dedicate myself
to scientific work, completely free of any
professional duties. I know to what a high
degree I am indebted to it. Only reluctantly
do I take leave of its circle, not least because of the stimulus and the good human
relationships that I enjoyed during this long
time as its member and always held in the
highest esteem.
However, I find my dependency on the
Prussian government resulting from my
position intolerable under the current
circumstances."
Einstein writes this on the return journey
from the United States to Europe aboard
the passenger ship Belgenland.

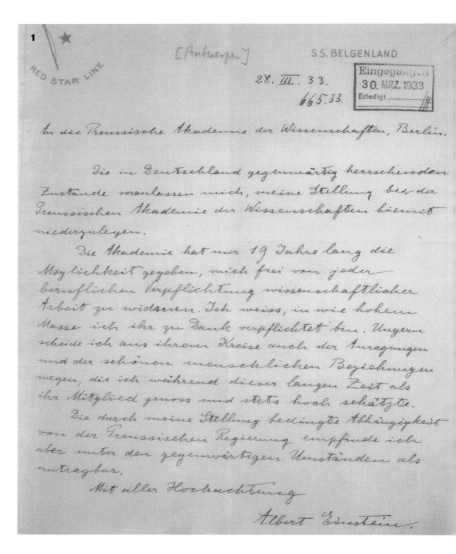

2 (Letter from the Office of the Führer's
Deputy Rudolf Hess to the Prussian Minister of Education Bernhard Rust
1 March 1934
After the death of Fritz Haber in January
1934, Einstein's friend Max von Laue writes
an obituary for the Kaiser Wilhelm Society's
journal *Die Naturwissenschaften*. In it he
writes: "Themistocles went down in history
not as an exile at the Persian court, but as
the victor of Salamis. Haber will go down
in history as the ingenious inventor of the
process of combining nitrogen and hydrogen, as the man [...] who extracted bread

from air and who achieved a triumph in
service to his country and all of mankind."
This passage is clearly directed at the
National Socialists who, as shown in this
letter to Rust, feel provoked. Haber was of
Jewish descent but had been tolerated by
the National Socialists as the director of
the Kaiser Wilhelm Institute for Physical
Chemistry and Electrochemistry because
he was a *Frontkämpfer* (veteran of World
War I). He resigned from the directorship
in protest against the anti-Semitic policies
in April 1933.

The report to the Minister of Culture showed here contains the transcription of a letter to the *Führer's* Deputy by Philipp Lenard, an anti-Semite and representative of so-called "German physics." In his letter Lenard invents the specter of an enemy "Einstein Circle." Along with Laue its members include Planck, Sommerfeld, and his student, Heisenberg. Lenard is already active campaigning against Einstein and his theories immediately after World War I. Together with Johannes Stark they try to expose the theory of relativity as a fraud. However both of them lose influence during the 1930s. Heisenberg, who is denounced by Stark in an SS publication, manages to rehabilitate himself with the help of influential colleagues. Ludwig Prandtl, director at the Kaiser Wilhelm Institute for Hydro- and Aerodynamics works out a compromise with Himmler, which he informs Heisenberg about in March 1938: "He [Himmler] hinted that you yourself were partly at fault for the difficulties. He said if you are convinced of the correctness of Einstein's theory then, in his opinion, there should be nothing in the least against you representing it in word and writing. But you should clearly distance yourself from the person and politician Einstein, who now really is blatantly opposed to today's government. I promised Herr Himmler to pass this advice on to you and hereby do so. It appears to me to be a tolerable way out of your difficulties."

The infamy of calling this proposal "tolerable" illustrates the moral collapse of German scientists during the Nazi era. Standards of justice and fairness were discarded for purely professional considerations.

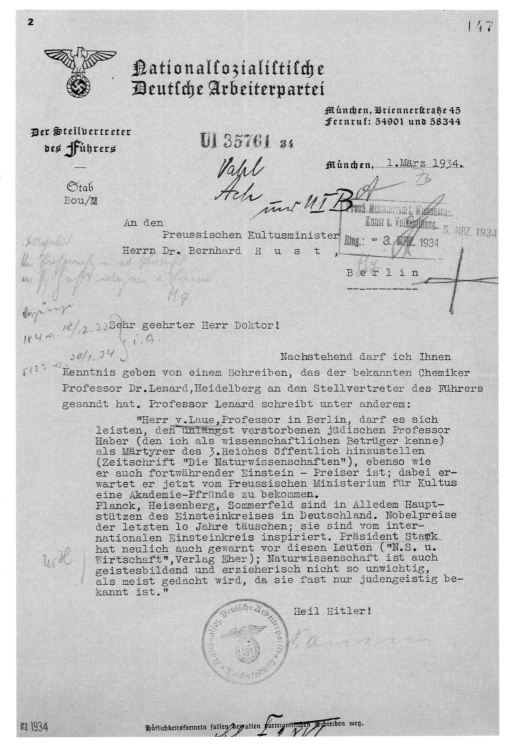

"Many terrible things"

heute auch in schlechter Lage. Die grossen jüdischen Geschäfte sind ja alle aufgeschmissen. Geld haben die Leute in ihrer Gewissenhaftigkeit nie herausgeschafft und so wird dieser alte, ererbte Reichtum beinahe in ein Nichts zerfliessen. Die Zustände in Deutschland sind grauenerregend. Ich weiss nicht, ob Sie wissen, was man uns zugefügt hat. Z.B. hat man unser gesamtes Vermögen, das wir dort hatten, konfisziert. Aber dies war nicht das schlimmste. Mehr Sorgen hatte man um die Kinder, die bis Ende August dort waren. Wenn Sie hier wären, könnte ich Ihnen stundenlang erzählen. Wenig Erfreuliches und viel Schreckliches.

1 (Elsa Einstein to Lili Petschnikoff
3 November 1933
During their visit to the United States in 1931, the Einsteins become friends with the Petschnikoff family. Lili Petschnikoff is a famous violinist who often meets Einstein to play music together and to whom Elsa Einstein confides during difficult times.
"The large Jewish stores are all in dire straits. In their conscientiousness people have not gotten their money out of the country and now the old, inherited wealth

2

will soon amount to next to nothing. The situation in Germany is horrifying. I don't know if you have heard what they have done to us. For example, they confiscated everything we owned there. But that is not the worst thing. We were more worried about the children, who were there until the end of August. If you were here, I could tell you about it for hours: hardly anything pleasant, and many terrible things."

2 (Einstein's Desk

Princeton

The desk is still in Einstein's apartment in May 1933 when the Nazi paramilitary SA ransack the place. With the help of the French ambassador, André François-Poncet, the desk and Einstein's papers are sent under diplomatic seal through France to the United States.

3 (Gestapo Report to the Prussian Minister of Finance about Seized Assets

16 December 1933, with seizure list

"In the attachment I respectfully send 4 lists of police seizures of shares and monies for the benefit of the Prussian state that were deposited in the police's main cash department in Berlin-Schöneberg, the office account of the Interior Ministry, and deposit account 90 of the *Dresdner Bank* with the request that upon receipt it be transferred to the appropriate administration."

This reproduction shows one page of the list mentioning Albert Einstein.

Lfd. Nr.	Tag der Niederlegung	Bezeichnung der niedergelegten Werte	Jahrgang	Nr.	Betrag RM	Niederleger	Bezeichnung der Masse
	11.10.33	4%Ungarische Renten-Anleihe	1892 D	156920	2,-	Dresdner Bank	Prof.Einstein
			1903 D	155615	4,-		
14.	11.10.33	Lübeckische Staatsanleihe von 1928	1928 C	2576	1000,-	Dresdner Bank	Prof.Dr. Einstein
			1928 C	2577	1000,-		
			1928 C	2578	1000,-		
			1928 C	2579	1000,-		
			1928 C	2580	1000,-		
15.	11.10.33	Pr.Pfandbriefbank Geld-Hypotheken-Pfandbriefe	1926 G	08145	50,-	Dresdner Bank	Prof.Einstein
16.	11.10.33	Geldpfandbrief-Zertifikate der Preuß. Pfandbrief-Bank	1926 A	3674	30,-	Dresdner Bank	Prof.Einstein
17.	11.10.33	Aktien der Commerz-u.Privat Bank	1932	168691	100,-	Dresdner Bank	Prof.Einstein
			1932	168692	100,-		
			1932	156504	100,-		
			1932	121489	100,-		
			1932	121490	100,-		
18.	11.10.33	Reichsbank-Anteilscheine	1930	675804	100,-	Dresdner Bank	Prof.Einstein
				675805	100,-		
				675806	100,-		
				675807	100,-		
				675808	100,-		
				675809	100,-		
				675810	100,-		
				675811	100,-		
				675812	100,-		
				675813	100,-		
				675814	100,-		
				675815	100,-		
				675816	100,-		
				675817	100,-		
				675818	100,-		
				675819	100,-		
				675820	100,-		
				675821	100,-		
19.	11.10.33	Aktien der Dresdner Bank	1932	230469	100,-	Dresdner Bank	Prof.Einstein
				230470	100,-		
				230471	100,-		
20.	11.10.33	Aktien der Schultheiß-Patzenhofer Brauerei A.G.	1932	66620	100,-	Dresdner Bank	Prof.Einstein
				66621	100,-		
				66655	100,-		
				66656	100,-		
				48065	100,-		
				48066	100,-		
				48067	100,-		
				48068	100,-		
				92994	20,-		

Einstein in America

Since 1933 the Institute for Advanced Study in Princeton provides Einstein with an ideal workplace. Here he can pursue his interests without distraction or financial worries. But the menacing situation in Europe also reaches this small idyllic American town. The German Reich is arming itself and is pursuing a ruthless foreign policy. Nuclear fission is discovered in Berlin in 1938. Less than a year later there is mounting evidence that Germany is considering its military use. At the urging of another émigré, the physicist Leo Szilard, Einstein signs a letter to the American president in 1939 warning him of the German plans. The letter does not lead directly to the building of the atomic bomb. This is prompted later by others. But the letter does show that Einstein, in the face of the mounting National Socialist threat, is willing to deviate from his pacifist views. Einstein is kept in the dark about the later project, as the authorities doubt his loyalty. He is only involved occasionally in smaller armament research projects.

Einstein on a skyscraper in New York

In America, Einstein is still a critical and uncomfortable contemporary. He remains fearless in the face of the paranoid outgrowths of heightened anti-communism in the 1950s. Einstein recommends everyone called to speak before Senator McCarthy's committee not show up in Washington and to refuse to submit to this ignoble procedure. This stance separates him from his younger colleague, J. Robert Oppenheimer. Formerly one of the directors of the atomic bomb project in Los Alamos, Oppenheimer has meanwhile been appointed director of the Institute in Princeton. He runs into difficulties because of his opposition to the development of the hydrogen bomb, as well as for his political views in the 1930s. Oppenheimer thinks that he can worm his way out of the committee's questioning, but is maneuvered into making statements about former colleagues, and ends up by losing the government's trust. He is not rehabilitated until the 1960s. Einstein's keen political awareness proves right not only in this case. While some may see his political idealism, which is object of public attention in America, as simple-minded, his pronounced aversion to every form of social injustice turns out to be in no way unrealistic. He speaks out against racial segregation and describes deep-seated prejudices as "degrading" and "fatal". He ascribes the difficult situation of the black population not to their lack of ability, but sees it as the result of hundreds of years of oppression. Toward the end of his life, Einstein is deeply troubled by the political state of the world. He warns of the dangers of a nuclear arms race and sees a form of a world government as the only solution that can prevent conflict between individual countries. Einstein's attitudes and appearance in the United States still shape his popular image today. He becomes a symbolic figure, even for protest movements that emerge after his death.

○ HISTORICAL CONTEXT

Institute for Advanced Study

The Institute for Advanced Study is founded in 1930 through a donation from the department store magnates Bamberger. Princeton, a university town, is chosen as the location for the new institute, not far from New York City. The American educator Abraham Flexner is appointed to realize its foundation. The most

distinguished theoretical scientists should dedicate themselves solely to research without the distraction of teaching or administrative duties. In 1933 the Institute begins its activities. Along with Einstein, the mathematicians John von Neumann and Hermann Weyl are among the first members. Almost all of the best physicists and mathematicians of the twentieth century spend some time at least as guests on the institute's campus. Today, along with the original mathematics department there are others for natural sciences, history, and social sciences.

The McCarthy Era

Joseph McCarthy (1908–1957) is a member of the US Senate from 1947 to 1957 at the height of the cold war. From 1953 to 1955 he directs the House Un-American Activities Committee, which is concerned with the dangers of communist agents infiltrating the United States. Based on the discovery of a few espionage cases, McCarthy conducts a vehement witch-hunt against anyone and everyone he deems is influenced by communism. Scientists – including Albert Einstein – writers, artists, trades union activists, anyone active on the left of the political spectrum, including those supporting peace, international understanding, and social rights all come under his scrutiny. This causes many people to lose their jobs or serve jail sentences although most of them have nothing whatsoever to do with the Communist Party. These witch-hunts poison the political climate inside the country and lead to an almost unprecedented persecution of critical positions in the United States.

○ PERSONALIA

Leopold Infeld

Son of a Polish-Jewish leather goods factory owner, Leopold Infeld (1898–1968) studies in his hometown of Krakow. In 1920 he continues his education in Berlin and consults with Einstein about his dissertation on the general theory of relativity, for which he receives his doctorate in Krakow in 1921.

Initially he works as a teacher and, starting in 1929, is an assistant at the University of Lemberg. From 1933 to 1935 he is a Rockefeller fellow in Cambridge, England. He works at Princeton together with Einstein from 1936 to 1938. In 1939 he is appointed professor at the University of Toronto, where he remains until 1950. In that year he becomes full professor for theoretical physics at the University of Warsaw and later on institute director. Infeld is concerned with the equations of motion in the general theory of relativity. Together with Einstein he writes a popular scientific book in 1937, *The Evolution of Physics*. In 1969 his memoir *Sketches from the Past* is published, in which he also recounts his collaboration with Einstein.

Leo Szilard

Leo Szilard (1898–1964), son of a Hungarian-Jewish civil engineer, initially studies electrotechnics in Budapest. He then studies physics in Berlin, where he gains his doctorate in 1922 supervised by von Laue. From 1922 to 1924 he works at the Kaiser Wilhelm Institute for Fiber Chemistry, from 1924 on at

Left: Einstein with Henry Wallace (left), presidential candidate of the Progressive Party, the journalist Frank Kingdon (second from right), and the singer and civil right's activist Paul Robeson, 1948

Right: Einstein in conversation with Institute's Director Robert Oppenheimer, 1947

the institute for theoretical physics of the University of Berlin. In 1933 he is forced to emigrate, first to England, then to the United States in 1938. He receives a position at Columbia University in 1940. He works in Chicago together with Enrico Fermi on the realization of the first atomic reactor. After the war he receives a professorship for biophysics in Chicago.

Szilard's technical interests lead to numerous patents, including one with Einstein in 1927 for a refrigerator and one for the idea of nuclear chain reactions in 1934 – before the discovery of atomic fission. When he learns of this discovery in 1939 the physical and political consequences are immediately clear to him. Using his acquaintance with Einstein, he instigates the famous letter to President Roosevelt.

Moral Courage

1 (Portrait of Einstein by Josef Scharl
1927
Einstein becomes acquainted with the
artist Joseph Scharl (1896–1954) in 1927
through the photographer Lotte Jacobi.
Scharl's paintings are stigmatized by the
National Socialists as "degenerate." As
in so many other cases Einstein acts as
guarantor for Scharl for his emigration
to the United States in 1938.

2 (A Plea for Help for the Violinist
Boris Schwarz
27 July 1936
Einstein writes to the American Jewish
Congress and refers to a promise from the
wife of Stephen Wise (1874–1949) to help
Boris Schwarz. As a Rabbi, Zionist, and prominent spokesman of American Jews, Wise
speaks out on behalf of the persecuted,
giving early and emphatic warnings about
the National Socialists to the political
leaders in Washington.

Saranac Lake N.Y. den 27.Juli 1936

An die Verwaltung des
Congress-House
50 West 68th Str.
New York City

Sehr geehrte Damen:
        Dieser Tage ist ein lieber junger
Freund von mir, der Geiger Boris Schwarz, aus Deutschland
angekommen, wo ihm und seinen Eltern als Ostjuden (Russen)
nicht nur die Existenz als Künstler untergraben, sondern
auch die Ausweispapiere weggenommen wurden. Frau Stephen
Wise hat uns freundlicherweise versprochen, dass für einen
solchen Fall die Hilfe des Congress-House offen stehe und
ich bitte Sie demgemäss, den feinen und bescheidenen jungen
Mann aufnehmen zu wollen.

        Mit ausgezeichneter Hochachtung
                A. Einstein.
        Professor Albert Einstein

A. Einstein Archive
28-296

3

3 (About Pacifism: Einstein's Answer to an Article by B. D. Allison
27 November 1934
"One thing is fundamentally clear to me: a real solution to the pacifist problem can only be brought about by the creation of a supranational arbitrational institution, which, in contrast to the current League of Nations in Genf, has the means to enforce its decisions – in short, an international court of justice with a permanent powerful military or police apparatus."
The impressions of the National Socialists' seizure of power and the German politics of war lead Einstein to change his attitude toward military conscientious objectors. The democratic states must have the means to oppose Nazi Germany.

EINSTEIN'S WORLD TODAY

"If you want to find anything out from the theoretical physicists about the methods they use, I advise you to stick closely to one principle: Don't listen to their words, fix your attention on their deeds!"

What worldview is science working on today, and what role does Einstein's legacy play within it? And what is science anyway: a guarantee of progress, or the modern form of superstition? In any case, the consequences of Einstein's work make apparent the long-term character of the development of knowledge. Not until decades later does the practical importance of many of his insights become clear, be it in the form of lasers or in satellite-based navigation.

Among the scientific challenges Einstein left behind is the search for gravitational waves, and for a unified theory of all forces of nature. Even today, borderline problems between the disciplines play a key role in scientific innovations, from biophysics to the question of the Big Bang.

More than ever, however, scientific questions are also questions for society at large. Which technologies should be our basis for generating energy? How can we prevent the proliferation of atomic weapons? Should scientific information be freely available in the Internet? Einstein's legacy also includes the demand for a responsible, free, and democratic culture of knowledge.

What is Science?

Nietzsche: Just look at the scholars, those worn out hens. All they can do is cluck around more than ever simply because they have to lay more and more eggs. What they don't realize is, the eggs are getting smaller and smaller by the day.

Einstein: Specialization gives the impression that the individual is "replaceable." Despite this, science can only be created by people who are filled with the desire to find the truth.

Nietzsche: That could also be a hidden desire to find death. What's the point of morals when life, nature, history are "immoral"?

Mach: Reality is immoral. But there are two ways to become reconciled with it. People become accustomed to its riddles. Or they learn through history to understand it, so they can then study it without hatred.

Einstein: Your *Science of Mechanics* shows that a historical study is not without effect. It shook the foundations of dogmatic belief in mechanics.

Mach: History has done everything, history can change everything.

Nietzsche: Nevertheless, there is a level of historical understanding that is detrimental to life.

Mach: A person who knows the whole historical process will be much freer to think about the significance of a current scientific development than a person who only perceives the momentary direction of the development.

Nietzsche: A time will come when people abstain from constructions of historical processes.

Perspectives

Lyotard: Quite right. From the very beginning science has been in conflict with the grand narratives of the modern age. When judged by their own criteria, they usually prove to be fables.

Habermas: But your postmodernist theories throw the baby out with the bath water and see crises not as the result of failing to exhaust the potential of reason, but as a logical result of a self-destructive program of rationalization.

Nietzsche: In future people will no longer study the masses, but again focus on the individuals who build a bridge across the turbulent river of development. One giant calls out to another through the desolate spaces of time, and undisturbed by the cacophony of the noisy dwarfish hoards that crawl around between them, the conversation between the higher minds continues.

Einstein: Restricting scientific knowledge to a small group of persons weakens the philosophical mind of a people and leads to its intellectual impoverishment.

Nietzsche: That is the notorious tailoring of the coat of science to fit the body of the "mixed audience."

Einstein: Concern for human beings must be our main focus of interest, so that the products of our intellect are of benefit rather than a curse for them.

Mach: Let's just imagine that we lose all our personal property. In that case we would recall that apart from ourselves there are still simple people, who in such a situation would be vastly superior to us in experience and skills. But what is science itself? It's simply the most economical summary of such experiences made by humankind!

Einstein: But your argument fails to recognize the free constructive element in the formation of concepts. All science is the refinement of everyday thought. No empirical methods can be created without speculative concept constructions; and there is no speculative thought whose concepts do not betray the empirical material to which they owe their origin.

○ PERSONALIA

Ernst Cassirer

According to Ernst Cassirer (1874–1945), human beings can only appropriate the outside world by constructing systems of meaning. Every meaning produced, starting from simple perception, is made by means of symbols. The human

being is an *animal symbolicum*. In his *Philosophy of Symbolic Forms*, Cassirer develops a systematic approach to comprehending the symbolic forms used in the construction of the world – for example, religion, art, and science. The neo-Kantian philosopher grapples intensively with the theory of relativity, in the end regarding it as a confirmation of his opinion that concrete thinking in substance categories should be developed further into relational thinking in function categories, most clearly expressed in mathematics. Cassirer and Einstein hold each other personally in high esteem. Cassirer, who sees freedom as a fundamental condition of culture and values democratic institutions, emigrates after the seizure of power by the National Socialists.

Above:
Ernst Cassirer,
around 1930

Thomas S. Kuhn

The physicist and philosopher of science Thomas S. Kuhn (1924–1996) presents a new theory of the development of science in his work *The Structure of Scientific Revolutions*. The history of science is not seen as the progressive accumulation of knowledge, but is characterized by deep disruptions, so-called paradigm shifts. A paradigm is a basic consensus about guiding theories, knowledge, and research problems shared by scientists in a field and taken for granted in their day-to-day work, "normal science." When unexpected results come to light, anomalies that cannot be made to fit into prevailing theories, the paradigm tilts into a crisis. After a phase of fundamental scientific controversy known as "extraordinary science," a new paradigm is established. The replacement of classical physics by modern physics represents for Kuhn just such a paradigm shift.

○ FOUNDATIONS

Logical Empiricism

The philosophical movement called "logical empiricism" originates in the so-called "Vienna Circle" after World War I with the appointment of Moritz Schlick (1882–1936) to the long-vacant chair of Ernst Mach (1838–1916) at the University of Vienna. Led by Otto Neurath (1882–1945) and Rudolf Carnap (1891–1970) and for a while known as the "Ernst Mach Club," the group around Schlick works on the rejuvenation of classical modern empiricism enriched by the latest developments in formal logic. The Vienna Circle is complemented by a Berlin Group around Hans Reichenbach (1891–1945) and Carl Gustav Hempel (1905–1997). Strongly influenced by the phenomenalism of Ernst Mach and the philosophy of Ludwig Wittgenstein (1889–1951) they try reduce all questions of philosophy, including ethics and aesthetics, to propositions about immediate experience and to clarify their relations by means of the new logical instruments. The source of such empirical propositions is increasingly sought in physics, which at the same time represents the model of the rational knowledge acquisition. The recent developments in physics after Einstein always play a central role. Only after the murder of Schlick in 1936 does the program of logical empiricism become increasingly confined to philosophy of science.

Mortiz Schlick,
around 1920

Postmodernism

In allusion to recent movements in art and architecture, some approaches to the philosophical analysis of science are called "postmodern." The postmodernists often appeal to Friedrich Nietzsche (1844–1900), Martin Heidegger (1889–1976) and post-structuralist French philosophy as shaped by Jacques Lacan (1901–1981), Michel Foucault (1926–1984), and Jacques Derrida (1930–2004). Since the notions of rationality in this philosophical movement are not derived from physics but come instead from art, the humanities, or other areas of human practice, the evaluation of the cognitive claims of science is not always positive – even in those cases where it is well informed. In particular science's claims to universality, necessity, objectivity, and truth are called in question and emphasis is placed on the incompleteness, uncertainty, contingency, and context dependency of our access to the world. Postmodernism, however, is less a philosophical investigation of the actual content of science than a critical reflection on the public image of scientific knowledge and on the systematic analysis of science pursued by contemporary philosophy of science in the aftermath of logical empiricism.

Work on Einstein

Research on Einstein and his work continues. And this includes the question: What is science? An important starting point for Einstein research is his estate, which is kept in the Albert Einstein Archives at the Hebrew University in Jerusalem. Time and again new documents are found that offer insights into his personality, his works, and their context. At the same time, the history of science constantly poses new questions, whether it be research into the structure of scientific revolutions or the relationships between science and its context. The great undertaking of an annotated edition of Einstein's writings is being carried out by Princeton University Press. It includes publications, manuscripts, and correspondence in a total of more than twenty volumes, nine of which have already appeared. This edition project is being complemented increasingly by the new possibilities that the Internet offers to make historical sources freely available and thus to make entire archives immediately accessible for both research as well as for a wider, less specialized audience.

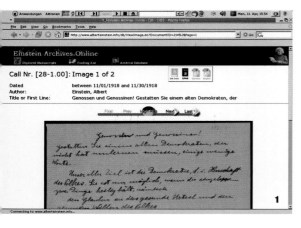

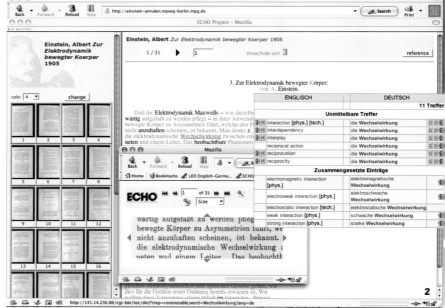

1 (Einstein Archives Online
*Hebrew University, Jerusalem/Einstein
Papers Project, Pasadena*
The Albert Einstein Archives of the Hebrew University together with the Einstein Papers Project are making important parts of Einstein's estate accessible to the public online. There are already more than 3,000 high-quality scans of Einstein manuscripts available, as well as a database detailing the entire archive of 43,000 documents.
http://www.alberteinstein.info/

2 (Einstein's Works in the Annalen der Physik
Wiley-VCH and Max Planck Institute for the History of Science, Berlin
Between 1901 and 1922 Einstein published forty-nine articles in the *Annalen der Physik*, including his most famous works on quantum theory, the special theory of relativity, and statistical physics. These articles are freely available over the Internet as part of the European Cultural Heritage Online (ECHO) initiative, together with many additional publications and research data. The documents are available as high-quality electronic facsimiles and can be searched in full-text mode. With the help of language technology they are linked to online dictionaries.
http://einstein-annalen.mpiwg-berlin.mpg.de/home

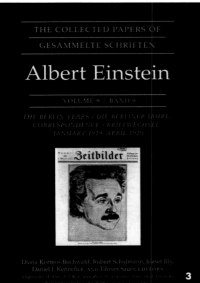

3 (Einstein Papers Project

California Institute of Technology,
Pasadena/Princeton University Press
The Einstein Papers Project has been
publishing the collected works of Albert
Einstein since 1986. More than 14,000
documents are planned to be published
in the original language with English translations. The texts are edited and annotated
by a team of Einstein researchers.
http://www.einstein.caltech.edu

4 (Gehrcke Estate

Max Planck Institute for the History
of Science, Berlin
In the 1920s the experimental physicist
and Einstein opponent Ernst Gehrcke
collects thousands of pamphlets and newspaper articles about Einstein. He tries to
show that the theory of relativity is merely
a scientific mass suggestion.
The Max Planck Institute for the History
of Science has acquired a large part of this
estate which allows a comprehensive look
at the public reception of Einstein and his
work. The documents are freely available
over the Internet as part of the ECHO
project.
http://echo.mpiwg-berlin.mpg.de/
content/relativityrevolution/gehrcke

Putting Relativity to the Test

For the special theory of relativity there is a direct pathway from its predictions to experimental testing: it rapidly becomes the everyday tool of nuclear and particle physicists. In combination with quantum theory it forms the foundation of modern particle physics and, in this context, is time and again subject to rigorous experimental tests.

In the case of the general theory of relativity, progress from concept to experimental verification is much slower: almost forty years elapse after publication before the heyday of experimental general relativity arrives. Progress is made in two major directions: first of all, there are tests involving the cosmological models and their predictions. Secondly, advances in observation methods, space technology, and atomic and quantum physics since the beginning of the 1960s allow for direct testing of the theory's predictions. One of the effects tested to great accuracy is the Shapiro effect – the time delay of radar and radio signals that pass through the solar system in the vicinity of the Sun – at first by measuring signals reflected from planets such as Mars, and later by using space probes such as Mariner 6 and Mariner 7, or recently, in summer 2002, the Cassini space probe. The influence of gravitation on time – as shown by clocks – can also be measured: Pound and Rebka first succeed in this in a terrestrial experiment, using methods from nuclear physics. Later, the effect is measured directly, for instance at the beginning of the 1970s in the context of the Gravity Probe A mission, the currently most exact direct measurements, for which an atomic clock is sent on a rocket trip. Also, the classical measurements that originally led to the breakthrough of the general theory of relativity are repeated with high precision. The relativistic perihelion shift — a minute deviation unexplainable in the framework of Newton's theory that was first demonstrated in Mercury's orbit – has meanwhile been confirmed also in the case of Venus, the Earth, and Mars. Measurements of the radio waves of distant quasars that sweep past the Sun confirm

Representation of the Cassini experiment

Einstein's predictions for the gravitational deflection of light with an accuracy of two parts in a thousand. Since the end of the 1970s it is possible to test Einstein's theory using gravitational fields that are far stronger than those in the solar system. This is achieved with the help of a binary star system consisting of two neutron stars. The most famous of such star systems points to future possibilities of testing Einstein's theory: over the years the orbital period of the binary neutron stars PSR 1913+16 becomes shorter and shorter - exactly in keeping with the general theory of relativity's prediction of how the emission of gravitational waves should influence an orbital movement. In addition to such cosmic laboratories, current research is carrying out new experiments with space probes: the Gravity Probe B satellite has been collecting data since autumn 2004 on the effects of spacetime distortions on the axes of rotation of free-falling gyroscopes, and the currently planned STEP experiment is designed to test the principle of equivalence with unprecedented precision.

○ HISTORICAL CONTEXT

Alternative Theories on Gravitation

Despite the impressive confirmation of the general theory of relativity, alternative theories play a role today in the search for a quantum theory of gravitation and for solutions to cosmological questions. They also play a role by providing a theoretical context for empirically testing Einstein's theory. Historically, the general theory of relativity originates in the intense interaction with alternative theories. These theories also pursue the aim of replacing Newton's theory of action at a distance with a field theory which maintains that gravity propagates at a finite speed. With Einstein's support, Nordström develops such a theory in 1913 within the framework of the special theory of relativity. Nordström's theory becomes the main competitor with Einstein's. It does not, however, predict light deflection. This development results in major insights into the structure of a field theory of gravitation that even without Einstein's contribution may have eventually led to a theory similar to the general theory of relativity, though admittedly a great deal later.

○ PERSONALIA

John Archibald Wheeler

John Archibald Wheeler (*1911) is one of the leading theoretical physicists of the twentieth century. He first gains wider recognition through his key work published jointly with Bohr in 1939 on the theory of nuclear fission. He then joins the Manhattan Project and later works on the development of the hydrogen bomb. One of his notable achievements is to regenerate interest among American theoretical physicists in the general theory of relativity during the 1960s, after it had been overshadowed by atomic, nuclear, and particle physics. He introduces new physical and mathematical concepts into general relativity research and creates the apt name "geometrodynamics" for the theory. The methodological key concept pursued by Wheeler together with his students is the idea that the essence of a theory is only revealed by taking it to all of its extremes. His name is associated with essential concepts of the modern general theory of relativity, such as black hole, worm hole, or quantum foam. His most important work is carried out at Princeton University and from 1977 on at the University of Texas.

○ FOUNDATIONS

Pulsars as Cosmic Laboratories

A binary star consisting of two neutron stars, one of which is a pulsar, is an ideal cosmic laboratory for testing the predictions of the general theory of relativity in a situation with powerful gravitational fields. Radio pulses reach us from the pulsar at extremely regular intervals, but the arrival times experience minute shifts due to relativistic effects. The effects that come into play depending on the orbital movements of the system are various. They include gravitational redshift, the transversal Doppler effect, and the influence of gravitational fields

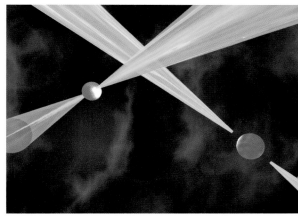

Left:
John Archibald Wheeler visits Black Hole in Nova Scotia, 1981

Above right:
Artist's impression of a binary pulsar

on the propagation of light. From observations of the pulse arrival time, one can calculate the orbital parameters of the system – the precondition for the first indirect proof of gravitational waves. In 2003 a binary neutron star was actually discovered consisting of two pulsars in close orbit, making measurements of unprecedented accuracy possible.

Current Experiments on the Theory of Relativity

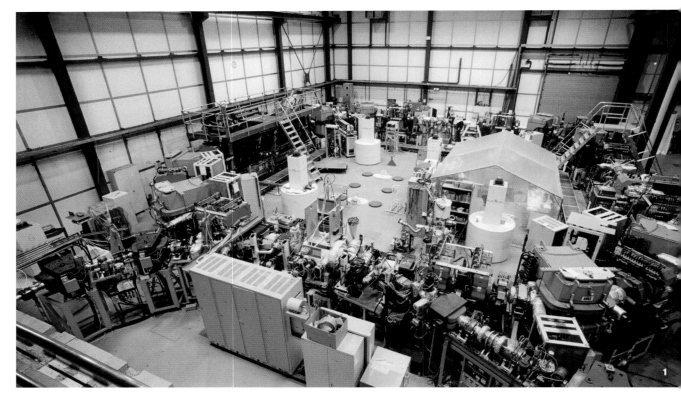

1 (Time Dilation Experiment
Max Planck Institute for Nuclear Physics, Heidelberg
As progress is made regarding experimental methods, new possibilities arise to test well-known effects, such as time dilation as predicted by the special theory of relativity, with ever-increasing accuracy. Currently the most precise tests use a particle accelerator at the Max Planck Institute for Nuclear Physics in Heidelberg. Lithium ions circulate at high speed in the accelerator ring. Like all atoms, they absorb or emit light at particular, characteristic frequencies. The value of the frequency – how fast

one light wave peak follows the next – is a direct gauge for how fast time passes in the reference system of the lithium ions. From the perspective of an external observer, clocks co-moving with the lithium ions must go slower in accordance with relativistic time dilation. Similarly the light emitted by the ions oscillates more slowly, corresponding to a lower frequency. Among other techniques the scientists in Heidelberg use laser technology to make sure only ions of a single, narrowly defined speed are excited. The time dilation for these ions can be tested with a relative precision of less than one part in a million.

2 (Gravity Probe B Satellite
NASA, Stanford University, Lockheed Martin, ZARM Bremen, scale model 1:5
Gravity Probe B is a space-born experiment to test two predictions of the general theory of relativity concerning the behavior of gyroscopes in a gravitational field. According to Newton's mechanics a gyroscope, free of any external influences, will maintain its axis of rotation within space. However, in curved spacetime the axis of rotation of such a gyroscope should chane its direction, compared with distant reference points: on the one hand because of curvature effects in the neighbourhood of a mass (de Sitter precession), on the other hand because spacetime around a rotating body exerts a "dragging" influence on objects in the vicinity (frame dragging or Lense-Thirring effect). In the Gravity Probe B

experiment, rotating spheres about the size of table tennis balls are put aboard a satellite and sent into an orbit that takes them over the Earth's poles. Measurements are made of the minutest changes in the spheres' axis of rotation in comparison with a distant guide star. In that situation, the de Sitter precession should lead to a change in the direction of the axis of rotation, of about two thousandths of a degree, while frame dragging should contribute about one hundred-thousandth of a degree per year. The Gravity Probe B gyroscopes are the most perfect spheres ever to be manufactured. If the Earth were such a perfect sphere none of the mountains would be more than two meters above the surface of the sphere. The whole experiment is contained in a cooling system that keeps the temperature at just 1.82 degrees above absolute zero. In this way the effects of superconductivity can be exploited to determine shifts in the axis of rotation. The first results are expected in 2006.

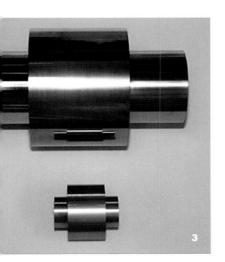

3 (Test Masses for the STEP Experiment

Irrespective of their material characteristics, all bodies fall in the gravitational field with the same acceleration – according to the principle of equivalence, the departure point of Einstein's relativistic theory of gravitation. The aim of the Satellite Test of the Equivalence Principle (STEP) experiment is to test this statement with extreme precision. Two test masses of different materials incapsulated in one another are brought into a special form of free fall: into orbit around the Earth. To achieve und isturbed free fall, they are placed inside a satellite constructed in such a way that all influences that could disturb the free fall are minimized, including the Earth's atmospheric friction, the solar magnetic field, and the drag exerted by particle currents from the Sun. If the test masses were to experience different free fall accelerations, they would over a longer period – in this case some thousands of seconds – shift in relation to each other, and these shifts could then be measured. The precision of previous experiments could be improved one hundred thousand times and the principle of equivalence could be confirmed with a precision of about 10^{-18}.

This is not just interesting as an amazing experimental achievement, but above all because some extensions of the standard model of elementary particle physics predict an additional fundamental force of nature that would lead to an apparent violation of the principle of equivalence that could be detected with the STEP experiment.

The Large Scale Structure of the World

The general theory of relativity permits a description of the large scale structure of the cosmos. The details depend not only on the properties of matter in the cosmos, but also on our knowledge of the dynamic processes in the universe. According to the cosmological standard model the universe originated some thirteen billion years ago from what is called the "big bang." Since then the universe has been expanding and cooling, and structures have evolved ranging from the galaxies to atoms that we know today. How convincing is this scenario? Although Hubble's 1929 discovery that galaxies are constantly moving away from each other indicates that the universe is expanding, the big bang model does not gain direct observation-

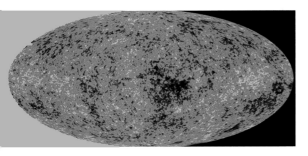

Microwave image of the sky

al support until the 1960s. At this time a great diversity of observations are found to correspond to the model's predictions, including the abundance of light elements in the cosmos and, most importantly, the existence and properties of the cosmic background radiation. In the beginning the universe consists of a hot gas composed of particles that can be described with the help of thermodynamics, the theory of relativity, and elementary particle physics. After more than 300,000 years the temperature has sunk to below 3,000 Kelvin so that charged particles form neutral atoms and the electromagnetic radiation can freely propagate through the cosmos: the universe becomes transparent. This primordial radiation has the characteristics of thermal radiation, as investigated by Planck. It cools down as the universe expands. The radiation that reaches us today from all directions as a relic of the big bang has a temperature of a mere 2.73 degrees above absolute zero. Minimal variations in the spectrum of this radiation are interpreted as the starting point for formation of cosmic structure. In the end, this leads to the universe as we know it today, with galaxies in clusters that are themselves part of super-clusters, which in turn are distributed along a network of filaments with huge gaping voids in between. While the standard model is to a great extent confirmed by astronomical observations, three major problems remain open: missing mass, the cause of the universe's accelerated expansion, and the transition from micro- to macrocosm in the very early universe. Evidence for the first problem is found, for example, through observations of the rotation of galaxies, allowing estimates of the mass they contain. This has led to the assumption of the existence of "dark matter" which accounts for about eighty-five per cent of matter. The second problem emerges in the mid-1990s from observations of distant supernovae which surprisingly indicate an acceleration of our universe's expansion. In conjunction with our knowledge that the geometry of space is flat (or Euclidean, as can be derived from properties of the background radiation) and with the known density of matter, it follows from the general relativity that there must exist a cosmic repulsion that causes this expansion. It can be described by the cosmological constant first introduced by Einstein, but later rejected. This repulsion corresponds to a "dark energy" that constitutes about seventy per cent of the energy in the universe. The third problem concerns the relationship between physical processes in the early cosmos and its present-day structure. This problem led to the assumption of an "inflationary" phase in which causally connected regions of the early universe are separated, as one small homogeneous region expands to a size much larger than what we can observe today. Also, the expansion explains the current flatness of space, while quantum fluctuations in the model account for the minute irregularities in the cosmic background radiation and are the starting point for the formation of cosmic structures. In order for a world such as ours to actually exist, the curvature and the density of the various constituents of the universe at the time of the big bang must have been tuned to each other with incredible precision.

History of Radio Astronomy

In 1931 employees of the telephone monopoly American Telephone and Telegraph test the possibility of setting up a transatlantic telephone service using radio waves with wavelengths of between 10 and 20 meters. The engineer Karl Jansky tackles the problem of atmospheric disturbances. He simply cannot get rid of, or explain, a particular hissing at wavelengths of around 15 meters. After taking detailed measurements he realizes that it is a signal coming from the center of our galaxy. This marks the beginning of radio astronomy, the observation of astronomical objects using the radio waves that they emit.

The most important radio astronomical measurement for Einstein's theory of relativity is also unplanned: in 1962 the physicists Arno Penzias and Robert Wilson start using a radio antenna, originally designed for telecommunications, to carry out radio astronomical investigations. At wavelengths of around 7.3 centimeters they observe a constant signal that seems to come from all directions, and at first they think it is noise. After excluding all possible other causes, including pigeon droppings, they start looking for a theoretical solution.

In 1965 a lecture by Jim Peebles starts spreading the idea of cosmic background radiation first introduced by Gamow, Dicke, and others. This observation is soon generally accepted as a confirmation of the big bang theory that then advances to become the standard theory of cosmology.

The Hunt for Neutrinos

Neutrinos, the electrically neutral elementary particles predicted in 1930 by Pauli and then discovered in 1955, interact extremely rarely with matter. Every square centimeter of our skin is penetrated every second by billions of neutrinos from the interior of the Sun. Only a material shield hundreds of millions times the diameter of the Earth would be able to stop them. In contrast to these, the electrically charged protons of cosmic radiation are even deflected by galactic magnetic fields and thus lose the information about their origins. This is why neutrinos make ideal probes for investigating cosmic catastrophes, such as supernova explosions, which are a source of highly energetic cosmic radiation. A few of these high-energy neutrinos, which presumably arrive for the most from precisely localizable directions in space, react with an atomic nucleus while passing through the Earth to produce a highly energetic muon, a heavier relative of the electron. In optical media, such as water or ice, its speed initially exceeds that of light. Similar to the sound cone behind a supersonic aircraft, the Čerenkov radiation is thus generated, seen as a flash of bluish light at the location of the observer. The largest neutrino telescopes, NT-200 in the depths of the Baikal Sea, and AMANDA buried in the Antarctic ice cap at the South Pole, lie in wait for such tell-tale flashes with their probes, ball-shaped photomultipliers. The world's biggest research project at the beginning of the twenty-first century is the extension of AMANDA into the far more sensitive, one cubic-kilometer-sized IceCube telescope.

Cosmic Background Radiation

So-called cosmic microwave background radiation reaches us from all parts of the sky. It corresponds to the thermal radiation emitted by a body at 2.73 Kelvin, colder than – 270 degrees centigrade, and is a clear confirmation of the big bang model. According to this model the universe became transparent about 300,000 years after the big bang: whilst the universe was filled with hot plasma before that time, the temperature then sank low enough for the formation of stable atoms. That marked the end of the direct interaction between matter and the electromagnetic radiation that filled the universe. From then on the latter remained unchanged, except for the fact that it became cooler and cooler with the expansion of the universe, until it reached its current temperature. Interestingly, the temperature of the background radiation fluctuates by tenths of thousandths of a degree, depending on which part of the sky the astronomers are watching. These fluctuations, detected by the COBE satellite telescope, provide informative direct images of the minute density fluctuations through which – starting with an initially very homogeneous universe – the cosmic structures as we see them today evolved. More recent, far more precise measurements with the Wilkinson Microwave Anisotropy Probe (WMAP) launched in 2001 have uncovered additional indications of an inflationary phase of accelerated expansion which the universe underwent shortly after the big bang.

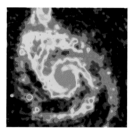

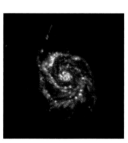

Images of the spiral galaxy M51 taken at different wavelengths

The Cosmic Messages of Radiation

1 (Effelsberg Radio Telescope
Max Planck Institute for Radio Astronomy, Bonn, model
The Effelsberg Radio Telescope of the Max Planck Institute for Radio Astronomy went into operation in 1972 and is the second largest fully steerable dish antenna in the world. It is used mainly for the observation of objects and phenomena in the Milky Way, such as star birth regions or super-nova remnants. The antenna's large surface is designed to pick up weak signals, while the large diameter of 100 meters is decisive for its high angular resolution

2 (Dual Horn Antenna for Microwaves
Jodrell Bank Observatory of the University of Manchester, Instituto de Astrofísica de Canarias
At the beginning of the 1990s the antenna serves to measure cosmic microwave background radiation. By comparing the measurements obtained with both antenna horns, which point in slightly different directions, it is possible for the first time to measure the temperature differences between background radiations reaching us from different directions in the sky.

3 (Infrared Speckle Camera SHARP I
Max Planck Institute for Extraterrestrial Physics, Garching
The camera takes many infrared photographs with short exposure times of a small section of the sky. Each picture looks slightly different because of atmospheric disturbances, a phenomenon that we see in daily life as the twinkling of stars. The effect of the disturbances can be corrected by combining the individual pictures in a computer. In 1997/98 this camera succeeds in making observations of stars at the center of our Milky Way leading to the conclusion that a black hole is located there.

(angular resolution being the smallest distance in the sky at which the telescope can distinguish between two different objects).

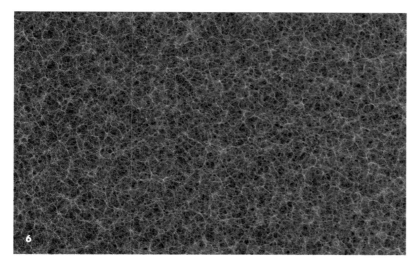

6 (Virgo Millennium Simulation
*Virgo Supercomputing Consortium:
Max Planck Institute for Astrophysics,
Garching / Edinburgh Parallel Computing
Centre*
Supercomputer simulation of structure
formation in the universe. The computer
follows the movement of over ten billion
particles under mutual gravitational attraction. The simulation shows the formation
and evolution of galaxies, galactic clusters,
and structures of the intergalactic medium
that correspond to current astronomical
observations.

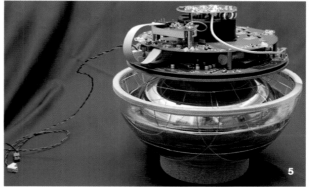

4 (European Photon Imaging Camera of
the XMM-Newton Satellite
*Max Planck Institute for Extraterrestrial
Physics, Garching*
Since 1999 a camera of this type is being
used in the XMM-Newton satellite of the
European Space Agency (ESA) for the observation of cosmic x-ray radiation. As the
x-rays emitted by astronomical objects are
absorbed by the Earth's atmosphere, they
can only be observed from space. With a
sensitive area of 36 square centimeters,
the camera has the largest x-ray detector
ever constructed. It consists of 12 charge-
coupled devices (CCD), electronic light
sensors allowing very high image resolution.

5 (IceCube Probe
*Deutsches Elektronen-Synchrotron DESY,
2005*
One of the 4,800 probes of the IceCube
neutrino telescope being built at the South
Pole. The probes are plunged into the ice
on cables to depths between 1,400 and
2,400 meters. At the extremely rare collisions between cosmic neutrinos and atoms
of the ice, elementary particles called
muons arise. The probe enables the observation of the blue light flashes caused by
the muons as they pass through the ice,
and so indirectly proves the entry of cosmic neutrinos. The first IceCube cable with
sixty probes was successfully lowered into
the ice at the beginning of 2005 and is now
in operation.

Gravitation as a Challenge

Since the end of the 1960s developments in the theory of relativity and breakthroughs in astronomy have led to a new image of the cosmos. Elements of this new picture are phenomena such as black holes, gravitational lenses, and gravitational waves. In all of them, gravitation is more than a force that binds masses together – it is a dynamic field, and it influences the propagation of light. One of the key challenges of gravitation is the nonlinearity of the general theory of relativity that makes it difficult to find solutions to its equations. Black holes are objects with such a strong gravitational field that not even light can escape. Already in the eighteenth century the assumption that light consists of particles that are influenced by the gravitational force leads to the idea of such objects. This assumption falls into oblivion with the

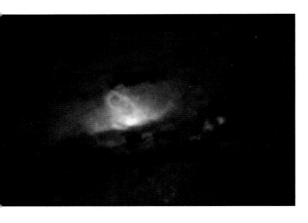

triumph of the wave theory of light. But when the theory of relativity shows there is a critical radius, below which neither light or anything else is able to leave a star, even Einstein has doubts about this conclusion and its significance in astronomy. Nevertheless, evidence for the existence of black holes has been mounting since the 1960s.

Gravitational lensing occurs when the light of a distant object is bent by a mass concentration in such a way that the mass acts like a lens for the observer on Earth, making the object appear closer and brighter, or even producing multiple images. Einstein predicts this effect as early as 1912, but thinks it is impossible to observe, and only publishes this hinsight in 1936 when prompted by an amateur scientist. Only in 1979 this prediction is confirmed after the discovery of quasars. Today, the observation of gravitational lensing helps in de-

Galaxy NGC 4438: At the brightly shining center is a black hole

termining cosmic structures. Like the electromagnetic waves generated by accelerated charges in electrodynamics, accelerated masses generate gravitational waves. As early as 1916 Einstein analyzes them within the framework of the theory of relativity. In 1974 Hulse and Taylor interpret the frequency change in a radio signal from a pulsar as a result of the fact that it is a binary star system that is losing energy through the radiation of gravitational waves. But there is no direct proof comparable to that of Hertz in the case of electromagnetic waves. The changes in distance by which these waves reveal their presence are extremely small. Even the distance between the Sun and the Earth would only fluctuate by the diameter of a hydrogen atom as a result of the waves from a supernova explosion. Gravitational wave astronomy promises insights into previously unobservable cosmic events ranging from the collision of black holes to the big bang.

The biggest challenge posed by gravitation is its relationship with other forces – a problem already investigated by Einstein. Attempts to find some kind of solution focus on borderline problems between microphysics, as described in the quantum theory, and macrophysics, as described by the general theory of relativity. In 1916 Einstein pointed out that, if quantum theory is not taken into consideration, an electron in an atom must also emit gravitational waves. The empirical inaccessibility of such borderline problems is a result of the fact that effects in which gravitation acts with a level of strength comparable to the other forces only occur at extremely high energies and small distances. Such promising borderline problems are represented by the big bang and black holes. The thermodynamics of black holes in particular – like the thermodynamics of radiation at the beginning of quantum era in early twentieth century – provides subtle hints as to what a quantum theory of gravitation might look like. The most important candidates for a theory of quantum gravity are string theory and loop quantum gravity.

◗ FOUNDATIONS

Gravitational Waves

One of the key predictions of the general theory of relativity is that distortions in the geometry of space spread in a wavelike manner at the speed of light. The aim of the researchers who attempt to directly detect gravitational waves with the

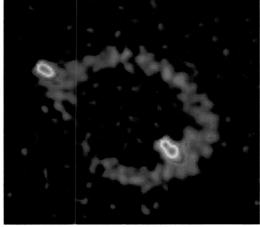

Radio image of an Einstein ring

help of highly sensitive detectors is the establishment of a new branch of astrophysics: gravitational wave astronomy. Gravitational waves convey information about some of the most powerful events in the universe, such as the merging of two black holes or the processes at the center of a supernova. The observation of very slowly vibrating gravitational waves would even make it possible to gain direct information about the events in the extremely hot early phase of our cosmos, mere fractions of a second after the big bang. This would open up areas of investigation previously inaccessible to researchers making traditional observations of light or electromagnetic waves.

Black Holes

According to Einstein's theory, when mass is extremely concentrated in a small region of space it builds a geometrical prison: a black hole, a region in which space and time are so distorted that light and matter can never escape once they

have crossed the "horizon" that separates the black hole from the outside world. Inside the black hole, the general theory of relativity reaches its limits: this is where the theory predicts a spacetime border, a "singularity" where Einstein's model for space and time loses its validity. In real black holes it is supposed that the quantum effects ignored by Einstein's theory are of prime importance.

For present-day astrophysics two types of black hole are of significance: stellar black holes, with several times the mass of our Sun, are the final stage of massive stars that died in a highly energetic supernova explosion. Supermassive black holes with more than a million times the Sun's mass are thought to exist in the central

regions of most galaxies. According to current scientific knowledge, the energy released when matter falls into a supermassive black hole is responsible for some of the most energetic astronomical phenomena, for instance the jet-like eruptions from radio galaxies.

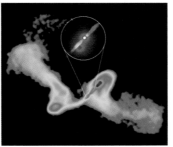

Above:
The Crab nebula recorded by the Hubble telescope

Below:
Radio galaxy NGC 326

Quantum Gravity

Modern physics is based on two pillars: general theory of relativity as a theory of gravitation, and quantum theory giving a microscopic description of matter and radiation. There are situations in which both theories should become important, for

instance close to the big bang during the extremely dense and energetic early phases of the universe, and inside black holes. However, attempts to transform Einstein's theory of gravitation into a quantum theory in the traditional way lead to problematic results. A quantum theory of gravitation must be approached differently. There are many promising approaches: string theory, for instance, a form of quantum mechanics in which the elementary particles are represented by one-dimensional strings, or loop quantum gravity. Admittedly, these approaches have to contend with huge mathematical difficulties. A great deal of research still remains to be done before a complete theory of quantum gravitation can be developed, which could provide reliable descriptions of the singular, extreme situations close to the big bang or in black holes.

The Cosmic Messages of Gravitation

The unusual phenomena predicted by Einstein's theory provide new insights into the structure of the universe.

Gravitational lensing has established itself as a tool of cosmology and enables scientists to detect dark matter. The greatest challenge is the first direct detection of gravitational waves. To this end groups of physicists throughout the world have set up extremely sensitive detectors, such as the GEO600 detector located in Ruthe near Hanover. It is being operated by the Max Planck Institute for Gravitational Physics (Albert Einstein Institute) in Golm/Potsdam, the University of Hanover, and British scientists from the Universities of Cardiff and Glasgow.

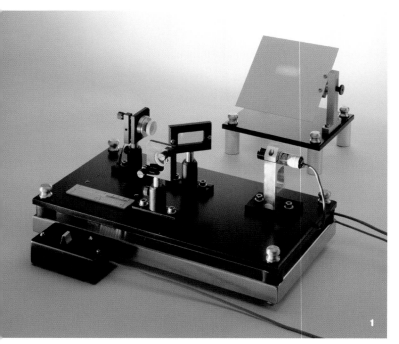

1 (Model of the Functional Principle of GEO600
Max Planck Institute for Gravitational Physics (Albert Einstein Institute), Potsdam/University of Hanover, 2004
GEO600 operates as a Michelson interferometer according to a principle that already played an important role in the creation of the special theory of relativity. Laser light is beamed along two different paths and then superposed. If, at the point of superposition, a wave crest coincides with another wave crest then amplification results; if a wave crest coincides with a wave trough the two partial beams can cancel each other. The model demonstrates how even the smallest changes in the length of the light path lead to marked changes in brightness. GEO600 is designed to detect in this way changes in length caused by gravitational waves.

2 (Model of the Mirror Mounting in GEO600
Max Planck Institute for Gravitational Physics (Albert Einstein Institute), Potsdam/University of Hanover, 2005
The detection of gravitational waves is only possible if the mirrors – between which the laser light needed for the measurement of the distances moves back and forth – are protected from external disturbances. The model shows the combination of pendulums, cantilever springs, and actuators designed to support the mirrors of GEO600.

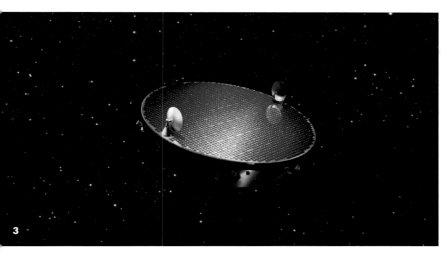

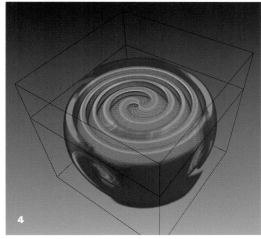

3 (Model of One of the LISA Satellites
Max Planck Institute for Gravitational
Physics (Albert Einstein Institute),
Potsdam/University of Hanover, 2005
Terrestrial detectors are unable to detect
slowly vibrating gravitational waves
because disturbances are too strong. The
Laser Interferometer Space Antenna (LISA)
is designed to remedy this problem. The
three satellites will fly in formation and
emit laser beams forming a triangle with
sides of five million kilometers in length.
This will help detect minute changes in
distance caused by gravitational waves.

4 (The Generation of Gravitational Waves
Simulation: Max Planck Institute for Gravi-
tational Physics (Albert Einstein Institute),
Potsdam; visualization: W. Benger, Max
Planck Institute for Gravitational Physics
(Albert Einstein Institute), Potsdam/
Konrad Zuse Institute, Berlin, 1998–2005;
computer simulation
Einstein's equations are so complex that
modeling realistic situations is only possible with the help of a computer. The results
are also important for the detection of
gra-vitational waves – they provide patterns for the analysis of detector data.

5 (Gravitational Lens
Max Planck Institute for Extraterrestrial
Physics/Max Planck Institute of Quantum
Optics, Garching, 2005, show model
The effect of a compact mass acting as
a gravitational lens that deflects the light
from objects behind it can be simulated
by means of a suitably formed glass lens.
Gravitational lens effects like the Einstein
ring or the Einstein cross then become
directly visible.

6 (Black Hole as a Gravitational Lens
Department of Theoretical Astrophysics,
University of Tübingen, 2004, interactive
simulation
Through simulation it is possible to inter-
actively investigate the gravitational lens
effect of a black hole that deflects light
from distant astronomical objects.

The Small Scale Structure of the World

In the mid-1930s the task of understanding the structure of matter seems to be essentially achieved. Based on Bohr's model of the atom it is possible to successfully describe the behavior of electrons in the electromagnetic field of atomic nuclei, both in quantitative and qualitative terms, using the newly developed theory of quantum mechanics. But some puzzles still remain, especially the question of what makes the nuclear components, the protons and neutrons, hold together as well as the cause of radioactive decay. It proves that apart from gravitation and electromagnetism, there are two other fundamental forces, the strong and the weak force. The strong force explains the cohesion of the atomic nucleus and the weak force explains certain kinds of radioactive decay.

In accelerator experiments numerous new particles are generated. In 1964 it turns out that it is possible to conceive a structural order of these particles if one assumes the existence of even smaller particles – which however cannot occur independently. These so-called quarks form the protons and neutrons. Their charges are measured in thirds of the elementary charge. It is not until 1995 that the last of the predicted quarks is experimentally detected.

A first completion of the standard model of elementary particle physics is achieved in the 1970s. It encompasses strong, weak, and electromagnetic interaction and explains the atomic structure of matter, especially nuclear binding and decay. The basis is the relativistic quantum field theory into which field theory, special relativity, and quantum physics merge. It leads to a new understanding of familiar concepts. Forces are interpreted as the exchange of quantum particles that represent the transmission of actions by means of fields. The photon transmits the electromagnetic force between two electrons,

Aerial view of CERN showing the location of the particle accelerator

just like a thrown ball can unleash a force of repulsion between two players. Quarks exchange "gluons," the agents of the strong interaction. Whether there is a "graviton," the quantum of the gravitational interaction, is still unclear. The vacuum is seen – like the aether in earlier times – as a medium capable of oscillations which, however, is unable to come entirely to rest, due to the uncertainty principle of quantum theory.

The unification of the different forces within the framework of the standard model is based primarily on insights into symmetry properties. Interactions of particles can be described by statements about quantities, such as the charge, which remain unchanged in these processes. Such conserved quantities, which generalize the classical concept of charge, correspond to symmetry properties of the underlying physical processes. Similarly, according to Piaget's developmental psychology, the understanding that the quantity of liquid remains the same when poured from one container into another, differently shaped one depends upon the fact that the operation can be reversed, meaning it is symmetrical with respect to the time direction. Indications of the connection between the different basic forces result from the fact that, in the framework of a quantum field theory, certain symmetries can be reproduced as "gauge invariances." The idea goes back to the unsuccessful attempt by the mathematician Weyl in 1918 to merge Einstein's theory of gravitation and Maxwell's theory of electromagnetism into a unified theory by introducing a variable length gauging. The major challenges of present-day particle physics include the unification of all forces and the explanation of the "particle zoo" systematics, especially the different masses of particles and the different strengths of the fundamental interaction forces (the so-called "hierarchy problem").

○ FOUNDATIONS

The Standard Model of Particle Physics

The standard model provides a unified description of strong, weak, and electromagnetic interaction. The elementary particles of matter are grouped into three families consisting of two quarks and two leptons each. (Electrons, for instance, are leptons.) The interactions between the particles of

matter are caused by the exchange of force particles: photons convey the electromagnetic interaction, W and Z bosons convey the weak interaction, and gluons the strong interaction. All of the postulated matter and force particles have meanwhile been detected. It is possible to describe the interactions of particles by means of conserved quantities which correspond to symmetry properties. If, however, one of two symmetry states still gains preference, this is referred to as "symmetry breaking." This mechanism is responsible for the generation of the different masses of the particles, and is mediated by another particle called the Higgs boson. The discovery of this boson would represent another milestone in the consolidation of the standard model.

Nuclear Fission and Nuclear Fusion

Atomic nuclei consist of protons and neutrons: the nucleons. The number of protons, which determines the chemical characteristics of the atom, increases from the hydrogen atom with just one proton as its nucleus, to the transuranics with more than 92 protons. The nucleons themselves are stable as a result of the strong interaction

that mediates their mutual binding within the atomic nucleus. This binding energy at first decreases with an increasing number of nucleons, reaches a minimum in iron, and increases again afterwards. If a heavy nucleus is split into two lighter ones (nuclear fission), or if two light nuclei fuse into a heavier one (nuclear fusion), the surplus binding energy is released. It corresponds to the Einstein's energy equivalent of a minute mass difference between the original and the final products. The fusion of four hydrogen nuclei to form helium is the main nuclear reaction taking place in the Sun's interior.

Plasma and Plasma Traps

An electrically charged atom – that is an atom whose shell does not possess the same number of electrons as the number of protons in the nucleus – is called an ion. And an ionized gas is called plasma. Most of the matter in the universe is in the plasma state. On the Earth we experience it in atmospheric phenomena such as lightning or

Left:
Tunnel of the Large Electron-Positron (LEP) collider at CERN

Right:
Segment of the twisted plasma vessel of the Stellarator Wendelstein 7-X

the Northern Lights. Inside the Sun electromagnetic fields and gravity work together to produce the plasma density and temperature needed for nuclear fusion. But in technical plants the confinement of plasma is a challenge. The favored technique is the magnetic cage. Two main types of construction prevail today, the Tokamak and the Stellarator. Their twisted magnetic field structures are generated in different ways.

Towards the X-Ray Laser

A microscope for atomic dimensions cannot work with visible light, because its wavelength exceeds atomic scales by far. Similarly, a long water wave passes over a small object, whereas a short wave is reflected making it "visible." An electron microscope is based on de Broglie's understanding of the wave nature of moving particles – an analogue of Einstein's wave-particle dualism of the light quantum (or photon). The higher the energy of the particles, the higher their frequency, and consequently the shorter their wavelength and the smaller the dimensions that a particle beam can image. In alternating electromagnetic fields electrons can be accelerated to almost the speed of light. If they are then forced through alternating magnetic fields to follow a slalom-like course, they emit photons that can be perfectly superimposed. Because of the relativistic length contraction the electrons "perceive" the external magnetic structures as extremely foreshortened, so that the wavelength of the emitted light is also shortened. A linear accelerator of several kilometers in length thus becomes an extremely powerful x-ray laser. Its extremely short and bright flashes of light are used for investigating biological and chemical structures and in materials research. This makes it possible for the first time to film molecular processes in biological and chemical reactions.

High Energy Physics: Concepts and Tools

In order to unify the electromagnetic and the weak force the standard model needs a super heavy particle, predicted in the 1960s but still to be discovered: the Higgs boson. Its interaction with other particles, for instance electrons, is supposed to impart them their mass. It is so heavy that it still cannot be generated with present-day accelerators. The Large Hadron Collider (LHC) at CERN will be the first accelerator with enough energy to make the discovery of the Higgs boson possible. The accelerator's proton-proton collisions will produce gluons that can fuse to form Higgs bosons.

The LHC program will also attempt to gather indications of the existence of a super-symmetric mirror world (SUSY) of the standard model, duplicating the whole of the "particle zoo." This would satisfy another symmetry idea: the unification of matter and force particles (fermions and bosons). SUSY particles are expected to be far heavier than those of the "standard" world. They are candidates for the explanation of "dark matter" in the universe. From the energy and directional distribution of the decay products of particle collisions as measured in detectors it is possible to draw conclusions about the reaction that has taken place. Alone in the inner section of the Large Hadron Collider's ATLAS detector one can expect to obtain some 40,000 measurement data for the reconstruction of a single event. In order to monitor even the most short-lived particles, detectors with a total of 140 million pixels are arranged in the form of cylindrical shells in the immediate vicinity of the proton-proton collision.

1 (Half Shell of a Silicon Detector
European Center for Nuclear Research CERN, Geneva, 1989–2000
This detector unit was situated close to the electron-positron collision zone in the OPAL detector of the Large Electron-Positron Collider (LEP). From 1989 to 2000 this experiment investigates the interaction particles (bosons) of the weak force and their decay. Positrons are the positively charged "antiparticles" of electrons that differ from them only by the sign of the charge.

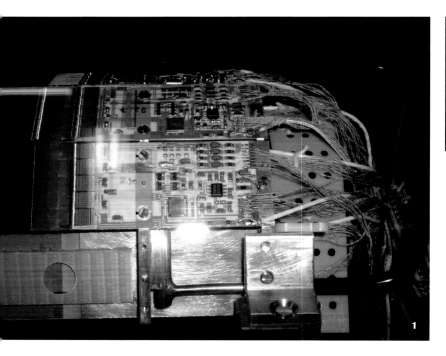

2 (Lead-Scintillator Calorimeter
Deutsches Elektronen-Synchrotron DESY, 1999
The purpose of the calorimeter is to detect the electromagnetic component in particle reactions. In the lead plates of the lead-scintillator sandwich high energy photons generate electron-positron pairs. Scintillators are made from a material that emits light when such charged particles pass. A photoelectron multiplier then transforms this light signal into an electrical signal.

Superconductivity in Research Technology

In order to determine their momentum in a detector, charged particles are deflected by powerful magnetic fields generated largely loss-free in superconducting coils. The principle of superconductivity, in which the electric conductor puts up virtually no resistance to the current flow, was discovered as early as in 1911 by Kammerlingh-Onnes at temperatures close to absolute zero (- 273 degrees Celsius), but it was not explained until 1957 by Bardeen, Cooper, and Schrieffer. It has become well-established in accelerator technology and is used in the construction of modern facilities for controlled nuclear fusion.

4 (Coils of the Stellarator Wendelstein 7-X
Max Planck Institute of Plasma Physics,
Greifswald, under construction 2005,
desk model
Fifty superconducting coils form the magnetic cage for the plasma at the fusion facility Wendelstein 7-X which is currently under construction. The model shows a section consisting of five coils to a scale of 1:20. With their complex shape they create the optimal structure for a magnetic cage of the Stellarator type. The aim of fusion research is to develop a power plant that – like the Sun – generates energy from the fusion of light atomic nuclei. Wendelstein 7-X is the world's largest fusion experiment of the Stellarator type. It is designed to show that the concept is feasible for power stations.

3 (TESLA Accelerator Cavity
Deutsches Elektronen-Synchrotron DESY,
2000, partial view
The hollow space resonator (cavity) made of the metal niobium, which becomes superconducting at -264 degrees Celsius, generates a high-frequency electromagnetic field by which particles are accelerated to almost the speed of light. This energy-saving superconductor technology has been developed at the *Deutsches Elektronen-Synchrotron* DESY within the framework of the international TESLA collaboration.

It is being used in the European x-ray laser project X-FEL and is planned for use in the future international linear accelerator ILC.

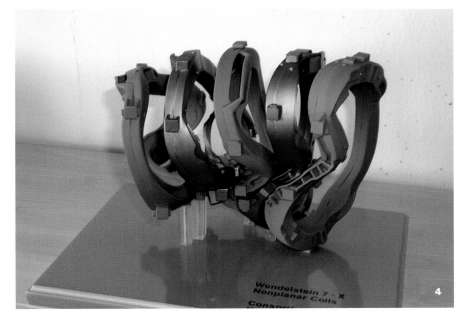

Strange Connections

In spite of the importance of atomistic concepts in present-day physics, the idea of a hierarchically structured world where its characteristics at a particular level of complexity can be derived from the laws governing the level immediately below has proved untenable. Nor can the world in general be interpreted as the result of individually acting separate parts. As Einstein's work and its consequences show, there are strange connections between the constituent parts of the microworld, as well as between the microworld and the macroworld.

His analysis of the Brownian molecular motion makes clear that between the microworld of the atoms and the macroworld of the bodies of our environment there is an intermediary "mesocosmos" with its own statistical laws. Einstein's analysis becomes the starting point not only for the development of new mathematical methods but also for research into physical, chemical, biological, and even economic systems reaching far beyond the horizons of classical physics. The new spheres range from thermal systems outside equilibrium, to soft, easily deformable matter, fluctuating exchange rates, genetic processes, and the spread of epidemics. Whilst Einstein explains the macroscopic diffusion of a substance (for instance ink in water) as the sum of the random movements of Brownian particles, current research now focuses on so-called Lévy processes. In these random processes the overall effect depends on the character of the individual steps, in the same way as the spread of a disease can nowadays depend on the intercontinental flights of infected individuals.

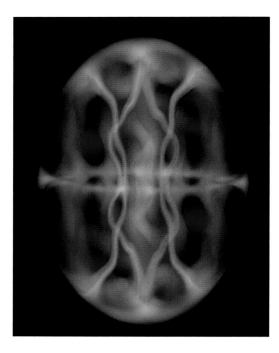

Rotating Bose-Einstein condensate with a vortex structure typical in superfluidity (simulation)

Strange connections also result from Einstein's heritage in the area of quantum physics. Works by Bose and Einstein in 1924 establish the idea of a gas with the property that its atoms are no longer individually distinguishable. Depending on the statistical properties, a distinction is made between bosons and fermions. The indistinguishability of atoms causes them to behave as if a strange connection existed between them – a consequence of the larger statistical probability of bosons to remain in the same state. Also the self-amplification of light in the laser is explained by the fact that light quanta are bosons. In the case of matter, one can predict from the statistical characteristics of bosons a new kind of state at low temperatures called the "Bose-Einstein condensate." In such a condensate all the atoms of a gas stay in the same quantum state. While a similar state, "superfluidity" or fluidity without viscosity, is discovered in helium already in 1938, Bose-Einstein condensates are not realized until 1995. Since then they have become a focus of research, also because of their optical properties which can result in non-linear effects, such as an extreme deceleration in the speed of light.

Finally, recent experiments on quantum mechanically "entangled," but spatially separate particles confirm that the results of measurements of their states are inevitably connected, even though the individual particles exhibited no quantum mechanically describable state before the measurement takes place. This refutes Einstein's hope that quantum mechanics may be completed to a classical theory. And so it seems that the entanglement which Einstein mocked as a "spooky action at a distance" is an inevitable part of our reality.

○ PERSONALIA

Paul Lévy

Paul Lévy (1886–1971) was born into a family of mathematicians in Paris. He is regarded as one of the most important mathematicians of the twentieth century. After studying at the *École Polytechnique* in Paris he teaches there from 1919 to 1959 as professor of mathematics. He is one of the founders of functional analysis and especially mod-

ern probability theory (1919). Lévy makes fundamental contributions to the theory of stochastic processes (1937), and his works on Brownian motion (1938/39) mark the beginnings of stochastic analysis. During his lifetime Lévy is considered as an outsider. His fundamental role becomes clear only with the development of fractal geometry by Benoit Mandelbrot and others.

Above:
Paul Lévy,
around 1960

○ FOUNDATIONS

Confinement of "Ultracold" Matter

Twenty-five years of research into techniques for the generation and confinement of extremely low-energy particles and particle clouds precede the realization of a strange state of matter very close to absolute zero temperature: the Bose-Einstein condensate. Prototype of these techniques is the Paul trap in which a quadrupole field generates a kind of saddle that is constantly being "rotated" (i.e., its polarity is reversed at high frequency) in order to prevent a charged particle from "falling down." In 1985 scientists succeed in trapping an electron for ten months. Although atoms are electrically neutral, it is in fact possible to use the weak residual external effect of the field between nucleus and electron shell to trap them electromagnetically. The "vessel" containing the atoms is then comparable to a very flat dish, and the atoms have to be actively stopped to prevent them from slipping over the edge. Scientists succeed in this with the help of laser radiation (laser cooling), also in 1985. What still has to be done on the way to absolute zero temperature is to learn how to gradually lower the rim of the dish to release the most energetic atoms (evaporative cooling). The combination of both techniques leads to the first Bose-Einstein condensate in 1995.

As a quantum computer, this fifty-ion crystal would provide the equivalent of the world's currently available computing capacity.

Quantum Computer

In contrast to traditional computers, information processing machines based on peculiarities of quantum physics are called quantum computers, whether they are real or only theoretical – as is mostly the case until today. The basic information unit in traditional computers is the bit which can have the logical value of true or false and is represented by physical systems that exhibit two states, for instance charged or uncharged. But if these systems are replaced by objects that obey the laws of quantum physics (like individual photons or atoms, for instance), these may exist not only in one of two states but also in a superposition of both. In this case one speaks of qubits.

In computers numbers are represented as sequences of bits. If sequences of qubits are created, the result are objects that can store combinations of states, that is, combinations of numbers. Calculations with such objects correspond to the simultaneous calculation of many single results to which however there is only limited access.

In order to make quantum computing possible, it will be necessary to develop new problem-solving algorithms that are compatible with the laws of quantum physics. As qubits tend to lose their superposition character when interacting with their environment, the realization of quantum computers poses serious problems. So far the most powerful quantum computers operate with only a few qubits and a single algorithm each.

Control of Photons

Canned light is a vision like from the *Tales from the Arabian Nights*. A top headline in 2005 says: "Photons under control." The part of physics called optics that describes the propagation of light and its interaction with matter changes radically due to quantum physics and the new concepts of light and matter associated with it. The quantum physical understanding of atoms and molecules, single or as a collective – for instance in crystals or in gases – makes it possible to precisely influence their interaction by means of light. It becomes possible to slow down light pulses in such media to the speed of a cyclist. It is even possible to capture light pulses completely and to release them again later in exactly the same form – that is, in a sense, to bring light to a standstill. All of this opens up completely new possibilities to understand the real nature of the photon and how it interacts with optical media. Nowadays it is also possible to give the wave form any desired time structure, and even to generate and control single photons with predetermined characteristics. As light rays are composed of innumerable photons, their temporal succession contributes to determine the character of the radiation. When photons are emitted in a regular pace, we speak of nonclassic light. It is of great interest as far as safe encoding and other optical information processing are concerned.

Particles and Light under Control

1 (Paul Trap
*Max Planck Institute of Quantum Optics,
Garching, 1987*
An exemplar of a Paul trap for storing
charged particles. A kind of rotating saddle
is created inside by a high-frequency

2 (Ring Trap
*Max Planck Institute of Quantum Optics,
Garching, 1987*
A variant of the Paul trap. It enables the
arrangement along a ring of the regular
structures, called ion crystals, produced

gases in the universe. This extreme state
of matter close to absolute zero temperature was predicted by Einstein in 1924.
It represents a matter wave in which the
single atoms give up their individuality in
favor of a completely synchronized move-

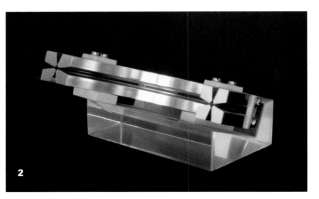

quadrupole field that prevents a particle
from moving in the unstable direction
(downhill), because a reverse in polarity
immediately reverses it into the stable
direction (uphill).

by the electrostatic repulsion among the
ions. This type of trap was initially used for
investigations into laser cooling and the
transition into the ordered state. Today it
is mainly used in basic research aiming at
the realization of a quantum computer.

3 (Apparatus for Producing
Bose-Einstein Condensates
University of Constance, 1997
An original apparatus for producing Bose-
Einstein condensates (BEC), the coldest

ment together with the others. As a result,
the macroscopic condensate behaves like
a single atom. In 1997 scientists at the
University of Constance used this apparatus to successfully produce a Bose-Einstein
condensate, which had been produced for
the first time in 1995 by Eric Cornell and
Carl Wieman at the University of Colorado.

4 (Photon Statistics
Max Planck Institute of Quantum Optics, Garching, 2005
The interactive animation of the photon sequence in either classical light (light bulb), laser light, or non-classical light was programmed by a student at the Gymnasium of Unterhaching. It describes a real photon-statistical experiment and shows the chronological sequence of photon emissions in the three types of light. Non-classical light sources are sources that emit a particularly regular photon sequence, such as the light from a single captured ion that is stimulated by a laser beam.

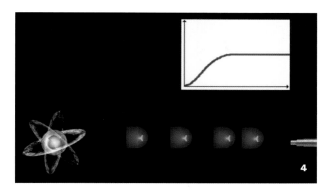

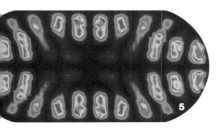

5 (Quantum Chaos
University of Marburg
The animation illustrates quantum chaos, the quantum mechanical equivalent of chaotic dynamic systems. The doubts expressed by Einstein in 1917 about the possibility of transferring Bohr's atomic model of hydrogen to more complex atoms form the starting point for the studies on which this particular animation is based. Periodic orbits play a special role in the quantum mechanics of chaotic systems. In order to illustrate them the wave function is represented in a situation called "stadium billiard" that was actually measured in microwave experiments. It is an example of the "whispering gallery" mode, where the waves move along the walls

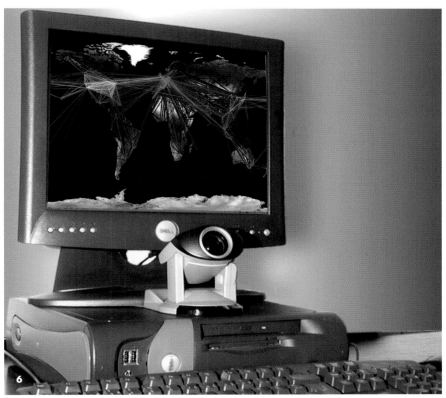

while the center remains untouched. The same effect can be found in sound waves in Romanesque domes.

6 (Eye Tracking System
SensoMotoric Instruments, Teltow
The stationary eye tracking system can be used to compare classical and non-classical statistics. Eye movements provoked by image contrasts are abrupt, as opposed to the Brownian molecular motion

investigated by Einstein. Today epidemics also spread in huge leaps via air traffic networks. Both of these are phenomena to which non-classical Lévy statistics apply. Other examples include such widely differing phenomena as the dynamics of financial markets, abrupt climatic changes in the Earth's history, and the foraging behavior of animals.

Applications

Basic research can yield far-reaching consequences, not only for the development of new worldviews, but also for industry and everyday life. Conversely, problems of application can be the starting point for deeper scientific insights. Both effects become apparent in Einstein's scientific heritage. He is not only the author of new theories of space and time, he is also a practical thinker with a feeling for riddles of everyday life ranging from the secret of flying to why rivers meander – themes to which he dedicates individual publications. He even holds patents for appliances including a refrigerator and a gyrocompass.

Certainly, Einstein cannot foresee the most significant technical consequences of his scientific work. They result from research and development that place his insights into new and surprising contexts. Even his boldest hypotheses, about the particle nature of light and the dependence of the lapse of time on motion and gravity, form the starting point for the development of technologies that nowadays determine our everyday life. The photoelectrical effect described in Einstein's work on the light quantum hypothesis now forms the basis of photovoltaics, that is, the generation of electricity by means of light-sensitive substances in solar cells. Photoelectron multipliers, in which a light quantum (or photon) triggers a cascade of electrons, are used in movement sensors and highly sensitive cameras.

The laser also has its origins in Einstein's contribution to basic research. In 1916 he obtains a new derivation of Planck's radiation law from the assumption that two kinds of light emission occur in the interaction between atoms and light: the spontaneous emission and the "induced" or "stimulated" emission triggered by irradiation. Induced emission opens up the possibility to use a light quantum to cause an ensemble of excited atoms to emit a cascade of similar light quanta, to which corresponds a wave that is exactly the same as that of the

GPS satellite   original photon. This is the basic principle of "light amplification by stimulated emission of radiation," which gave the laser its name.

The experimental demonstration of induced emission is carried out at the Kaiser Wilhelm Institute for Physical Chemistry in 1928, but it is not until 1954 that the laser principle is first realized with the development of the so-called maser in the microwave range. The laser itself is developed in 1960 and now is an integral part of everyday life. It is used at supermarket check outs for scanning barcodes and it is behind the technology for playing CDs or DVDs. One single microscopically fine glass fiber, the light conductor through which a laser beam is sent, enables millions of users to make simultaneous phone calls. In medical technology the laser enables minimal invasive surgery.

And special as well as general theory of relativity have also found their way into technological applications that are part of modern everyday life. According to the general theory of relativity, the atomic clock in a satellite situated in a weaker gravitational field than the observer on Earth will go faster than the observer's atomic clock. On the other hand, according to the special theory of relativity, the clock in a fast orbiting satellite goes somewhat slower than the stationary clock on Earth. But the effect of the general theory of relativity dominates, so that the on-board clocks go faster. Were these effects to be neglected, then satellite navigation would be unreliable.

◯ HISTORICAL CONTEXT

Telecommunications

During the nineteenth century public postal systems gradually replace messengers on foot or horseback and carrier pigeons as the means of transmission of news. In response to political and military needs, parallel advances are made in telegraphy, initially with optical means. With his pointer telegraph, which translates letters and digits into electrical impulses, Werner

Siemens wins in 1848 the contract for the telegraphy line between Berlin and Frankfurt am Main, the seat of the German National Assembly. When the Kaiser is elected in 1849 the real sensation is that the news takes less than an hour to reach Berlin. At the end of the 1870s the telephone starts competing with the telegraph and eventually supersedes it after the range of telephony is substantially increased by Michael Pupin's amplifying coils. The 600-kilometer-long Rhineland Cable (1912–1921) becomes the precursor of the European telephone network.

◉ FOUNDATIONS

Laser Principle

When an atom changes from an excited state to a lower energy state it emits a photon. The opposite process, in which a photon encounters an atom, is called absorption. In 1916 Einstein describes, besides spontaneous emission and absorption, also a third kind of interaction between light and atom, the induced emission. In this case a photon strikes an excited atom and

Left:
Mounting a Pupin coil near Baumgartenbrück, Germany, around 1920

Right:
From solar radiation to electricity

triggers the emission of another photon with identical characteristics. The more atoms there are in an excited state, the greater the probability that a photon will induce the emission of another one, rather than being absorbed itself. This effect is utilized in the laser. In a process called pumping, an above-average number of atoms inside a resonator composed of two mirrors are brought into an excited state by means of an external energy source. When photons bounce back and forth between the mirrors the induced emission produces a landslide effect. Part of the light amplified in this way can be decoupled as a laser beam.

Satellite-Based Navigation

The Global Positioning System (GPS) operated by the USA since 1995 consists of twenty-four satellites positioned at about 20,000 kilometers altitude of which at least four are simultaneously visible from almost every point on Earth. The system has become indispensable for commercial and military purposes and is being increasingly used for private navigation. Each satellite has an atomic clock on board and transmits time and position at regular intervals. With three readings its distance from the receiver can be determined while one reading shows the local time of the receiver. Einstein's theories of relativity come into play with the problem of synchronizing the atomic clocks. A deceleration in the on-board clock, caused by the high speed in orbit (compared with the rotating Earth), contrasts with an acceleration effect due to reduced gravity. All in all on-board clocks go on average around 0.04 milliseconds faster per day, which would result in a positioning error of ten kilometers per day. A simple trick to get around this problem is to "detune" the satellite clocks so that they go slower by the same time difference. The European Galileo system planned for 2008 will operate thirty satellites at 24,000 kilometers altitude and offer meter-perfect precision, and above all, advanced failure safety.

Photovoltaics

Already in 1839, Alexandre Becquerel detects the "internal" photoeffect in the semi-conducting material selenium and uses it to produce weak voltages. Semi-conductors are solids that become electrically conductive under the influence of light (or heat) so that electrons can move

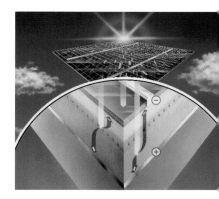

almost freely within the atomic union. In 1886 Hertz discovers the "external" photoeffect, in which electrons are set completely free from metal surfaces, and lets Halbwachs to investigate it experimentally. After Hertz's early death, Lenard, another of his former students, takes up the topic and discovers contradictions to the generally accepted wave nature of light. This then leads Einstein to the light quantum hypothesis in 1905. The energy of the electrons released from (or activated within) the atomic union depends only on the frequency, and not on the intensity, of the incident light. With his explanation, Einstein contributes substantially to the theoretical foundations of modern photovoltaics, the direct transformation of sunlight into electrical current. Today the most important semi-conductor used in solar cells is silicon, the second-most common element in the Earth's crust.

Applications

From the Rhineland Cable to Laser Free Transmission

1 (Navigation Console
MILANO medien GmbH, Frankfurt am Main, 2005
Siemens VDO Automotive AG, Regensburg
Design for a console to demonstrate modern forms of satellite-based navigation. The console conveys complex multimedial

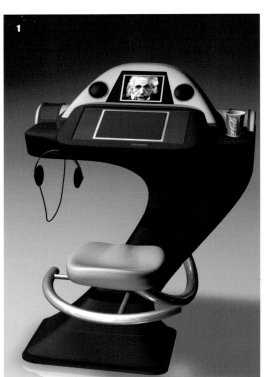

As an exemplification of the idea of augmented reality, that is, the integration of navigational details and other informations into a continually updated camera image of the vehicle's surroundings, audiovisual background information is conveyed when approaching the destination.

2 (Rhineland Cable
Siemens & Halske, Munich, around 1920
Prepared end of the segment Dortmund-Düsseldorf of the 600-kilometer-long Rhineland Cable, the telephone connection between Berlin and Cologne completed in 1921 by the company *Siemens & Halske*. This cable was the precursor of the Euro-

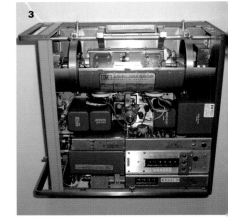

information about routes and destinations pertaining to a selection of locations where Einstein lived and worked. The "head-up display" system transfers navigation details from the on-board computer directly onto the windshield. An animation illustrating satellite-based navigation is shown as an example of information provided in a networked car.

pean telephone network. With modern optical fiber it is possible to transmit an average of 12.5 million simultaneous conversations, while the material needed for the total length of 600 kilometer, using a 120-micron-thin glass fiber, reduces to an amount of quartz sand that corres-ponds in volume to less than one meter of Rhineland Cable.

3 (Atomic Clock HP 5061
Hewlett Packard Company, Palo Alto
The HP 5061A is a frequency standard (an extremely stable pacemaker designed for local timekeeping such as in GPS satellites) based on a cesium atomic beam. It is produced by *Hewlett Packard* of Palo Alto. The instrument utilizes the transition frequency of 9.192.631.770 MHz within the hyperfine structure of the ground state of the cesium-133 atom. This is the basis of the current internationally recognized time unit. Fine and hyperfine structures in atomic energy levels result from minute systematic variations in the motion of the shell electrons and in their interaction with the atomic nucleus.

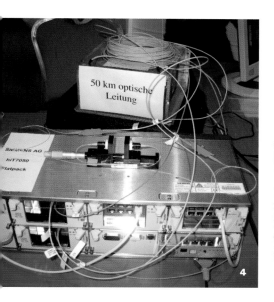

4 (Laser-Based Video Transmission
Siemens AG, IC Networks, Berlin, 2005
Internet emulation via the digital telephone network. The central device in the installation is the modular platform SURPASS hiT 7050, an affordable network router in optical technology with multi-service functionality, suited for different types of nets and different data transfer rates. Video transmission via infrared laser is demonstrated in a fifty-kilometer-long fiber optic cable and in a laser free transmission line. Laser free transmission also serves in communications between satellites in orbit. A further example of its application is its use in optical components to split the beam.

5 (Laser Lithotripsy
Clyxon Laser GmbH, Berlin, 2005
Demonstration console for lithotripsy (from the Greek, meaning "disintegration of stones") in hollow bodily organs (kidney, bladder, gall bladder) by means of a shock wave generated with the help of a laser focus. The laser device U100 plus generates

pulses of a power of just 1.2 watts and of such a short duration that it is technically impossible to accidentally damage soft tissue.

6 (Laser Cutting: Grids for Transmission Tubes
Siemens AG, CT MM, Munich
Grids of pyrographite for transmission tubes up to 1 MW transmission power. Their filigree structures are cut from the moulding blank by means of a laser. The largest of the depicted tube grids is about 50 cm high and takes about ten hours to produce. A typical size of the mesh of this material is 0.2 mm. In metals such as molybdenum a mesh of around 0.1 mm size can be achieved without any burr formation. In order to heat the material as little as possible short laser pulses are used. A typical cutting speed is 2 mm/s. Laser cutting, just one example of a broad spectrum of applications of laser material processing, also finds usage in medicine and surgery in the shape of the laser scalpel.

The Theory of Relativity under Suspicion

Karl Vogtherr,
Where is the Theory of Relativity Leading?, 1923

Ever since the 1920s, the relativization of time and space has come under fierce public attack in the name of "common sense." Common sense stands for the demand that a physical theory be clear and intelligible to all; accordingly, the only correct description of reality is one which all can comprehend from everyday experience. To this day, the theory of relativity has been criticized on the basis of these assumptions. Has anyone ever seen a moving clock run more slowly, or a moving measuring stick grow shorter? Are space and time not objective foundations of reality which cannot be defined at will by physics? On top of that, Einstein introduces the relativity of simultaneity to fulfill another postulate of the relativity theory which seems just as dubious to everyday thinking, namely the constancy of the speed of light independent of the motion of the light source or the observer. Many of Einstein's critics regard the abolition of the classical addition of velocities as a breach of logic. They try, for example, to explain the constant speed of light by positing an aether that moves along with the Earth. While relativity theory manages without the assumption of an aether, opponents of the theory continue to posit the necessary existence of the aether as a medium for the propagation of light waves and as a material intermediary for physical processes in outer space. How can an electromagnetic field have physical properties if it is immaterial? The same holds for the question about the nature of gravitation and its description as a warping of the space-time continuum. Einstein is accused of failing to find gravitation's true physical cause and indeed of neglecting to seek it in his geometrical description. While many of Einstein's opponents hold to Newton's gravitational force, others offer alternative explanations. For example they attempt to explain gravitation as the thrusts of tiny aether atoms pushing bodies toward one another. Often Einstein's opponents advocate amateur scientific theories of the world based on mystical assumptions, such as the elemental force of the aether, which have no place in modern physics.

Einstein's contemporaries see the theory of relativity as more than just a new physical theory. Newspapers report a revolution in physics which has changed the entire current worldview. Parallels are drawn with political and artistic upheavals. "Everything is relative" becomes a catchphrase, and for some, the overcoming of classical notions of space and time is tantamount to the relativization of reality and the undermining of traditional values and morals. These arguments are also reflected in anti-Semitic attacks against Einstein in the1920s and 1930s. The theory of relativity is denounced as "Jewish physics," abstract and mathematical, in contrast to the concreteness of German natural sciences. But even for those without such prejudices, the theory remains an intellectual provocation.

"A generally understandable defense of common sense against Einstein's attacks!"

◯ HISTORICAL CONTEXT

Amateur Science

At the beginning of the twentieth century popular science is an important aspect of bourgeois public life. Numerous scientific amateurs botanize, experiment, and discuss current scientific questions in natural science societies. The theory of relativity, with its new concepts of space, time, and matter, presents

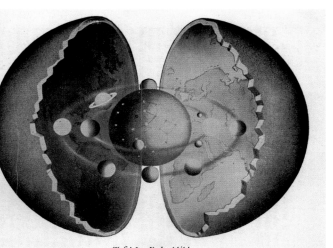

Tafel I · Erdweltbild
(Schema, Himmel überzeichnet, alle Gestirne nur Punkte)

amateur scientists with a challenge. The 1920s bring hundreds of publications attacking relativity theory, written by doctors, engineers, and chemists who expect physics to provide a clear and easily understandable explanation of the world. They often propound unconventional theories in an attempt to refute the theory of relativity. Some amateur theories such as the "world ice theory," which claims that the universe arose from the dualism between fire and ice, and the "hollow Earth theory," which states that the Earth is a hollow sphere with humanity living on the inside surface, find numerous adherents.

◉ FOUNDATIONS

Philosophy and the Theory of Relativity

Changed notions of space, time, and matter in physics force philosophy to thoroughly reexamine its premises as well. Philosophers of all schools grapple with the theory of relativity. In 1926 Bertrand Russell notes, "There has been a tendency [...] for every philosopher to interpret the work of Einstein in accordance with his own metaphysical system." Two influential philosophical schools of the time, neo-Kantianism and vitalism, have particular reservations about relativistic concepts of space and time. In the view of neo-Kantians such as Paul Natorp (1854–1924) and Richard Hönigswald (1875–1947), the theory of relativity does not affect the foundations of transcendental philosophy; rather, transcendental philosophy is the prerequisite of all science. Space and time as purely intuitive forms in the sense of Kant must be strictly separated from perceptible times and spaces that can be measured by physics. Hönigswald argues that the determination of relative time

periods and distances in relativity theory presupposes absolute time and absolute space as a point of reference for the relative. Vitalism draws a fundamental distinction between life and matter, making it necessary to use two difference concepts of time. Thus Henri Bergson (1859–1941) distinguishes the "temps scientifique," which is measurable with mechanical clocks and which can be affected by the relativistic dilation of time, from the time that is directly experienced by every living being. This "durée réelle" (real duration) is an absolute time which cannot be measured scientifically, only experienced intuitively. Vitalists criticize the fact that Einstein also applies the relativization of time to moving, organic clocks, i.e. organisms.

Representation of the hollow Earth theory, from: Karl Neupert, *Umsturz des Weltalls* (The Overturning of the Universe), 1929

The Twin Paradox

"To stay young, keep moving." In a way, this is backed up by the special theory of relativity: someone who flies across the Atlantic from Europe is a few quadrillionths of a second younger upon arrival than those who stayed at home. The faster the airplane, the younger the person becomes. Many versions of this illustration of the consequences of time dilation describe a set of twins who age at different rates, but one could also use Albert and his sister Maja, who was two years younger than he. If Albert were to fly from the Earth to a star four light years away, traveling at eighty per cent the speed of light, and then return to Earth, ten years would have passed for Maja in the meantime. By contrast, Albert would have aged by only six years. He would now be Maja's little brother. This is startling, but is it a paradox? The paradox lies in the fact that one could argue that Maja has moved relative to Albert, meaning that she should actually be six years younger than her big brother. This argument often plays a central role in attempts to refute the theory of relativity.

The apparent contradiction vanishes, however, when one takes into account the fact that Albert accelerates and slows down during his journey. This destroys the physical equivalence of the two siblings' locations.

Pamphlets

1 (Gusztáv Pécsi: The Liquidation of the Theory of Relativity: The Calculation of the Speed of the Sun
Regensburg, 1925
Gusztáv Pécsi, a Hungarian professor of natural philosophy, believes that absolute motions can be detected. "Even if relativity were to run rampant for a thousand years, it would not shake humanity's conviction that a train or a ship whose engine or locomotive is being greased and stoked by the engineer is in motion, rather than the surroundings or the shore." For Pécsi, these mechanical motions are absolute motions which require no system of reference. According to him, the theory of relativity is based upon a false interpretation of motion phenomena and has devastating consequences for the proud edifice of classical physics: "But now darkness has fallen again. Everything has been undermined and made uncertain by relativity." Pécsi attempts to use "astromechanical methods" to prove that the Sun's velocity is absolute, thus providing the breakthrough for his new system of mechanics.

2 (Hans Israel, Erich Ruckhaber, Rudolf Weinmann (eds.): 100 Authors against Einstein
Leipzig, 1931
100 Autoren gegen Einstein is a collection of twenty-eight statements solicited from Einstein's opponents along with excerpts from previously published writings attacking Einstein. It is initiated and published in 1931 by the engineer Hans Israel, the actor Rudolf Weinmann, and the writer Erich Ruckhaber, the last attempt during the Weimar era to take concerted action against the theory of relativity. The Einstein

Liquidierung der
Relativitätstheorie

Berechnung der
Sonnengeschwindigkeit

Von Dr. Gusztáv Pécsi

Erste und zweite Auflage

Regensburg 1925. Verlagsanstalt vorm. G. J. Manz
Buch- und Kunstdruckerei A.-G., München-Regensburg

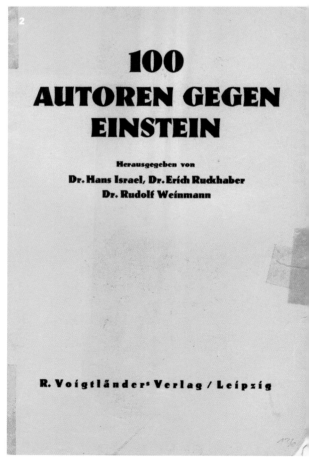

100
AUTOREN GEGEN
EINSTEIN

Herausgegeben von
Dr. Hans Israel, Dr. Erich Ruckhaber
Dr. Rudolf Weinmann

R. Voigtländer-Verlag / Leipzig

opponents Gehrcke and Lenard, who spoke out publicly in the early 1920s, do not contribute new statements: Gehrcke claims to have nothing new to add to his previous well-known writings. The philosopher Hugo Dingler declines to participate as well, stating that the refutation can be achieved only through sound arguments – which he claims to have made already – and not by majority decision. In fact the "majority" that speaks out against Einstein here is quite heterogeneous. The editors give this a positive spin: "We cannot present a unified front [...] when the opposing front is so ambiguous and unclear. But surely a revealing counterargument will be found for each argument put forth by the Einstein side." *100 Autoren gegen Einstein* repeats familiar arguments from the anti-Einstein polemics of the 1920s: it is absurd to suppose that the speed of light is constant, the aether is a conceptual scientific necessity, the equivalence principle is nothing but a fallacy, and the theory of relativity is ultimately physical relativism.

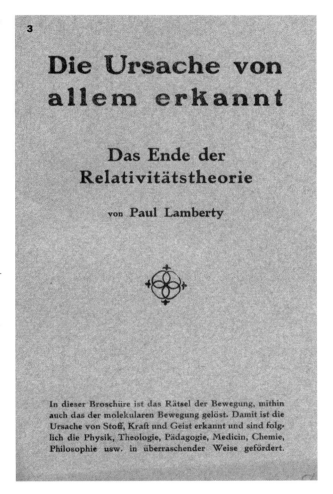

3 (Paul Lamberty: The Cause of Everything Discovered: The End of the Theory of Relativity
Haag, 1925
"This pamphlet solves the riddle of motion, including molecular motion. This reveals the cause of matter, energy, and spirit and thus gives an astonishing boost to physics, theology, education, medicine, chemistry, philosophy, etc."

Paul Lamberty propounds an unconventional aether theory as a basis for attacking the theory of relativity: while Einstein's theory manages without the concept of an aether, according to Lamberty the aether is the "world stuff," the eternal force in the universe which determines all events from the propagation of light to the social behavior of human beings.

Einstein under Suspicion

Scientists often come under suspicion – in Einstein's day as now. Even the political refugees who emigrated to the United States in the 1930s are regarded distrustfully by the authorities. The mission of the FBI becomes politicized, and fear of communist infiltration encourages informers and overzealous bureaucrats to spy on prominent intellectuals. The FBI creates an extensive file on Einstein, containing more than a thousand pages. As early as World War I Einstein is kept under surveillance because of his pacifist views. State agencies also keep a watchful eye on him during the Weimar Republic because of his prominent political role.

The US passport of Paul Weyland

Einstein's American file shows that he was also excluded from the atomic bomb project (Manhattan Project) for political reasons. Due to his public statements on issues of global politics and especially his support of American civil rights activists, the FBI uses targeted misinformation in an attempt to socially isolate and discredit the famous scholar. They even make use of informants who vilified Einstein back in the 1920s. Paul Weyland, already prominent in the anti-Semitic anti-Einstein campaign in Germany, emigrates to the United States in 1948, becomes an FBI informer in 1953, and denounces Einstein as a communist.

During the Cold War the FBI compiles the most preposterous information about Einstein – for example that his Berlin apartment served as a contact office for Soviet agents with his knowledge and approval. His employees, in particular his secretary, Helen Dukas, are accused of having been members of the Communist Party. Furthermore, Einstein supposedly smuggled the Soviet atomic spy Klaus Fuchs, whom he never in fact met, into the Manhattan Project. There are even conjectures that Einstein experimented on secret weapons together with other scientists. The head of the FBI, J. Edgar Hoover, is obsessed with documenting Einstein's supposedly un-American views and depicts him as the key figure in an intellectual network of supposedly communist organizations. His file lists more than seventy such organizations, including a dozen civil rights organizations, to which Einstein is supposed to have connections.

Still, the FBI does not dare to take action against a famous scientist who is one of the most prominent of world figures. Unlike other victims of the FBI and the McCarthy era, Einstein is too popular. Any attempt to publicly humiliate Einstein would create a scandal and a wave of solidarity, threatening the hoped-for anti-communist propaganda coup and the intended intimidation and disciplining of the American public. The material compiled is too implausible and absurd. It is of little use for an espionage story let alone as documentation of subversive activities. Yet to this day a sense of suspicion remains against unpredictable scientists who refuse to fall into line.

○ PERSONALIA

J. Edgar Hoover

John Edgar Hoover (1895–1972) takes a degree in law and joins the American Justice Department in 1917. After working as an assistant for the Attorney General, in 1924 he is named director of the Bureau of Investigation, which becomes the

FBI in 1935. Initially Hoover distinguishes himself in the fight against organized crime. In 1936 the FBI is entrusted with tasks relating to domestic security and becomes more of an anti-espionage institution. After World War II Hoover ultimately turns it into a political instrument that primarily targets left-wing political groups. Material is collected on politically active intellectuals, including Einstein. In some cases, surveillance is followed by direct intimidation. Under John F. Kennedy's presidency, Hoover's influence begins to wane. At the time of his death and up until today, his methods and his political legacy are highly controversial.

● FOUNDATIONS

Einstein's German Files

In March 1915, Einstein joins the *Bund Neues Vaterland* (New Fatherland League), a pacifist organization which is banned in February 1916. This brings him to the attention of the imperial military and police authorities. In March 1916 the military headquarters in Berlin demands information about him from the Academy of Sciences, of which he is a member, and in January 1918 Einstein's name is included in a police list of prominent pacifists in Berlin and the surrounding areas. Beginning in 1916 Einstein is even placed under surveillance. The informer can only report that Einstein is indeed a member of the *Bund Neues Vaterland*, but is not particularly prominent there – and, in addition, enjoys "the best moral reputation imaginable." In the Weimar Republic many political functionaries, especially the ministerial bureaucracy largely retained from the *Kaiser*'s day, continue to regard Einstein with suspicion. The Foreign Ministry collects reports about Einstein's journeys and his appearances abroad. Einstein's file grows.

Left:
J. Edgar Hoover, 1939

Right:
Einstein's files from the Political Archive of the German Foreign Ministry

Einstein's FBI Files

The FBI's files on Albert Einstein contain an extensive collection of informers' reports, newspaper articles, and denunciatory letters. While overflowing with half-truths and rumors, they also document Einstein's social activism and his championing of civil rights in the United States. In the 1950s, Einstein is a provocative figure because of his refusal to be intimidated by the anti-communist witch-hunt. Einstein's activities are monitored in an attempt to smear his reputation, but today the documentation of this enterprise produces the opposite effect. The Freedom of Information Act requires that the files be made available to the public. They can be read online.
http://foia.fbi.gov/foiaindex/einstein.htm

1 (Sheriff Howard Durley, Ventura County, California, to J. Edgar Hoover
24 April 1934
"At a recent speech in this County Dr. Einstein was referred to as being a communist and to be furtherer of communist activities in this country."

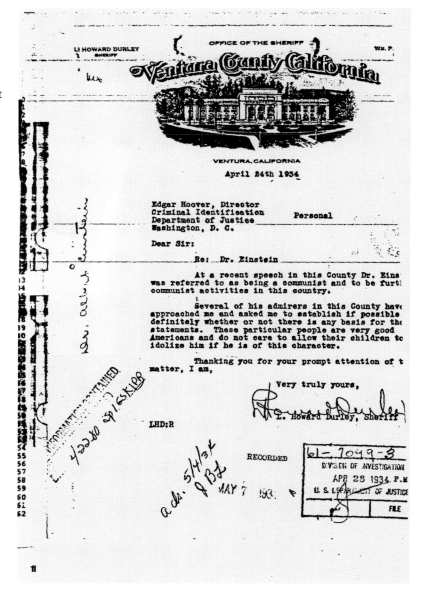

2 (Report on Einstein's Contacts with the Veterans of the Abraham Lincoln Brigade
23 November 1953

'Subject on 5/15/45 sent telegram to Veterans of Abraham Lincoln Brigade meeting praising Veterans of Abraham Lincoln Brigade fight against Fascism.
[...]
Morning Freiheit' editorial of 6/15/53 lauds EINSTEIN's opposition to Internal Security investigations, and states that EINSTEIN orders the American intellectuals to 'refuse to testify'."

The Abraham Lincoln Brigade consists of American volunteers fighting during the Spanish Civil War on the side of the legitimate republican government against the insurgent Franco's forces.

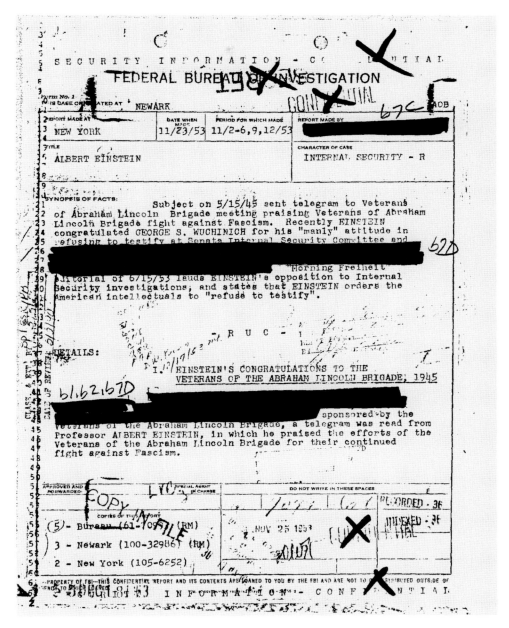

The Atomic Bomb

The wars of the twentieth century are waged using all available resources and with a system of organization capable of selectively harnessing scientific innovations for technical applications under strict deadlines. For this purpose, political players create new structures that link the military, science, and industry. The development of the first atomic bombs by the Manhattan Project sets precedents for the effective integration of these spheres, comparable at the time only to Germany's rocket research program. An army of experts uses the most sophisticated technology to release enormous amounts of energy, enabling the military to destroy entire cities in a single blow. However, the projects also develop a momentum of their own which is very difficult for individuals to control. Shortly before finishing their work, many of the participating scientists realize that they will have no political control over the decision to use the new weapon. Their research becomes a political challenge for them.

Preparations for the first detonation of a nuclear bomb, New Mexico, 1945

When the atomic bombs are dropped on Japan in 1945, public perceptions of the relationship between science and war also shift. The physicists involved now symbolize a profession in the limelight.

Despite the devastating effects of the atomic bomb, Einstein continues to regard the willingness to go to war, not the means with which it is fought, as the basic evil. He sees the necessity of international negotiation as a political problem which the new weapons have not necessarily changed but simply made more urgent. Nonetheless, he joins his colleagues in their efforts to reach out to the public and explain the new dangers of war waged with atomic bombs.

In Germany scientists only begin to take a stand against atomic armament in the mid-1950s, among them scientists who worked on weapons research under the National Socialist regime. Here, too, there is a discussion about the extent of scientists' responsibility and the ways in which the future "misuse" of scientific discoveries can be prevented. Yet it is problematic and somewhat misleading to speak of the "misuse" or the "lost innocence" of science. This can retroactively shift responsibility entirely onto the political sphere. In the case of the German physicists, it is ultimately also a way of compensating for their own past. In any event, the development of nuclear physics in the twentieth century makes it clear that even basic research raises the issue of responsibility. There seems to be no way for scientists to avoid reflecting on their contributions in the light of societal developments.

○ HISTORICAL CONTEXT

Cold War Manifestos

The perils of the atomic arms race makes many scientists realize that they will have to face up to their political responsibility. Some (such as Niels Bohr and J. Robert Oppenheimer) choose to use their personal influence on politicians, while

Left:
Pascual Jordan,
1930s

others make public statements against the use of atomic weapons. The first appeal against the atomic bomb, the *Franck Report*, appears in June 1945, while the Manhattan Project is still underway. It is followed by the *Stockholm Peace Appeal* (1950), in which Frédéric Joliot-Curie plays a crucial role, the *Russell-Einstein Manifesto* (1955), the *Mainau Declaration* by Nobel laureates, initiated principally by Otto Hahn, and the *Göttingen Declaration* (1957) by eighteen German atomic scientists (the "Göttingen 18"). In 1962 the American winner of the Nobel Prize for chemistry, Linus Pauling, is awarded the Nobel Peace Prize for his efforts to stop the further testing of nuclear weapons and for his initiation of the appeal against atomic testing which is submitted to the United Nations in 1957.

○ PERSONALIA

Max Born and Pascual Jordan

In 1925 Max Born (1882–1970) and his students Werner Heisenberg (1901–1976) and Pascual Jordan (1902–1980) develop a theory of atomic processes based on the "matrix mechanics" originated by Heisenberg ("Three-Man Work"). In 1933 teacher and disciples go their separate ways. Born, of Jewish descent, is forced to emigrate to Great Britain, while Heisenberg remains. Under the National Socialists, with whom he sympathizes, Jordan pursues a career as a professor in Rostock and Berlin. From 1957 to 1961 he is a Member of the German Federal Parliament for the Christian Democratic Union (CDU). In 1953 Max Born returns to Germany. He and Jordan take opposing positions on the issue of nuclear armament, with Jordan speaking out against the "Göttingen 18." To the outrage of his old teacher, Jordan calls his opponents' political judgment into question.

Leslie R. Groves

Leslie R. Groves (1896–1970), engineer, is the brigade general who serves as military leader of the Manhattan Project starting in 1942. He studies at the University of Washington and at MIT and graduates from the Military Academy at West Point in 1918. A crucial figure in the construction of the first atomic bombs, he recruits the scientists involved in the project, organizes the procurement and production of fissile material, and chooses the research facilities and testing ranges. When several members of the project, led by James Franck and Leo Szilard, try to prevent the use of the weapon shortly before the end of the war, Groves firmly overrules them and encourages the President of the United States to drop the bomb on Japan. After the war he heads an army department for "special weapons" for several years before being promoted to lieutenant general in 1948 and retiring shortly thereafter.

J. Robert Oppenheimer and Edward Teller

J. Robert Oppenheimer (1904–1967) receives his doctorate in 1927 under Max Born. Beginning in 1929 he teaches at the University of California in Berkeley and at the California Institute of Technology. In 1942 he is appointed scientific director of the Manhattan Project in Los Alamos. Later he becomes director of the Institute for Advanced Study in Princeton and an influential figure at the Atomic Energy Commission. However, in 1954 his security clearance is revoked after a public hearing on his political convictions, and he retires from public life. The Hungarian-born Edward Teller (1908–2003) receives his doctorate in 1930 under Werner Heisenberg. In 1935 he emigrates to the United States, where he teaches at George Washington University and Columbia University. During the war he heads a section of the Manhattan Project and subsequently works on the development of the hydrogen bomb. In contrast to Oppenheimer, who is skeptical of this new weapon, Teller's deep-seated anti-Soviet views lead him to become one of its most staunch supporters.

Right:
J. Robert Oppenheimer (left) and Leslie R. Groves (right) visiting the remains of a test tower after a detonation, 1945

From Armaments Research to Peace Activism

the Russell-Einstein manifesto

In the tragic situation which confronts humanity, we feel that scientists should assemble in conference to appraise the perils that have arisen as a result of the development of weapons of mass destruction, and to discuss a resolution in the spirit of the appended draft.

We are speaking on this occasion, not as members of this or that nation, continent or creed, but as human beings, members of the species Man, whose continued existence is in doubt. The world is full of conflict; and, overshadowing all minor conflicts, the titanic struggle between Communism and anti-Communism.

Almost everybody who is politically conscious has strong feelings about one or more of these issues; but we want you, if you can, to set aside such feelings and consider yourselves only as members of a biological species which has had a remarkable history, and whose disappearance none of us can desire.

We shall try to say no single word which should appeal to one group rather than to another. All, equally, are in peril, and, if the peril is understood, there is hope that they may collectively avert it.

We have to learn to think in a new way. We have to learn to ask ourselves, not what steps can be taken to give military victory to whatever group we prefer, for there no longer are such steps; the question we have to ask ourselves is: what steps can be taken to prevent a military contest of which the issue must be disastrous to all parties?

The general public, and even many men in position of authority, have not realised what would be involved in a war with nuclear bombs. The general public still thinks in terms of the obliteration of cities. It is understood that the new bombs are more powerful than the old, and that, while one A-bomb could obliterate Hiroshima, one H-bomb could obliterate the largest cities, such as London, New York and Moscow.

No doubt in an H-bomb war great cities would be obliterated. But this is one of the minor disasters that would have to be faced. If everybody in London, New York and Moscow, were exterminated, the world might, in the course of a few centuries, recover from the blow. But we now know, especially since the Bikini test, that nuclear bombs can gradually spread destruction over a very much wider area than had been supposed.

It is stated on very good authority that a bomb can now be manufactured which will be 2,500 times as powerful as that which destroyed Hiroshima. Such a bomb, if exploded near the ground or under water, sends radioactive particles into the upper air. They sink gradually and reach the surface of the earth in the form of a deadly dust or rain. It was this dust which infected the Japanese fishermen and their catch of fish.

No one knows how widely such lethal radioactive particles might be diffused, but the best authorities are unanimous in saying that a war with H-bombs might quite possibly put an end to the human race. It is feared that if many H-bombs are used there will be universal death—sudden only for a minority, but for the majority a slow torture of disease and disintegration.

Many warnings have been uttered by eminent men of science and by authorities in military strategy. None of them will say that the worst results are certain. What they do say is that these results are possible, and no one can be sure that they will not be realised. We have not yet found that the views of experts on this question depend in any degree upon their politics or prejudices. They depend only, so far as our researches have revealed, upon the extent of the particular expert's knowledge. We have found that the men who know most are the most gloomy.

Here, then, is the problem which we present to you, stark and dreadful, and inescapable: Shall we put an end to the human race; or shall mankind renounce war? People will not face this alternative because it is so difficult to abolish war.

The abolition of war will demand distasteful limitations of national sovereignty. But what perhaps impedes understanding of the situation more than anything else is that the term "mankind" feels vague and abstract. People scarcely realise in imagination that the danger is to themselves and their children and their grandchildren, and not only to a dimly apprehended humanity. They can scarcely bring themselves to grasp that they, individually, and those whom they love are in imminent danger of perishing agonisingly. And so they hope that perhaps war may be allowed to continue provided modern weapons are prohibited.

This hope is illusory. Whatever agreements not to use H-bombs had been reached in time of peace, they would no longer be considered binding in time of war, and both sides would set to work to manufacture H-bombs as soon as war broke out, for, if one side manufactured the bombs and the other did not, the side that manufactured them would inevitably be victorious.

Although an agreement to renounce nuclear weapons as part of a general reduction of armaments would not afford an ultimate solution, it would serve certain important purposes. First: any agreement between East and West is to the good in so far as it tends to diminish tension. Second: the abolition of thermonuclear weapons, if each side believed that the other had carried it out sincerely, would lessen the fear of a sudden attack in the style of Pearl Harbour, which at present keeps both sides in a state of nervous apprehension. We should, therefore, welcome such an agreement, though only as a first step.

Most of us are not neutral in feeling, but, as human beings, we have to remember that, if the issues between East and West are to be decided in any manner that can give any possible satisfaction to anybody, whether Communist or anti-Communist, whether Asian or European or American, whether White or Black, then these issues must not be decided by war. We should wish this to be understood, both in the East and in the West.

There lies before us, if we choose, continual progress in happiness, knowledge and wisdom. Shall we, instead, choose death, because we cannot forget our quarrels? We appeal, as human beings, to human beings: Remember your humanity, and forget the rest. If you can do so, the way lies open to a new Paradise; if you cannot, there lies before you the risk of universal death.

9TH JULY 1955

1

Bertrand Russell
1872–1970

Albert Einstein
1875–1955

resolution

We invite this Congress, and through it the scientists of the world and the general public, to subscribe to the following resolution:

"In view of the fact that in any future world war nuclear weapons will certainly be employed, and that such weapons threaten the continued existence of mankind, we urge the Governments of the world to realise, and to acknowledge publicly, that their purposes cannot be furthered by a world war, and we urge them, consequently, to find peaceful means for the settlement of all matters of dispute between them".

Professor Max Born
Professor of Theoretical Physics at Göttingen; Nobel Prize in Physics

Professor P.W. Bridgman
Professor of Physics, Harvard University; Foreign Member of the Royal Society; Nobel Prize in Physics

Albert Einstein

Professor L. Infeld
Professor of Theoretical Physics, University of Warsaw; Member of the Polish Academy of Sciences

Professor J.F. Joliot-Curie
Professor of Physics at the College de France; Nobel Prize in Chemistry

Professor H.J. Muller
Professor of Zoology, University of Indiana; Nobel Prize in Physiology or Medicine

Professor L. Pauling
Professor of Chemistry, California Institute of Technology; Nobel Prize in Chemistry

Professor C.F. Powell
Professor of Physics, Bristol University; Nobel Prize in Physics

Professor J. Rotblat
Professor of Physics in the University of London, at St. Bartholomew's Hospital Medical College

Bertrand Russell

Professor Hideki Yukawa
Professor of Theoretical Physics, Kyoto University; Nobel Prize in Physics

Following the release of the Russell-Einstein Manifesto on 9 July 1955, efforts were begun to convene an international conference of scientists for a more in-depth exchange of views on ways to avert a nuclear catastrophe. With the support of Cyrus Eaton, the first Pugwash Conference was held at the Eaton summer home in Pugwash, Nova Scotia, from 7–10 July 1957. A total of 22 participants from 10 countries attended and issued conference reports on nuclear radiation hazards, control of nuclear weapons, and the social responsibilities of scientists. From this first meeting the Pugwash Conferences have evolved into an international organization with national groups in more than 50 countries, which by the summer of 2001 had organized 265 meetings, involving more than 3,500 individual scientists, academics, and policy specialists. In recognition of its efforts to eliminate the nuclear threat, Pugwash and its then President, Joseph Rotblat, were jointly awarded the 1995 Nobel Peace Prize.

Bertrand Russell
and Joseph Rotblat

1 (Russell-Einstein Manifesto

London, 1955

At a press conference in London on 9 July 1955, the philosopher and mathematician Bertrand Russell presents the famous peace manifesto demanding nuclear disarmament and a stop to the development of the hydrogen bomb. The disarmament docu-ment is signed by eleven prominent scientists, including nine Nobel laureates. Russell initiates the document with Einstein's active support, and Einstein signs it a few days before his death. The *Russell-Einstein Manifesto* sets precedents for scientists' commitment to peace and disarmament.

2 (Göttingen Declaration

Göttingen, 1957

The *Göttingen Declaration* against the planned nuclear armament of the West German army is published on 12 April 1957. It is signed by eighteen of Germany's most prominent atomic scientists, including four Nobel laureates. The signatories emphasize their unwillingness to participate in the production, testing, or use of atomic weapons.

The declaration creates a sensation because it leads to a controversy between prominent scientists and the German government. It marks the beginning of the discussion about scientists' political commitment in West Germany. However, it can also be seen in light of the fact that the signatories' role in armaments research under the National Socialists still has not been clarified at the time.

GÖTTINGER ERKLÄRUNG   (12. April 1957)

Die Pläne einer atomaren Bewaffnung der Bundeswehr erfüllen die unterzeichneten Atomforscher mit tiefer Sorge. Einige von ihnen haben den zuständigen Bundesministern ihre Bedenken schon vor mehreren Monaten mitgeteilt. Heute ist die Debatte über diese Frage allgemein geworden. Die Unterzeichneten fühlen sich daher verpflichtet, öffentlich auf einige Tatsachen hinzuweisen, die alle Fachleute wissen, die aber der Öffentlichkeit noch nicht hinreichend bekannt zu sein scheinen.

1. *Taktische Atomwaffen haben die zerstörende Wirkung normaler Atombomben.* Als "taktisch" bezeichnet man sie, um auszudrücken, daß sie nicht nur gegen menschliche Siedlungen, sondern auch gegen Truppen im Erdkampf eingesetzt werden sollen. Jede einzelne taktische Atombombe oder -granate hat eine ähnliche Wirkung wie die erste Atombombe, die Hiroshima zerstört hat. Da die taktischen Atomwaffen heute in großer Zahl vorhanden sind, würde ihre zerstörende Wirkung im ganzen sehr viel größer sein. Als "klein" bezeichnet man diese Bomben nur im Vergleich zur Wirkung der inzwischen entwickelten "strategischen" Bomben, vor allem der Wasserstoffbomben.

2. *Für die Entwicklungsmöglichkeit der lebensausrottenden Wirkung der strategischen Atomwaffen ist keine natürliche Grenze bekannt.* Heute kann eine taktische Atombombe eine kleinere Stadt zerstören, eine Wasserstoffbombe aber einen Landstrich von der Größe des Ruhrgebiets zeitweilig unbewohnbar machen. Durch Verbreitung von Radioaktivität könnte man mit Wasserstoffbomben die Bevölkerung der Bundesrepublik wahrscheinlich heute schon ausrotten. Wir kennen keine technische Möglichkeit, große Bevölkerungsmengen vor dieser Gefahr sicher zu schützen.

Wir wissen, wie schwer es ist, aus diesen Tatsachen die politischen Konsequenzen zu ziehen. Uns als Nichtpolitikern wird man die Berechtigung dazu abstreiten wollen; unsere Tätigkeit, die der reinen Wissenschaft und ihrer Anwendung gilt und bei der wir viele junge Menschen unserem Gebiet zuführen, belädt uns aber mit einer Verantwortung für die möglichen Folgen dieser Tätigkeit. Deshalb können wir nicht zu allen politischen Fragen schweigen. Wir bekennen uns zur Freiheit, wie sie heute die westliche Welt gegen den Kommunismus vertritt. Wir leugnen nicht, daß die gegenseitige Angst vor den Wasserstoffbomben heute einen wesentlichen Beitrag zur Erhaltung des Friedens in der ganzen Welt und der Freiheit in einem Teil der Welt leistet. Wir halten aber diese Art, den Frieden und die Freiheit zu sichern, auf die Dauer für unzuverlässig, und wir halten die Gefahr im Falle des Versagens für tödlich.

Wir fühlen keine Kompetenz, konkrete Vorschläge für die Politik der Großmächte zu machen. Für ein kleines Land wie die Bundesrepublik glauben wir, daß es sich heute noch am besten schützt und den Weltfrieden noch am ehesten fördert, wenn es ausdrücklich und freiwillig auf den Besitz von Atomwaffen jeder Art verzichtet. Jedenfalls wäre keiner der Unterzeichneten bereit, sich an der Herstellung, der Erprobung oder dem Einsatz von Atomwaffen in irgendeiner Weise zu beteiligen.

Gleichzeitig betonen wir, daß es äußerst wichtig ist, die friedliche Verwendung der Atomenergie mit allen Mitteln zu fördern, und wir wollen an dieser Aufgabe wie bisher mitwirken.

Fritz Bopp, Max Born, Rudolf Fleischmann, Walther Gerlach, Otto Hahn, Otto Haxel, Werner Heisenberg, Hans Kopfermann, Max von Laue, Heinz Maier-Leibnitz, Josef Mattauch, Friedrich-Adolf Paneth, Wolfgang Paul, Wolfgang Riezler, Fritz Strassmann, Wilhelm Walcher, Carl Friedrich Frhr. von Weizsäcker, Karl Wirtz.

2

Science as a Challenge

After World War II, against the background of Nazi crimes and the development of weapons of mass destruction, the question of scientists' responsibility becomes an increasing focus of public awareness. The dissolution of moral boundaries in scientific research at the service of criminal regimes is an especially disturbing phenomenon, leading to the formulation of new ethical guidelines in many spheres. In the twenty-first century, fifty years after Einstein's death, science still confronts political challenges, especially those posed by the industrialization of science. As a rule, scientific applications such as nuclear technology, stem cell research, and nanotechnology harbor both risks and opportunities. Genetic technology offers the chance to combat hunger and insidious diseases. But at the same time, it raises the issue of the perils posed by unprecedented human intervention into biological evolution. The political challenges faced by science include the tensions between humanitarian, military, political, and commercial interests when it comes to using the results produced. When medicines are expensive due to the high costs involved in producing them, how can they still be made available to all who need them? To take another example, how can the dangerous results of armaments research be prevented from spreading throughout the globe?

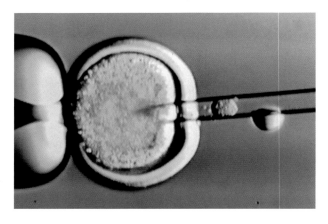

Cloning technology: An embryo cell is injected into the ovum of a sheep

But even basic research raises societal issues. For example, the growing material demands of science raise the question of society's priorities in research funding. Large-scale research projects for exploring outer space or elementary particles, or for developing new methods of producing energy, are costly. What returns are expected from these investments? In open societies, the sciences are under constant pressure to legitimize themselves. Sponsors usually take a result-oriented approach, in terms of political prestige as well. The increasing complexity of the scientific research system and its dependency on a web of particular societal interests conflicts with the demand for transparency in an open society. The interrelation between basic research and technical applications also contributes to this complexity. Thus, mechanisms of institutional control and self-control play a key role in processes of mediation between the values of scientific freedom and the necessity of democratically legitimating publicly-funded science. Without public awareness of the value and nature of science, however, these mechanisms lack social support, as does science itself. But how can the public form an opinion on the social and political challenges of science, and how can it intervene? The prerequisite for a productive discussion is the willingness to inform oneself adequately about the controversial issues. Here the Internet has created new opportunities and helped close the gap between inaccessible expert knowledge and the knowledge which the public needs for orientation. Science journalism and popular science can also contribute to developing a public culture of science, as can the disciplines engaged in reflecting upon science, such as the history and philosophy of science. They can create the foundations upon which a well-informed public can regard science neither as a per*petuum mobile* of progress nor as a threat, but rather as a chance for shaping the future.

○ PERSONALIA

Joseph Rotblat

The physicist Joseph Rotblat is born in Poland in 1909 and studies in Warsaw, where he receives his doctorate in 1938. After emigrating, he participates in the development of the American atomic bomb. By 1944, convinced that Germany would not succeed in building the bomb, he leaves the project, and after the war he devotes himself to

the medical uses of nuclear physics. The dropping of the atomic bombs on Japan in August 1945 makes him a resolute opponent of nuclear weapons. He is one of the signers of the *Russell-Einstein Manifesto* and the founder of the Pugwash Conferences on Science and World Affairs, serving as its president since 1988. In 1995 he and the Pugwash Conferences are awarded the Nobel Peace Prize for their campaigns against the perils of nuclear technology and for promoting social responsibility among scientists.

○ FOUNDATIONS

Whistleblowing

"Whistleblower" is the term used for conscientious research project employees who call attention to violations of ethical standards of science of which they have acquired inside knowledge. In the United States they enjoy legal protection. In Germany, too, attempts are being made to provide institutional safeguards for this

practice, protecting whistleblowers against reprisals by fellow employees and superiors.

As demonstrated by major scandals involving faked research results, whistleblowers can serve as a check, helping to maintain scientific standards and ensure the responsible use of resources. Furthermore, they can also raise fundamental questions about the moral viability of certain methods and fields of research. This keeps the public informed about problems specific to certain fields whose societal implications are not yet foreseeable and which have not yet received adequate public discussion.

Nanotechnology

Nanotechnology studies structures that are smaller than one hundred nanometers and develops their applications. A nanometer, one millionth of a millimeter, is about the size of ten hydrogen atoms placed side by side. In this range molecular forces and quantum effects, as well as the fact that nanoparticles have a higher ratio of surface to volume, begin to radically influence the particles' mechanical, electronic, chemical, and optical qualities. Nanotechnological research on the border zone between the disciplines of physics, chemistry, biology, and computer science, has produced a large number of new products. The qualities of ordinary materials can be purposefully altered by embedding nanoparticles. For example, nanoparticular titanium dioxide absorbs ultraviolet radiation in sunscreens, but its small particle size renders it invisible to the naked eye. Surfaces coated with nanoparticles have special qualities. They can be scratch-resistant, antiseptic, or, as in the case of the lotus effect, self-cleaning. Tiny motors can be constructed using nanotechnological components. New applications are anticipated especially in information processing and medicine. Visionary concepts concerning the potential of nanotechnology, such as the idea of self-reproducing nanorobots, must, however, be subjected to critical scrutiny not only with regard to their feasibility but also with regard to their potential economic, social, ecological, and military implications.

Open Access

The Internet revolution has been compared to the upheavals created by the invention of writing and the printing press. For the first time in history, the Web enables free, global, interactive access to scientific and cultural information and thus to humanity's cultural legacy as a whole. However, the question of how best to use this potential is still being debated. How can the technical, institutional, political, and economic prerequisites be provided to ensure that information is made available in a lasting, sustainable manner? Publishers are transferring traditional forms, such as scientific journals, to the new medium. Meanwhile, scientific institutions and the custodians of cultural resources are increasing efforts to make information and research findings freely available on the Internet to society, following the principle of open access. These efforts led to the formulation of the *Berlin Declaration on Open Access to Knowledge in the Sciences and Humanities*. Initiated in 2003 by the Max Planck Society, it is supported by major research and cultural institutions throughout the world.

Left: Joseph Rotblat, around 2004

Right:
Soldiers using the Global Positioning System (GPS)

Current Global Challenges

Globalization has not only affected trade in goods and services. The natural sciences and technology have created the means for monitoring, changing, and controlling the world on a global scale.

Satellites can record every inch of the globe, enabling precise navigation and positioning. The process of urbanization accelerated by science and technology is changing the biosphere on a planetary scale. The use of nuclear energy makes it possible to use weapons systems, such as submarines and aircraft carriers, on a constant basis and around the globe. Scientists are confronted with global challenges, in terms of the perception of problems as well as approaches to solutions. Only in an international context can epidemics and climate change be brought under control. Setting a course in biotechnology also has implications that transcend the national context – the biosphere does not respect political borders. Current challenges for science policy also include the well-considered creation of global networks for education and research, the use of scientific knowledge to solve global problems, and, more generally, the responsible use of technological potentials. It is evident that international authorities must play a more active role in these processes – very much in the spirit of Einstein, who was convinced of the possibility of a peaceful world order in which science would occupy a prominent place.

1 (Military Use of Space Travel
18 December 2004
An Ariane-5 rocket lifts off from the ESA Space Center in French Guayana, carrying a French surveillance satellite into space.

2 (Global Epidemics
Globalization causes epidemics to spread with unprecedented speed, meaning that regional and national relief measures no longer suffice.

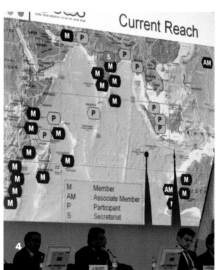

3 (An Homage to Nuclear Power
The crew of the USS Enterprise draws
a bold connection between Einstein's
famous formula and the force powering
their ship. The formula refers to the forty
years since the introduction of nuclear-
powered aircraft carriers in the American
navy. The picture dates from November
2001, when the ship was returning
from Operation "Enduring Freedom"
(the Afghanistan conflict).

4 (Plan for a Tsunami Early
Warning System
The responsible use of scientific knowledge also entails making it available and
putting it to prompt use in economically
disadvantaged regions. Had there been
a functional tsunami early warning system
in the Indian Ocean, it is certain that the
tsunami on 26 December 2004 would
have claimed far fewer victims.

Science in the Media

Since the beginning of the 1920s, when the international press made Albert Einstein the first scientist to be celebrated by the mass media, scientific success stories and breakthroughs have become major social events. However, this popularization sometimes goes along with a tendency to overestimate the possibilities of science, and often an exaggerated fear of its supposedly uncontrollable consequences.

In the twentieth century scientists must learn for the first time how to cope with the mass media. Initially they shy away from playing too public a role. In Germany the educated establishment's fear of "trivialization" and "commercialization" is still quite an issue, even though the culture of popularization has existed for some time, encouraged in part by the scientists themselves. The use of the atomic bombs and the technologization of warfare alter the public perception of science. A belief in heroic progress on the one hand and apocalyptic social criticism on the other reflect the extreme perceptions that result from public anxiety.

Einstein surrounded by journalists, 1934

Meanwhile, scientists use the new avenues provided by the mass media to advance their own scientific interests or those of the political authorities. During the Cold War the media becomes a forum for debates on nuclear energy and atomic armament. Ultimately, the triumph of television creates new forms of conveying knowledge and a new type of scientific media star. In Germany Heinz Haber explains the "blue planet" to millions of viewers, while in the USA Carl Sagan takes his audience along on a journey into the cosmos. As the author of an interstellar "message in a bottle" on board the Voyager space probe, Carl Sagan even takes on the role of intermediary between human beings and extraterrestrial beings, the search for which he so vehemently advocates.

Yet these triumphs of popularization are accompanied by a fundamental problem: the sciences have developed a highly technical jargon, branching out into so many subdisciplines that even for educated laypeople it is virtually impossible to follow developments in their specifics. The media are faced with the challenge of mediating between current scientific problems and the public's need to know. Despite the extreme specialization and formalization of scientific information, the social consequences do not affect the experts alone. And scientists themselves are increasingly faced with the task of "translating" their findings into everyday language so that politicians and the public can hold a meaningful discussion on the costs and benefits of scientific research. These "translations" can also increase understanding between different branches of science, thus fostering a culture of interdisciplinarity.

○ PERSONALIA

Carl Sagan

Carl Sagan studies physics, astrophysics, and astronomy in Chicago. From 1968 as a professor at Harvard University and from 1971 at Cornell University he performs research in astronomy and exobiology – the search for extraterrestrial life.

port his heroine quickly to a remote part of the universe, and asks his friend, the astrophysicist Kip Thorne, for advice. This inspires Thorne to explore theories about the existence of wormholes – tunnels in spacetime that could connect distant points of the universe.

◉ FOUNDATIONS

Science Journalism

On television, in newspapers and in magazines, science journalists regularly report on science news and explain complex research findings intelligibly. However, the mass media report primarily on isolated, emotionally charged, spectacular phenomena in the natural sciences rather than explaining their background and placing them in a larger context. This can lead to a one-sided, uncritical public image of the sciences. Public discourse is often affected by the fact that even scientific news has the character of a commodity. Many scientists are unable to resist the temptation to seek publicity by occasionally exaggerating the significance of their achievements. There is, in any case, too little discussion of the fact that research takes a great deal of time and effort which can sometimes be in vain.

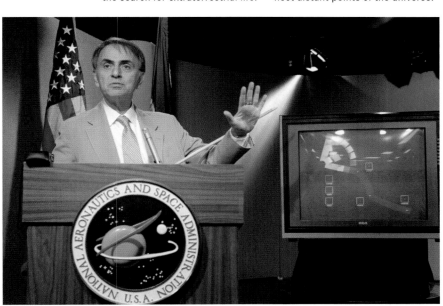

In 1978 he is awarded the Pulitzer Prize. With more than 200 articles and over twenty books, he has significantly contributed toward popularizing science and making it comprehensible for the public. His television show *Our Cosmos* has over 500 million viewers in over 60 countries. While writing the science fiction novel *Contact*, Sagan looks for a scientifically sound way to trans-

Carl Sagan at a NASA press conference, 1990

Science in Film

As the entertainment culture begins to take an interest in science, a new film hero is discovered: the scientist. Different types develop: a lonely, often unrecognized genius undertakes a daring self-experiment; an equally lonely Cassandra foretells impending disaster; a researcher blinded by hubris creates machines or beings that get out of control and turn on their creator. For all the hype and special effects, the stories expose underlying hopes and fears evoked by modern science. The film caricature of the scientist reflects the high level of anxiety and resentment on the part of the audience. The superior intellect must suffer: he is bumbling, lonely, unworldly, he is persecuted or killed in fanciful ways by the very products of his work. The balance is restored when the striver is forced to pay a high price for his vision.

All this has little to do with Albert Einstein. Though he experiences personal setbacks and is also persecuted politically, he is anything but a tragic figure. Yet there is something dubious about the occasional smug references to his personal life in the media. They parallel the schadenfreude with which films show how geniuses supposedly fail at everyday life, demonstrating an urge which is only satisfied when the person raised to the status of legend has been toppled from his pedestal again.

Science in One Day's Newspapers

Public discourse about science follows intellectual trends. People with enough time to regularly comb all of one day's newspapers for science news can keep track of the shifts in
fashionable topics. Some of these "waves of discourse" can almost take on the character of
campaigns. Increasingly, journalists assume the role not only of reporters, but of actors.
They influence the public mood with regard to current research issues, by no means
always to the scientists' advantage.

Einstein, the Legend

The confirmation of the general theory of relativity in 1919 makes headlines. The *Berliner Illustrirte* calls Einstein a "new celebrity of world history." What is it about the creator of relativity theory that fascinates the masses?

In the period of economic, political, and ideological uncertainty following World War I, relativity theory is mystified – "everything is relative" becomes a catchphrase. At the beginning of the 1930s, Einstein's colleague Ernest Rutherford calls the relativity theory a "great work of art." But even Einstein himself becomes a work of art and a legend in his lifetime, a cult figure of public life, inspiring photographs, drawings, and paintings. His portraits become icons of pop culture. The photo of Einstein sticking his tongue out becomes one of the best-known pictures in the history of the mass media.

"Cosmoclast Einstein," *Time*, 1 July 1946

COSMOCLAST EINSTEIN
All matter is speed and flame.

Some celebrate Einstein as a hero for liberating the human mind, while others accuse him of nihilism. The development of the atomic bomb adds a new dimension to the Einstein legend: the so-called "father of the bomb" returns penitently to his pacifist ideals and naively champions the reconciliation of all humanity. Other than that, Einstein is presented to the public as an absent-minded genius with wild hair and without socks who reveals the ultimate secrets of the universe. The bohemian professor occasionally flirts with this personality cult, but remains ever conscious of the moral responsibility his fame brings with it. He makes selective use of it for the humanitarian causes he supports. In recent years journalists and biographers have shown an increased interest in Einstein's private life. New legends are created, for example that of the chauvinistic sociopath for whom physics is mainly a way to escape the demands of the everyday world. In this version, too, the legend of Einstein conceals from view that his moral example and his brand of open-eyed science is still a provocation today.

○ HISTORICAL CONTEXT

The Marketing of Einstein

Posters, postcards, and coffee mugs flaunt the unconventional professor as an idol of non-conformism. The marketing patterns typical of cult and commerce apply to him as well – a limited stock of images and statements is comprehensively and internationally exploited. In the advertising branch Einstein's face has even been referred to as a "logo for genius." His name, his signature, and his face are now subject to systematic commercial use, appearing alongside the stylized images of Che Guevara, Marilyn Monroe, and John Lennon. Hollywood films such as Nicolas Roeg's *Insignificance* (1985), which tells of a fictitious encounter between Einstein and Marilyn Monroe, keep the legend alive today.

The 1999 New Year's edition of *Time* features Einstein as "Person of the Century."

Einstein's Brain

In 1957 Roland Barthes publishes his *Mythologies*, a book containing the short essay "Einstein's Brain," which reduces the materialism of the genius cult to a simple formula: "Mythologically, Einstein is matter."

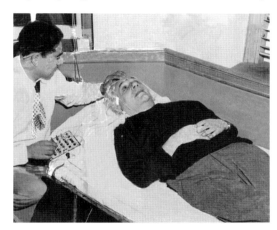

head shape and character traits. In the mid-twentieth century attempts are made to locate genius in measurable brain functions. Referring to the photograph of Einstein with electrodes on his head, Barthes writes: "The mythology of Einstein

Albert Einstein undergoing a physiological examination of his brain, 1951

Barthes assumes that Einstein had left his brain to science. The truth is that the pathologist Thomas Harvey illegally removed Einstein's brain from his corpse and then dissected and preserved it.

The assumption that there is an obvious connection between the histology of the brain and its intellectual performance has proved as untenable as the assumption that there is some relationship between depicts him as a genius so lacking in magic that people speak about his thought as if it were some functional procedure analogous to the mechanical production of sausages, the grinding of corn, or the crushing of ore." The notion that Einstein's genius can be found in the biological substance of his brain culminates in the grotesque practice of treating his remains, even today, as relics.

Above:
Andy Warhol,
portrait of Albert
Einstein from the
series *Ten Portraits
of Jews of the
Twentieth Century*,
1980

Below:
Still from the film
Insignificance by
Nicolas Roeg, 1985

Einstein and Art

Even as Albert Einstein is making his discoveries, the European continent sees the emergence of new artistic styles that confront issues of space, time, motion, and velocity. The language of the artistic avant-garde differs from that of physics, but there are clear parallels between the issues dealt with by physics and art. Einstein breaks with traditional notions of physics, while the artistic avant-garde breaks with the representational traditions of the nineteenth century. Their experiments test the frontiers of the representational just as Einstein's theories test the frontiers of the conceivable.

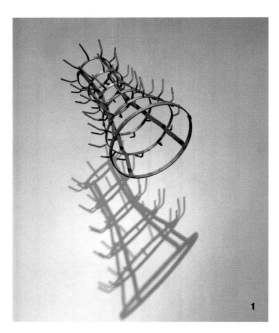

1

2 (El Lissitzky "Proun Composition"
around 1919/1920
The *Proun* compositions by the Russian artist El Lissitzky are innovative space concepts and architectural fantasies which he refers to as a "transfer stations" between painting and architecture. In 1925 Lissitzky writes in his essay *A. and Pangeometry*: "We see that suprematism has created the illusion [...] of irrational space with its infinite extensibility into the background and foreground. [...] With a mobile approach to surfaces, reference is made to the most modern scientific theories [...] (multidimensional spaces, relativity theory, Minkowski world, etc.)."

1 (Marcel Duchamp: "Bottle Dryer"
1914/1964
The French artist Marcel Duchamp studies intensively publications on mathematics and physics of the early twentieth century. His ready-made *Bottle Dryer* is an attempt to illustrate a thought experiment: the shadows of three-dimensional bodies are two-dimensional. Accordingly, the shadows of four-dimensional bodies must be three-dimensional. When the bottle dryer is hung in an undefined space, it can no longer be determined whether the bottle dryer is a shadow or an object.

2

3

4 (Giacomo Balla:
"Little Girl Running on the Balcony"
1912/13
"When the natural sciences of today deny
their past, they are in agreement with the
material needs of our time. In the same
way, art must deny its past to fulfill the
intellectual needs of our time," proclaim
the Futurists in their manifesto of 1910.
In this painting the Italian artist Giacomo
Balla sets a figure in motion by depicting
a series of identical images overlapping
one another and breaking the light down
into countless color units. In this way temporal succession is translated into spatial
juxtaposition.

3 (Theo van Doesburg:
"Contra-Construction Maison Particulière"
1923
In Theo van Doesburg's opinion, the task of
the architect is to overcome three-dimensional space through the proper division of
the "spatial body." The new architect must
construct a "time-space." In his architectural drawings he takes an axonometric
view which relativizes the subjective stand-
point, rather than a central perspective
which imitates the perspective of the observer. His use of axonometry makes the
architectural sketches seem to float. Doesburg emphasizes this floating and the
openness of the space by omitting architectural connecting elements.

4

Einstein on Postage Stamps

Einstein is a popular motif among stamp collectors. Postal administrations around the world produce stamps with his portrait or with motifs commemorating his achievements. By 1978 a good dozen such stamps had appeared. In 1979, the centenary of his birth, the number rocketed. Since then new stamps appear almost every year, bringing the current total to well over one hundred. In July 2005 the German Postal Administration will release a fifty-five-cent stamp for the Einstein Year.

1 (Postage Stamp Commemorating Einstein's 100th Birthday
German Democratic Republic, 1979

2 (Postage Stamp Commemorating Einstein's 100th Birthday
People's Republic of China, 1979

3 (Postage Stamp Commemorating Einstein's 125th Birthday
Serbia and Montenegro, 2004

4 (First Day Cover Commemorating "The Nobel Prize: 100th Anniversary"
United States of America, 2001
The two American stamps with Einstein portraits are from the years 1966 and 1979, while the stamp with the portrait of Alfred Nobel is from 2001.

Einstein on Coins

In contrast to the large number of postage stamps, only ten coins commemorating Einstein have appeared to date. In 2004 the German Ministry of Finance invited fifteen artists to submit designs for a ten-euro commemorative coin entitled *Albert Einstein – 100 Years of Relativity Atoms Quanta*. The coin will be released in the Einstein Year 2005.

1 (Winning Design by Heinz Hoyer (Berlin)

2 (Second Prize by Victor Huster (Baden-Baden)

3 (Third Prize by František Chochola (Hamburg)

WORLDVIEW AND KNOWLEDGE ACQUISITION

On the Shoulders of Giants and Dwarfs

"Dialogue on Mount Olympus"
Film by Jürgen Renn and Michael Dörfler, 2005
Max Planck Institute for the History of Science and
produktion2/cineplus, Berlin

"Images of the World" Part 1
Film on children's workshops by Katja Bödeker and
Henning Vierck, 2004
Comenius-Garten and Max Planck Institute
for the History of Science, Berlin

Invisible Forces

Electrostatic machine
Second half of the 18th century
Deutsches Museum, Munich

Table with Otto Hahn's experimental appliances
Berlin, 1938, reconstruction
Deutsches Museum, Munich

Gravity and Inertia

Double balance
Philipp von Jolly, around 1880
Deutsches Museum, Munich

Maxwell's wheel
Eberhard Gieseler, 1905
Deutsches Museum, Munich

Ernst Mach: *Die Mechanik in ihrer Entwicklung his-
torisch-kritisch dargestellt* (The Science of Mechanics:
A Critical and Historical Account of Its Development)
Leipzig, 1883
Max Planck Institute for the History of Science, Berlin

Albert Einstein: "Ernst Mach"
(Obituary for Ernst Mach)
Physikalische Zeitschrift, vol. 17, 1916
Max Planck Institute for the History of Science, Berlin

Transmutation

Earthen melting crucible
16th–18th century
Deutsches Museum, Munich

Iron mortar and pestel
16th–18th century
Deutsches Museum, Munich

Earthen alembic
16th century
Deutsches Museum, Munich

Equipment for distillation
16th century, reproduction
Deutsches Museum, Munich

Glass alembic
18th–19th century
Deutsches Museum, Munich

Destillery vessel
16th century
Deutsches Museum, Munich

Florentine bottle for aqua regia
17th–18th century
Deutsches Museum, Munich

Florentine bottle for aqua fortis
17th–18th century
Deutsches Museum, Munich

Ampulla
17th–18th century
Deutsches Museum, Munich

Alchemic instruments
Sammlung zur Laborgerätetechnik –
Rainer Friedrich, Mahlow

Instruments for fire assaying
Deutsches Museum, Munich;
Sammlung zur Laborgerätetechnik –
Rainer Friedrich, Mahlow

Apothecary instruments
Deutsches Museum, Munich;
Sammlung zur Laborgerätetechnik –
Rainer Friedrich, Mahlow

Chemical Changes

Blowpipe with accessories in a box
1907
Sammlung zur Laborgerätetechnik –
Rainer Friedrich, Mahlow

Friedrich Mohr: "Titration Method"
Brunswick, 1855
Sammlung zur Laborgerätetechnik –
Rainer Friedrich, Mahlow

Burette
Second half of the 19th century
Sammlung zur Laborgerätetechnik –
Rainer Friedrich, Mahlow

Instruments for metal analysis
Sammlung zur Laborgerätetechnik –
Rainer Friedrich, Mahlow

Mohr-Westphal density balance
Baird&Tratlock, London, 1920
Sammlung zur Laborgerätetechnik –
Rainer Friedrich, Mahlow

Instruments for volumetry
Sammlung zur Laborgerätetechnik –
Rainer Friedrich, Mahlow

Ebullioscope
1890
Sammlung zur Laborgerätetechnik –
Rainer Friedrich, Mahlow

Instruments from a chemical laboratory
of the 20th century
Sammlung zur Laborgerätetechnik –
Rainer Friedrich, Mahlow

Mysterious Effects

Geiger-Müller tube
Phywe Company, Göttingen

Cathode ray tube with Helmholtz coils
Phywe Company, Göttingen

Cathode rays tube with fluorescent screen
and magnet
Phywe Company, Göttingen

Light mill
Phywe Company, Göttingen

Light, Heat, Radiation

Albert Einstein: "Über das Boltzmann'sche Prinzip
und einige unmittelbar aus demselben fliessende
Folgerungen" (On the Boltzmann Principle and a
Number of Conclusions Directly Following from It)
Manuscript, around 1910, reproduction
Zentralbibliothek, Zurich
NL. Zangger, Schachtel 1a

Albert Einstein: "Zur Theorie der Radiometerkräfte"
(On the Theory of Radiometer Forces)
Manuscript, Berlin, 1924, facsimile
The Hebrew University of Jerusalem – Einstein Archives
Call Nr. 1–042

Albert Einstein: Abstract of an article by Emil Rupp
Manuscript, Berlin, 1926, facsimile
The Hebrew University of Jerusalem – Einstein Archives
Call Nr. 1–053

Cathode ray tube
Eugen Goldstein, around 1880
Deutsches Museum, Munich

Concave mirrors for reflecting heat radiation
Around 1810
Museo per la Storia dell'Università di Pavia

Aeolipile
Model, J. B. Haas&Co., London, 1790
Museo per la Storia dell'Università di Pavia

Electric radiometer after Puluj
Robert Goetze Company, Leipzig, around 1900
Deutsches Museum, Munich

Steam engine
Designed by Henry Maudslay, 19th century
Museo per la Storia dell'Università di Pavia

Dilatometer
Around 1904
Museo per la Storia dell'Università di Pavia

Cathode ray tube with Maltese cross
Around 1893
Museo per la Storia dell'Università di Pavia

Cathode ray tube with fluorescent mineral
Around 1893
Museo per la Storia dell'Università di Pavia

Parabolic mirror for Hertzian waves
Around 1900
Museo per la Storia dell'Università di Pavia

Geiger-Müller tube
Otto Pressler Company, Leipzig, 1938
Deutsches Museum, Munich

Projection apparatus with polarization filter
Jules Duboscq, Paris, around 1875
Museo per la Storia dell'Università di Pavia

Photoelectric selenium cell
Around 1904
Museo per la Storia dell'Università di Pavia

Flint glass prism
Joseph Fraunhofer, beginning of the 19th century
Deutsches Museum, Munich

Thermometer
Johann Georg Repsold, Hamburg, around 1826
Deutsches Museum, Munich

X-ray tube
C. H. F. Müller Company, around 1922
Deutsches Museum, Munich

Magnetism, Electricity, Induction

Albert Einstein: Notes on H. F. Weber's physics lecture
Manuscript, Zurich, winter semester 1897/98
Facsimile of page concerning the Volta element
The Hebrew University of Jerusalem – Einstein Archives
Call Nr. 3-001

Albert Einstein: "Grundgedanken und Methoden der
Relativitätstheorie in ihrer Entwicklung dargestellt"
(Fundamental Ideas and Methods of the Theory of
Relativity Presented in their Development)
Manuscript, January 1920, reproduction
The Pierpont Morgan Library, New York
Misc. Heinemann MAH 278

Spherical gyrocompass
Anschütz-Kaempfe and Einstein, around 1926
Deutsches Museum, Munich

Armed loadstone
1781
Museo per la Storia dell'Università di Pavia

Mechanical model of induction
After Hermann Ebert, Erlangen, 1905
Deutsches Museum, Munich

Electromagnetic motor
C. A. Grüel, Berlin, around 1858
Museo per la Storia dell'Università di Pavia

Electric motor
Matthias Hipp, late 19th century
Museo per la Storia dell'Università di Pavia

Leyden jar
Second half of the 18th century
Deutsches Museum, Munich

Electric pile
Alessandro Volta, 1800
Deutsches Museum, Munich

Electrodynamic current balance
William Thomson (Lord Kelvin), Glasgow, around 1887
Deutsches Museum, Munich

Rotating bar magnet for demonstrating the effect
of a magnetic field
E. Leybold's Nachfolger, Cologne
Deutsches Museum, Munich

Device for investigating the Peltier effect
Heinrich Daniel Rühmkorff, Paris, around 1877
Museo per la Storia dell'Università di Pavia

Electroscope with plate capacitor
Alessandro Volta, late 18th century
Museo per la Storia dell'Università di Pavia

Electrometer
Alessandro Volta, late 18th century
Museo per la Storia dell'Università di Pavia

Device for demonstrating the Seebeck effect
Max Kohl Company, Chemnitz, around 1900
Deutsches Museum, Munich

Demons and Ghosts

Letter from Albert Einstein to Heinrich Zangger
Berlin, [March 1930]
Zentralbibliothek, Zurich
Nl. Zangger, Schachtel 1a

Letter from Eduard Einstein to Heinrich Zangger
Berlin, 12 March 1930
Zentralbibliothek, Zurich
Nl. Zangger, Schachtel 1a

Flyer of the Deutsche Okkultistische Gesellschaft
(German Occultist Society)
Around 1920
Max Planck Institute for the History of Science, Berlin

Modern divining rod
Private collection

Automatic spring finder
Adolf Schmid, Bern, around 1911
Deutsches Museum, Munich

Labyrinth of the Microworlds

Glass bell jar with vacuum pump
Phänomenta Company

Models of the Cosmos

Mechanical model of the geogentric world system
Michael Sendtner Company, Munich, 1912
Deutsches Museum, Munich

Celestia. Computer program for virtual travel through
outer space
Chris Laurel, 2001–2005
http://www.shatters.net/celestia

Models of the Earth

Washington Irving: *Life and Voyages of Christopher
Columbus*
New York, 1892
Max Planck Institute for the History of Science, Berlin

Ptolemaic world map
Cod. Neap. lat. VF. 32, f 71v/72r
Around 1466, reproduction
Max Planck Institute for the History of Science, Berlin

Ebstorf world map
13th century, destroyed during World War II,
reconstruction
Rechen- und Medienzentrum der Universität Lüneburg,
Fach Kulturinformatik

Globe after Martin Behaim 1492
Reproduction
Mathematisch-Physikalischer Salon, Dresden

Terrella, model of the Earth as a magnet
after William Gilbert, late 18th century
Museo per la Storia dell'Università di Pavia

Model of the Earth as an electromagnet after P. Barlow
C. Dell'Acqua, Milan, around 1840
Museo per la Storia dell'Università di Pavia

Declination compass
Cristoforo Scalvino, Milan, early 19th century
Museo per la Storia dell'Università di Pavia

Inclination compass
Georg Friedrich Brander, around 1779
Deutsches Museum, Munich

Letter from Albert Einstein to Erwin Schrödinger
Princeton, 10 September 1943
Österreichische Zentralbibliothek für Physik, Vienna
Nl. Schrödinger

Advancing Towards Infinity

Astrolabe
Regiomontan school, Nuremberg, 15th century
Deutsches Museum, Munich

Quadrant
16th century
Deutsches Museum, Munich

Telescope
Galileo Galilei, around 1620, replica
Istituto e Museo di Storia della Scienza, Florence

Sights with celestial globe
Johann Gabriel Doppelmayr, 1730 (globe)
Mathematisch-Physikalischer Salon, Dresden

Heliometer
Utzschneider and Fraunhofer, 1815
Deutsches Museum, Munich

Spectrograph
Erich Neubauer Company, Berlin-Karow, around 1940
Astrophysikalisches Institut, Potsdam

Models of the Heavens

Galileo Galilei: *Dialogo sopra i due massimi sistemi del mondo* (Dialog on the Two Chief World Systems)
Second Italian edition, Florence, 1710
Max Planck Institute for the History of Science, Berlin

Athanasius Kircher: *Iter extaticum coeleste*
Würzburg, 1660
Max Planck Institute for the History of Science, Berlin

René Descartes: *Principia philosophiae*
Amsterdam, 1644
Max Planck Institute for the History of Science, Berlin

Robert Fludd: *Utriusque cosmi maioris scilicet et minoris metaphysica, physica atque technica historia*
Oppenheim, 1617/1618
Max Planck Institute for the History of Science, Berlin

Pamphlets about the appearance of the comet in the year 1664
Anonymus: *Kurzverfasster historischer Bericht*
Augsburg, 1664
Gabriel Kepler: *Kurtzer jedoch gründlicher Bericht*
s.l., 1665
Geminiano Montanari: *Cometes Bononiae Observatus*
Bologna, 1665
Johannes Praetorius: *Ivdiciolvm Asteriae*
Leipzig, 1664
Johann Matthias Schneuber: *Umständliche Beschreibung des grossen Cometen*
Strassburg, 1665
Christian Theophilus: *Cometen, Propheten*
Nuremberg, 1665
Max Planck Institute for the History of Science, Berlin

Armillary sphere
17th century
Istituto e Museo di Storia della Scienza, Florence

Tellurium
Charles François Delamarche, around 1800, replica
Istituto e Museo di Storia della Scienza, Florence

Andreas Cellarius: *Harmonia macrocosmica seu Atlas universali*
Second edition, Amsterdam, 1708
Staatsbibliothek, Berlin

Wilhelm F. Herschel's planetarium
William and Samuel Jones, London, around 1800
Mathematisch-Physikalischer Salon, Dresden

Curved Spaces

Karl Schwarzschild: "Über das zulässige Krümmungsmaß des Raumes" (On the Permissible Degree of Curvature of Space)
Vierteljahrsschrift der Astronomischen Gesellschaft, vol. 35, 1900
Staatsbibliothek, Berlin

Albert Einstein: "Zurich Notebook"
Manuscript, 1912–1913, facsimile of page 41
The Hebrew University of Jerusalem – Einstein Archives
Call Nr. 3–006

Albert Einstein: *Geometrie und Erfahrung*
(Geometry and Experience)
Berlin, 1921
Max-Planck-Institut für Wissenschaftsgeschichte, Berlin

Illustration of curved spaces by means of two-dimensional models
Euclidean surface, elliptic surface, hyperbolic surface
Museumstechnik, Berlin

EINSTEIN – HIS LIFE'S PATH

Treasure Chamber – Original documents on Einstein's life

Postcard from Albert Einstein to Carl Paalzow
Milan, 12 April 1901
Archiv zur Geschichte der Max-Planck-Gesellschaft, Berlin
Va. Abt., Rep. 2, Nr. 2

Caricature of Albert Einstein by Maurice Solovine
1903
The Hebrew University of Jerusalem – Einstein Archives
Call Nr. 36–621

Postcard from Albert and Mileva Einstein to Conrad Habicht
Munich, n.d.
Bibliothek der Eidgenössischen Technischen Hochschule, Zurich
Hs 1457:40

Letter from Albert Einstein to the Council of Education, Cantons of Zurich
Application for a job as a teacher
Bern, 20 January 1908
Staatsarchiv des Kantons Zürich
U 84 d.2

Albert Einstein: "Meine Meinung über den Krieg" (My Opinion about the War)
Draft of the contribution to *Das Land Goethes 1914–1916*
Manuscript, 1915
Staatsbibliothek, Berlin
Slg. Darmstädter, F 1e 1908(7), Bl. 9–10

Postcard from Albert Einstein to Paul Ehrenfest
[Berlin], 4 February 1917
The Hebrew University of Jerusalem – Einstein Archives
Call Nr. 9–396

Letter from Romain Rolland to Albert Einstein
[Villeneuve], 23 August 1917
Laurent Besso, Lausanne

Letter from Adolf von Harnack to Albert Einstein,
Berlin, 12 September 1917
Archiv zur Geschichte der Max-Planck-Gesellschaft, Berlin
I. Abt., Rep. 34, Nr. 2

Draft of letter from Wilhelm von Siemens to Albert Einstein
Berlin, 21 January 1918
Archiv zur Geschichte der Max-Planck-Gesellschaft, Berlin
I. Abt., Rep. 1A, Nr. 1656, Bl. 31

Postcard from Albert Einstein to
Max and Hedwig Born
Ahrenshoop, 2 August 1918
Staatsbibliothek, Berlin
Nl. Born, Nr. 188, Bl. 4

Postcard from Albert Einstein to Pauline Einstein
[Berlin], 11 November [1918]
The Hebrew University of Jerusalem – Einstein Archives
Call Nr. 29–352

Albert Einstein: On the Need for a Legislative National
Assembly
Draft of a speech held at a meeting of the Bund Neues
Vaterland (New Fatherland League)
[Berlin, 13 November 1918]
The Hebrew University of Jerusalem –
Einstein Archives
Call Nr. 28–001

District Court of Zurich: Decree of divorce between
Mileva and Albert Einstein
14 February 1919
Staatsarchiv des Kantons Zürich
BEZ Zch 6314.43

Letter from Albert Einstein to Paul Ehrenfest
[Berlin], 22 March 1919
The Hebrew University of Jerusalem – Einstein Archives
Call Nr. 9–427

Letter from Albert Einstein to Heinrich Zangger
[Berlin], 18 June [1919]
Zentralbibliothek, Zurich
Nl. Zangger, Schachtel 1a

Letter from Albert Einstein to Fritz Haber
Berlin, 9 March 1921
The Hebrew University of Jerusalem – Einstein Archives
Call Nr. 12–332

Albert Einstein: "Über ein jüdisches Palästina"
(On a Jewish Palestine)
Manuscript for a speech, Berlin, June 1921
The Hebrew University of Jerusalem – Einstein Archives
Call Nr. 28–010

Albert Einstein: Diary of the journey to South America
Manuscript, 5–11 March 1925
The Hebrew University of Jerusalem – Einstein Archives
Call Nr. 29–132

Albert Einstein: Declaration on international under-
standing for the British magazine No More War
Manuscript, April 1929
The Hebrew University of Jerusalem – Einstein Archives
Call Nr. 28–076

Max von Laue's car guest book
Around 1930
Archiv zur Geschichte der Max-Planck-Gesellschaft,
Berlin
III. Abt., Rep. 50, Nachtrag 1

Letter from Albert Einstein to the Prussian Academy
of Sciences
S. S. Belgenland, 28 March 1933
Archiv der Berlin-Brandenburgischen Akademie der
Wissenschaften, Berlin
II–III, Bd. 57, Bl. 6

Letter from Elsa Einstein to Antonina Vallentin
Scheveningen, 11 April 1933
Archiv zur Geschichte der Max-Planck-Gesellschaft,
Berlin
Va. Abt., Rep. 2, Bl. 105/17

Letter from Albert Einstein to Fritz Haber
19 May 1933
The Hebrew University of Jerusalem – Einstein Archives
Call Nr. 12–378

Letter from Albert Einstein to Max von Laue
Oxford, 7 June 1933
Archiv zur Geschichte der Max-Planck-Gesellschaft,
Berlin
III. Abt., Rep. 50, Nr. 536, Bl. 1

Letter from Albert Einstein to Heinrich Zangger
Le Coq sur Mer, 1 October 1933
Zentralbibliothek, Zurich
Nl. Zangger, Schachtel 1a

Albert Einstein: "Antwort auf einen Artikel von B. D.
Allinson" (Reply to an Article by B. D. Allinson
[recte: Allison])
Manuscript, 27 November 1934,
Published under the title "A Reexamination of
Pacifism", Polity, 1935
The Hebrew University of Jerusalem – Einstein Archives
Call Nr. 28–296

Letter from Max von Laue to Albert Einstein
Berlin, 10 January 1939
The Hebrew University of Jerusalem – Einstein Archives
Call Nr. 16–114

Albert Einstein: Poem for Melanie Serbu
[Princeton], 20 February 1939
Schweizerische Landesbibliothek, Bern
Schweizerisches Literaturarchiv
Schachtel Nr. 223/SSLB-24-Nr. 8

Albert Einstein: Presentation of The Black Book:
The Nazi Crime Against the Jewish People
Manuscript for a broadcast recording, 27 March 1946
The Hebrew University of Jerusalem – Einstein Archives
Call Nr. 28–692

Telegram from the President of the German Academy
of Sciences Johannes Stroux to Albert Einstein
Berlin, 26 July 1946
The Hebrew University of Jerusalem – Einstein Archives
Call Nr. 36–074

Albert Einstein: Address to the conference of the
National Urban League
Manuscript, 16 September 1946
The Hebrew University of Jerusalem – Einstein Archives
Call Nr. 28–707

Albert Einstein: "Der Ausweg" (The Way Out)
Manuscript of the contribution to
D. Masters and K. Way (eds.) One World or None,
New York, 1946
The Hebrew University of Jerusalem – Einstein Archives
Call Nr. 28–684

Letter from Albert Einstein to Otto Hahn
Princeton, 28 January 1949
Archiv zur Geschichte der Max-Planck-Gesellschaft,
Berlin
III. Abt., Rep. 14A, Nr. 814, Bl. 5-6

Draft of letter from Albert Einstein to
William Frauenglass
16 May 1953
The Hebrew University of Jerusalem – Einstein Archives
Call Nr. 34–636

Letter from Albert Einstein to the District Mayor of
Berlin-Neukölln Kurt Exner
Princeton, 15 May 1954
Albert-Einstein-Gymnasium, Berlin-Neukölln

Letter from Albert Einstein to Max von Laue
Princeton, 3 February 1955
Archiv der Johann-Wolfgang-Goethe-Universität,
Frankfurt am Main
Nl. von Laue

Albert Einstein: Declaration on Israel's
Independence Day
Manuscript, 1955
The Hebrew University of Jerusalem – Einstein Archives
Call Nr. 28–1098

The Einstein Papers Project

The Collected Papers of Albert Einstein
Volumes 1–9
Princeton, NJ, 1987–2004
Max Planck Institute for the History of Science, Berlin

Borderline Problems of Classical Physics

"Dispute about Borderline Problems of Classical
Physics"
Film by Jürgen Renn and Michael Dörfler, 2005
Max Planck Institute for the History of Science and
produktion2/cineplus, Berlin

Flat ring dynamo
Schuckert & Co., Nuremberg, around 1893
Deutsches Museum, Munich

Galvanometer with astatic needle pair
A researcher's construction, around 1830
Deutsches Museum, Munich

Exhibits

Hendrik Antoon Lorentz: *Versuch einer Theorie der electrischen und optischen Erscheinungen in bewegten Körpern* (Attempt of a Theory of Electric and Optical Phenomena in Moving Bodies)
Leiden, 1895
Max Planck Institute for the History of Science, Berlin

Mechanical models after Franz Reuleaux
Berlin, around 1880
Deutsches Museum, Munich

Atwood's free-fall machine
Milan, around 1865
Museo per la Storia dell'Università di Pavia

Ludwig Boltzmann: *Vorlesungen über die Principe der Mechanik* (Lectures on the Principles of Mechanics)
Leipzig, 1897
Max Planck Institute for the History of Science, Berlin

Hot-air engine
Goetze Company, Leipzig, around 1925
Deutsches Museum, Munich

Ice calorimeter after Lavoisier and Laplace
E. Leybold's Nachfolger, Cologne, 1911
Deutsches Museum, Munich

Max Planck: *Vorlesungen über Thermodynamik* (Lectures on Thermodynamics)
Leipzig, 1897
Max Planck Institute for the History of Science, Berlin

Challenges in the Laboratory

Black Body and Photoelectric Effect

Vacuum pump with an oil seal
Max Kohl Company, Chemnitz, 1935
Deutsches Museum, Munich

Moving–coil galvanometer
Hartmann & Braun, around 1894
Deutsches Museum, Munich

One-filament electrometer after Theodor Wulf
Günther & Tegetmeyer, Brunswick, around 1930
Deutsches Museum, Munich

Price list no. 56: Measuring instruments for laboratories and installation work
Siemens & Halske AG, Berlin, 1912
Max Planck Institute for the History of Science, Berlin

Regulating resistance
Around 1926
Deutsches Museum, Munich

Photocell after J. Elster and H. Geitel
Otto Pressler Company, Leipzig, 1930
Deutsches Museum, Munich

Precision balance
Hugo Schickert Company, Dresden, around 1860
Deutsches Museum, Munich

Battery
Heinrich Hertz, 1882
Deutsches Museum, Munich

Black body
Around 1950
Physikalisch-Technische Bundesanstalt, Berlin

Radiation measurement laboratory at the Physikalisch-Technische Reichsanstalt (National Institute for Physics and Technical Standards), Berlin
Photograph, around 1900
Max Planck Institute for the History of Science, Berlin

Rheostat
Edelmann & Sohn, Munich, 1935
Deutsches Museum, Munich

Quadrant electrometer
Wilhelm Hallwachs, 1895
Deutsches Museum, Munich

Philipp Lenard's laboratory journal
1904
Deutsches Museum, Munich

Tube for studying the photoelectric effect
Philipp Lenard, 1904
Deutsches Museum, Munich

Potential amplifier
Wilhelm Hallwachs, 1886
Deutsches Museum, Munich

Influence machine after James Wimshurst
J. Robert Voss Company, Berlin, 1906
Deutsches Museum, Munich

Induction device
Max Kohl Company, Chemnitz, 1906
Deutsches Museum, Munich

Discharge tube
Johann Wilhelm Hittorf, 1869 (reconstruction, 1968)
Deutsches Museum, Munich

Anode radiation tube
Ernst Gehrcke and Otto Reichenheim, 1906
Deutsches Museum, Munich

Electrostatic voltmeter
Hartmann & Braun, Frankfurt am Main, around 1930
Institut für Energie- und Automatisierungstechnik, Technische Universität, Berlin

Normal ammeter
Hartmann & Braun, Frankfurt am Main, 1911
Institut für Energie- und Automatisierungstechnik, Technische Universität, Berlin

Mirror galvanometer
Hartmann & Braun, Frankfurt am Main, around 1920
Institut für Energie- und Automatisierungstechnik, Technische Universität, Berlin

Electromagnetic Induction

Induction device
E. Leybold's Nachfolger, Cologne, 1911
Deutsches Museum, Munich

Universal measuring bridge after Kohlrausch
Hartmann & Braun, 1888
Deutsches Museum, Munich

Hans Landolt and Richard Börnstein: *Physikalisch-Chemische Tabellen* (Physical-Chemical Tables)
Berlin, 1905
Max Planck Institute for the History of Science, Berlin

Spectroscope
Reinfelder & Hertel, around 1870
Deutsches Museum, Munich

Brownian Motion

Induction machine after Hippolyte Pixii
Show model, 1908
Deutsches Museum, Munich

Device for measuring surface brightness
Optisches Institut A. Krüss, Hamburg, around 1902
Deutsches Museum, Munich

Magnetoelectric machines after Joseph Saxton
Around 1900
Deutsches Museum, Munich

Multiplier after Schweiger
Around 1827
Museo per la Storia dell'Università di Pavia

Device for experiments on secondary rays
Ernst Dorn, 1900
Deutsches Museum, Munich

Ernst Abbe's device for studying optical diffraction phenomena
Zeiss Company, Jena, around 1873
Deutsches Museum, Munich

Robert Brown: "Kurze Nachricht von den mikroscopis-chen Beobachtungen über die in dem Saamenstaube der Pflanzen enthaltenen Partikelchen" (Short Notice about the Microscopical Observations Concerning the Particles Contained in the Pollen of Plants)
Notizen aus dem Gebiet der Natur- und Heilkunde, October 1828, reproduction
Max Planck Institute for the History of Science, Berlin

Experiment on Brownian motion
TILL Photonics, Munich, 2005

Milieu of a Childhood

Birth certificate of Albert Einstein
Ulm, 15 March 1879, reproduction
Standesamt, Ulm
Geburtsregister 1879

Hermann Einstein
Photograph
SV-Bilderdienst, DIZ, Munich

Pauline Einstein, née Koch
Photograph
SV-Bilderdienst, DIZ, Munich

Anker stone building set
F. Ad. Richter & Co., Rudolstad, 1885
Deutsches Museum, Munich

Plan for an extension of the property at Rengeweg 14,
belonging to Mr. Einstein and Cie.
Munich, June 1886
Stadtarchiv, Munich

Compass
19th century
Deutsches Museum, Munich

Max Talmey: *The Relativity Theory Simplified and the
Formative Period of its Inventor*
New York, 1932
Max Planck Institute for the History of Science, Berlin

Aaron Bernstein: *Naturwissenschaftliche Volksbücher*
(Popular Books on Natural Science)
5th edition, Berlin, 1897
Max Planck Institute for the History of Science, Berlin

Ludwig Büchner: *Kraft und Stoff* (Force and Matter)
Frankfurt am Main, 1894
Max Planck Institute for the History of Science, Berlin

Theodor Spieker: *Lehrbuch der ebenen Geometrie mit
Übungsaufgaben für Höhere Lehranstalten* (Textbook
of Plane Geometry with Exercises for Institutes of
Higher Education)
Potsdam, 1886
Max Planck Institute for the History of Science, Berlin

Einstein's school class
Photograph, Canton School of Aarau, 1896
Bibliothek der Eidgenössischen Technischen
Hochschule, Zurich

Jost Winteler
Photograph
Schweizerische Landesbibliothek, Bern
Schweizerisches Literaturarchiv

Conrad Wüest: *Abriss der Geschichte der Elektrizität*
(Synopsis of the History of Electricity)
[Aarau, 1886], reproduction
Universitätbibliothek, Kassel

Albert Einstein: "Über die Untersuchung des
Aetherzustandes im magnetischen Felde"
(On the Investigation of the State of the Aether
in a Magnetic Field)
Manuscript, 1895, reproduction
Kaller Historical Documents, Malboro, NJ

Report about the instrumental music examination
Canton School of Aarau, [31 March 1896],
reproduction
Staatsarchiv des Kantons Aargau, Aarau
Erziehungsdirektion 1896, Mappe Ks. 4, no. 1049

Letter from Albert Einstein to the Education
Authority, Canton of Aargau
Registration for the Matura (secondary-school
leaving examination)
Aarau, 7 September 1896, reproduction
Staatsarchiv des Kantons Aargau, Aarau
Erziehungsdirektion 1896, Mappe Ks. 5, no. 2270

Albert Einstein: "Mes projets d'avenir"
(My Plans for the Future)
Matura French examination
Manuscript, 18 September 1896, reproduction
Staatsarchiv des Kantons Aargau, Aarau
Erziehungsdirektion 1896, Mappe Ks. 5, no. 2270

Results of the Matura examination
September 1896, reproduction
Staatsarchiv des Kantons Aargau, Aarau
Erziehungsdirektion 1896, Mappe Ks. 5, no. 2270

The Electrical Factory J. Einstein & Comp.

Advertisement by the J. Einstein'schen Fabrik für
Wasserförderung & Central-Heizungen
Adreßbuch von München, 1878
Deutsches Museum, Munich

The first electrical lighting of the Oktoberfest
Postcard, Munich, 1885
Münchener Stadtmuseum, Munich

*Officieller Katalog der Kraft- und Arbeitsmaschinen-
Ausstellung für das Deutsche Reich*
(Official Catalog of the German National Power
Machine and Engine Exhibition)
Munich, 1888
Deutsches Museum, Munich

Program for the opening of the Schwabing power
station
Schwabing, 26 February 1889
Stadtarchiv, Munich
Schwabig 499

Illustration of the Schwabing power station
Centralblatt für Elektrotechnik, 12, 1889
Bibliothek der Technischen Universität, Berlin

Invoice from J. Einstein & Comp. to the Magistrate
of the City of Schwabing
Munich, 28 February 1889
Stadtarchiv, Munich
Schwabig 499

Letter from J. Einstein & Comp. to the Main Committee
of the 7th German Gymnastics Festival
Munich, 26 July 1889
Stadtarchiv, Munich
Amt für Leibesübungen 55

Letter from J. Einstein & Comp. to the Chairman of the
Association of German Engineers Theodor Peters
Munich, 11 April 1890
Staatsbibliothek, Berlin
Slg. Darmstädter, F 1e 1909

Illustration of the interior of the J. Einstein & Comp.
electrical factory
Julius Kahn: *Münchens Großindustrie und Großhandel*
(Munich Large Industrial and Wholsale Companies)
Munich, 1891
Deutsches Museum, Munich

Advertisement by the J. Einstein & Cie electrical
factory
*Elektrizität. Offizielle Zeitung der Internationalen
Elektrotechnischen Ausstellung* (Electricity. Official
Publication of the International Electrotechnical
Exhibition)
Frankfurt am Main, 1891
Deutsches Museum, Munich

*Offizieller Bericht über die Internationale Elektro-
technische Ausstellung in Frankfurt am Main 1891*
(Official Report about the International
Electrotechnical Exhibition)
Frankfurt am Main, 1893
Max-Planck-Institut für Wissenschaftsgeschichte,
Berlin

Generator
Deutsche Elektrizitätswerke Garbe, Lahmeyer & Co,
Aachen, 1890
Deutsches Museum, Munich

Arc lamp
Hugo Bremer Company, Neheim, around 1900
Deutsches Museum, Munich

The Electrical Exhibitions

Pictures from the International Electrotechnical
Exhibition in Frankfurt am Main 1891
Das Buch für Alle, vol. 28, 1892, p. 68
Max Planck Institute for the History of Science, Berlin

Exhibits

Map of the International Electrotechnical Exhibition
*Offizieller Bericht über die Internationale Elektro-
technische Ausstellung in Frankfurt am Main 1891*
(Official Report about the International Electro-
technical Exhibition), vol. 1
Frankfurt am Main, 1893
Max Planck Institute for the History of Science, Berlin

Illustration of the Helios AG stand
*Offizieller Bericht über die Internationale Elektro-
technische Ausstellung in Frankfurt am Main 1891*
Vol. 1, Frankfurt am Main, 1893
Max Planck Institute for the History of Science, Berlin

Illustration of the great hall of machines
*Offizieller Bericht über die Internationale Elektro-
technische Ausstellung in Frankfurt am Main 1891*
Vol. 1, Frankfurt am Main, 1893
Max Planck Institute for the History of Science, Berlin

Small arc lamp
C. Hoffmann Company, around 1890
Deutsches Museum, Munich

Wall telephone
Kgl. Bayerische Posten und Telegraphen, 1899
Deutsches Museum, Munich

James Clerk Maxwell: *Die Elektrizität in elementarer
Behandlung* (An Elementary Treatise on Electricity)
Brunswick, 1883
Private collection

Leo Graetz: *Die Elektrizität und ihre Anwendungen*
(Electricity and its Applications)
Stuttgart, 1912
Private collection

Nernst lamp
1906
Deutsches Museum, Munich

Direct current switchboard
Schuckert&Co, Nuremberg, 1890
Deutsches Museum, Munich

*Weltausstellung in Paris 1900. Amtlicher Katalog der
Ausstellung des Deutschen Reichs*
(Paris World's Fair 1900: Official Catalog for the
Exhibition of the German Reich)
Berlin, 1900
Max Planck Institute for the History of Science, Berlin

Price lists no. 1–7b
Siemens-und-Halske-A.G., Berlin, 1897
Max Planck Institute for the History of Science, Berlin

Catalog for machines and fittings
Siemens-Schuckert Werke G.m.b.H., 1909
Max Planck Institute for the History of Science, Berlin

Generator for manual operation
C.&E. Fein, Stuttgart, 1881
Deutsches Museum, Munich

Flat sealing ring generator
Sigmund Schuckert Company, Nuremberg, 1881
Deutsches Museum, Munich

The Electrical Factory Einstein, Garrone & C.

Façade of the Einstein family home in Pavia
Photograph
F. and G. Torti, Pavia, 2005

The Einstein, Garrone & C. factory in Pavia
Lithograph, around 1900
Museo per la Storia dell'Università di Pavia

Ground plan of the Einstein, Garrone & C. factory
Pavia, 1894
The Einstein Papers Project, Pasadena, CA

Catalog of goods from Einstein, Garrone & C.
Pavia, 1896
Camera di Commercio, Pavia

Concession for Hermann Einstein to erect an electrical
power station in Canneto sull'Oglio
1899
Archivo Storico Comunale di Canneto sull'Oglio

Two invoices from the H. Einstein Company to the
township of Canneto sull'Oglio for electrical lighting
April and June 1900
Archivo Storico Comunale di Canneto sull'Oglio

Pelton turbine
Ing. A. Riva, Monneret&C., Milan, 1901
ASM Voghera s.p.a., Voghera

Letter from Albert Einstein to Ernesta Marangoni
Princeton, 16 August 1946
Museo per la Storia dell'Università di Pavia

Milieu of a Revolution

"Meeting of the Olympia Academy"
Film by Jürgen Renn and Michael Dörfler, 2005
Max Planck Institute for the History of Science and
produktion 2 / cineplus, Berlin

Student registry book of Albert Einstein
Zurich, 1896–1900, reproduction
Bibliothek der Eidgenössischen Technischen
Hochschule, Zurich
Rektoratsarchiv, Matrikel No. 82

Letter from Michele Besso to Aurel Stodola
Geneva, 22 August 1941, reproduction
Bibliothek der Eidgenössischen Technischen
Hochschule, Zurich
Hs 496:5

Michele Besso and his wife Anna Winteler
Photograph, around 1900
Laurent Besso, Lausanne

Albert Einstein with his mother Pauline
Photograph
Schweizerische Landesbibliothek, Bern
Schweizerisches Literaturarchiv

Results of Albert Einstein's physics examination
Zurich, 27 July 1900, reproduction
Bibliothek der Eidgenössischen Technischen
Hochschule, Zurich
Schulratsarchiv, 1900, 532

Bestowal of the Swiss citizenship to Albert Einstein
Politisches Departement der Schweizerischen
Eidgenossenschaft
Bern, 8 March 1900
Schweizerisches Bundesarchiv, Bern
E 21/23560

Postcard from Albert Einstein to Carl Paalzow
Milan, 12 April 1901, reproduction
Archiv zur Geschichte der Max-Planck-Gesellschaft,
Berlin
Va. Abt., Rep. 2, Nr. 2

Letter from Hermann Einstein to Wilhelm Ostwald
Milan, 13 April 1901, reproduction
Archiv der Berlin-Brandenburgischen Akademie der
Wissenschaften, Berlin
Nl. Ostwald, 677/1

Application by Albert Einstein to the Swiss Federal
Patent Office
Schaffahausen, 18 December 1901, reproduction
Schweizerisches Bundesarchiv, Bern
E 22/2338

Swiss Federal Patent Office
Photograph, Bern, around 1900
Schweizerische Landesbibliothek, Bern
Schweizerisches Literaturarchiv

Albert Einstein at the Swiss Federal Patent Office
Photograph, Bern, around 1905
Schweizerische Landesbibliothek, Bern
Schweizerisches Literaturarchiv

Einstein's first room in Bern, Gerechtigkeitsgasse 32
Reproduction of a drawing from a letter from Albert
Einstein to Mileva Marić, Bern, 4 February 1902
Max Planck Institute for the History of Science, Berlin

Mileva Einstein with son Hans Albert
Photograph, Bern, around 1908
Schweizerische Landesbibliothek, Bern
Schweizerisches Literaturarchiv

Directive to the District Court of Zurich concerning
the rent claim Gentner versus Einstein
Zurich, 26 August 1913, reproduction
Staatsarchiv des Kantons Zürich
BEZ Zch 6314.32

Einstein's advertisement for private tutoring
Anzeiger für die Stadt Bern, 19 February 1902
Schweizerische Landesbibliothek, Bern

Lucien Chavan: Notes on Albert Einstein's tutorial
on the theory of electricity
Bern, 1903–1906, reproduction
Schweizerische Landesbibliothek, Bern
Albert-Einstein-Archiv-Nr. 102

The members of the Olympia Academy: Conrad
Habicht, Maurice Solovine and Albert Einstein
Photograph, Bern, around 1903
Bibliothek der Eidgenössischen Technischen
Hochschule, Zurich

Caricature of Albert Einstein by Maurice Solovine
1903, facsimile
The Hebrew University of Jerusalem – Einstein Archives
Call Nr. 36–621

Postcard from Albert Einstein to Conrad Habicht
Bern, 6 August 1904
Bibliothek der Eidgenössischen Technischen
Hochschule, Zurich
Hs 1457:16

Letter from Albert Einstein to Maurice Solovine
3 April 1953, reproduction
Harry Ransom Humanities Research
Center, Austin (TX)
Estate Solovine: 12-486.00

Albert Einstein: *Eine neue Bestimmung der
Moleküldimensionen* (A New Determination
of Molecular Dimensions)
University of Zurich, Dissertation, 1905
Staatsbibliothek, Berlin

Letter from Albert Einstein to Conrad Habicht
Bern, May 1905, reproduction
Bibliothek der Eidgenössischen Technischen
Hochschule, Zurich
Hs 1457:20

Letter from Paul Habicht to Albert Einstein
Basel, 4 July 1908, facsimile
The Hebrew University of Jerusalem –
Einstein Archives
Call Nr. 35-546

Potential multiplier
Paul Habicht, Schaffhausen, around 1920
Physikalisches Institut, Eberhard-Karls-Universitat,
Tübingen

Natural History Societies

Memorandum about Albert Einstein's joining the
Natural History Society of Bern
Minutes of the meeting of the Naturforschende
Gesellschaft Bern, 2 May 1903
Burgerbibliothek, Bern
GA BNG Protokoll-Bd. 10, 987. Sitzung

Memorandum about Albert Einstein's first lecture
"Über die Theorie der elektromagnetischen Wellen"
(On the Theory of Electromagnetic Waves)
Minutes of the meeting of the Naturforschende
Gesellschaft Bern, 5 December 1903
Burgerbibliothek, Bern
GA BNG Protokoll-Bd. 10, 992. Sitzung

Mountain excursion by the
Natural History Society of Bern
Photograph, 1919
Burgerbibliothek, Bern

Albert Einstein's participation ticket to the 96th
annual convention of the Swiss Natural History
Society
Frauenfeld, 7–10 September 1913
Bibliothek der Eidgenössischen Technischen
Hochschule, Zurich
Hs 1457:73

Poincaré and the Problem of Simultaneity

Letter from Heinrich Hertz to Henri Poincaré
Bonn, 7 December 1890
François Poincaré, Paris

Letter from Robert Bourgeois to Henri Poincaré
Quito, 4 June 1902
François Poincaré, Paris

Electrical mother clock with three supplementary
clocks integrated
Peyer, Favarger & Cie, Neuchâtel, 1906
Deutsches Museum, Munich

Henri Poincaré: Popular science writings
Wissenschaft und Hypothese (Science and Hypothesis)
Leipzig, 1914
Der Wert der Wissenschaft (The Value of Science)
Leipzig, 1921
Wissenschaft und Methode (Science and Method)
Leipzig, 1914
Letzte Gedanken (Last Essays)
Leipzig, 1913
Private collection

The *annus mirabilis* 1905

annus mirabilis
Film by Jürgen Renn, Wolfram Deutschmann
and Michael Feser, 2005
Max Planck Institute for the History of Science and
Kick-Film, Berlin

Albert Einstein: "Über einen die Erzeugung und
Verwandlung des Lichtes betreffenden heuristischen
Gesichtspunkt" (On a Heuristic Point of View Concern-
ing the Production and Transformation of Light)
Annalen der Physik, vol. 17, 1905, reproduction
Max Planck Institute for the History of Science, Berlin

Albert Einstein: "Über die von der molekularkinetis-
chen Theorie der Wärme geforderte Bewegung von
in ruhenden Flüssigkeiten suspendierten Teilchen"
(On the Movement of Small Particles Suspended in
Stationary Liquids Required by the Molecular-Kinetic
Theory of Heat)
Annalen der Physik, vol. 17, 1905, reproduction
Schweizerische Landesbibliothek, Bern

Albert Einstein: "Zur Elektrodynamik bewegter
Körper" (On the Electrodynamics of Moving Bodies)
Annalen der Physik, vol. 17, 1905, reproduction
Bibliothek der Eidgenössischen Technischen
Hochschule, Zurich

Albert Einstein: "Ist die Trägheit eines Körpers von
seinem Energieinhalt abhängig?" (Does the Inertia
of a Body Depend on its Energy Content?)
Annalen der Physik, vol. 17, 1905, reproduction
Schweizerische Landesbibliothek, Bern

First Experimental Confirmations

Letter from Hendrik A. Lorentz to Henri Poincaré
Leiden, 8 March 1906, reproduction
François Poincaré, Paris

Part of an experimental setup by Walter Kaufmann
1906
Deutsches Museum, Munich

Detail of Walter Kaufmann's experimental setup
Illustration from Albert Einstein: "Über das Relativ-
itätsprinzip und die aus demselben gezogenen
Folgerungen" (On the Relativity Principle and the
Conclusions Drawn from It)
Jahrbuch der Radioaktivität und Elektronik, vol. 4, 1907,
reproduction
Max Planck Institute for the History of Science, Berlin

Part of an experimental setup by Alfred Bucherer
1908
Deutsches Museum, Munich

Laboratory of Alfred Bucherer
Photograph, Bonn, 1906
Deutsches Museum, Munich

Exhibits

Letter from Alfred Bucherer to Albert Einstein
Bonn, 7 September 1908, reproduction
The Hebrew University of Jerusalem – Einstein Archives
Call Nr. 6–188

Albert Einstein: *Einstein's 1912 Manuscript on the Special Theory of Relativity. A Facsimile*
New York, 1996
Max Planck Institute for the History of Science, Berlin

Albert Einstein: "Entwurf zu einer Vorlesung in Leiden" (Notes for a Lecture in Leiden)
Manuscript, 1911, facsimile
The Hebrew University of Jerusalem – Einstein Archives
Call Nr. 2–081

Jean Perrin: *Atoms*
Dresden, 1914
Max Planck Institute for the History of Science, Berlin

Ultramicroscope
Carl Zeiss Company, Jena, around 1910
Carl Zeiss AG, Oberkochen

Spark chamber
Deutsches Elektronen-Synchrotron DESY, Hamburg and Zeuthen, 1990

Bubble chamber
Photograph

Decay chain of elementary particles
Bubble chamber photograph
European Organization for Nuclear Research CERN, Geneva

Cloud chamber
Phywe Company, 1990
Deutsches Elektronen-Synchrotron DESY, Hamburg and Zeuthen

Academic Career

Letter from Albert Einstein to the Council of Education, Canton of Zurich
Application for a job as a teacher
Bern, 20 January 1908, reproduction
Staatsarchiv des Kantons Zürich
U 84 d.2

Letter from Albert Einstein to the Education Authority, Canton of Zurich
Application for reimbursement of removal costs
Zurich, 1 November 1909, reproduction
Staatsarchiv des Kantons Zürich
U 110 b.2 (44)

Letter from Zurich students to the Education Authority, Canton of Zurich
Petition to retain Albert Einstein at the University of Zurich
Zurich, 23 June 1910, reproduction
Staatsarchiv des Kantons Zürich
U 110 b.2 (44)

Excerpt from the minutes of the session of the Swiss Federal Council of Ministers
30 January 1912, reproduction
Bibliothek der Eidgenössischen Technischen Hochschule, Zurich
Schulratsarchiv 1912, Akten, No. 116

Proposal for Albert Einstein's membership in the Prussian Academy of Sciences
Manuscript in the hand of Max Planck, Berlin, 12 June 1913, reproduction
Archiv der Berlin-Brandenburgischen Akademie der Wissenschaften, Berlin
II–III, Bd. 36, Bl. 36–37

Ballotage balls of the Prussian Academy of Sciences
Archiv der Berlin-Brandenburgischen Akademie der Wissenschaften, Berlin

Letter from Albert Einstein to the President of the Swiss School Council Robert Gnehm
Zurich, 30 November 1913, reproduction
Bibliothek der Eidgenössischen Technischen Hochschule, Zurich
Schulratsarchiv 1913, Akten, No. 1289

Letter from Albert Einstein to the Prussian Academy of Sciences
Zurich, 7 December 1913, reproduction
Archiv der Berlin-Brandenburgischen Akademie der Wissenschaften, Berlin
II–III, Bd. 36, Bl. 54

The Riddle of Gravitation

Letter from Albert Einstein to Arnold Sommerfeld
Bern, 5 January 1908, reproduction
Deutsches Museum, Munich
NL 89, 007

Letter from Albert Einstein to Wilhelm Wien
Prag, 10 July [1912], reproduction
Paul Siebertz, Munich

Letter from Albert Einstein to Arnold Sommerfeld
[Zurich], 29 October [1912], facsimile
Niedersächsische Staats- und Universitätsbibliothek, Göttingen
Cod. Ms. Hilbert 92b, Bl. 2

Albert Einstein: "Zurich Notebook"
Winter 1912/13, facsimile
The Hebrew University of Jerusalem – Einstein Archives
Call Nr. 3–006

Albert Einstein and Marcel Grossmann: *Entwurf einer verallgemeinerten Relativitätstheorie und einer Theorie der Gravitation* (Outline of a Generalized Theory of Relativity)
Leipzig, 1913
Max Planck Institute for the History of Science, Berlin

Letter from Albert Einstein to Ernst Mach
Zurich, 25 June 1913, reproduction
Deutsches Museum, Munich
Nl. Mach

"The Einstein-Besso Manuscript"
Berlin and Gorizia, 1913/14, reproduction of a page
Aristophil, Paris

Michele Besso: Notes about the "hole argument"
Manuscript, 28 August 1913, reproduction of a page
Laurent Besso, Lausanne

Eötvös balance
Replica, 1930
Magyar Állami Eötvös Loránd Geofizikai Intézet, Budapest

Drop tower
Functional model, machtWissen.de, Bremen, 2005
Max Planck Institute for the History of Science, Berlin

The Completion of the Relativity Revolution

Letter from Max Planck to Wilhelm Wien
Grunewald, 29 June 1913, reproduction
Staatsbibliothek, Berlin
Nl. Wien, Nr. 70

Max Planck
Photograph, Berlin, around 1906
Archiv zur Geschichte der Max-Planck-Gesellschaft, Berlin

Albert Einstein's fountain pen
1912–1921
Museum Boerhaave, Leiden

"The Einstein-Besso Manuscript"
Berlin and Gorizia, 1913/14, reproduction of a page
Aristophil, Paris

Postcard from Albert Einstein to David Hilbert
[Berlin, 18 November 1915], reproduction
Niedersächsische Staats- und Universitätsbibliothek, Göttingen
Cod. Ms. Hilbert 92b:2

Postcard from David Hilbert to Albert Einstein
[Göttingen, 19 November 1915], facsimile
The Hebrew University of Jerusalem – Einstein Archives
Call Nr. 13–054

David Hilbert
Photograph
Deutsches Museum, Munich

Albert Einstein: "Die Feldgleichungen der Gravitation" (The Field Equations of Gravitation)
Sitzungsberichte der Königlich Preußischen Akademie der Wissenschaften, 25 November 1915
Bibliothek der Berlin-Brandenburgischen Akademie der Wissenschaften, Berlin

Letter from Albert Einstein to Arnold Sommerfeld
Berlin, 28 November 1915, reproduction
Deutsches Museum, Munich
NL 89, 1977–28/A, 78(2)

David Hilbert: "Die Grundlagen der Physik
(Erste Mitteilung)" (The Principles of Physics
(First Announcement))
Galley proof with handwritten corrections by Hilbert,
stamped 6 December 1915
Niedersächsische Staats- und Universitätsbibliothek,
Göttingen
Cod. Ms. Hilbert 634

Letter from Karl Schwarzschild to Arnold Sommerfeld
22 December 1915, reproduction
Deutsches Museum, Munich
NL 89, 059

Draft of letter from Karl Schwarzschild to
Albert Einstein
22 December 1915, reproduction
Niedersächsische Staats- und Universitätsbibliothek,
Göttingen
Cod. Ms. Schwarzschild 2:2, Bl. 2–3

David Hilbert: "Die Grundlagen der Physik
(Erste Mitteilung)" (The Principles of Physics
(First Announcement))
*Nachrichten von der Königlichen Gesellschaft
der Wissenschaften zu Göttingen. Mathematisch-
physikalische Klasse*
1915 [published in 1916], reproduction
Staatsbibliothek, Berlin

Letter from Albert Einstein to Hugo Andres Krüss
[Berlin], 10 January 1918, reproduction
Staatsbibliothek, Berlin
Autogr. I/4180, 1–3

Albert Einstein: *Äther und Relativitätstheorie*
(Aether and Relativity Theory)
Berlin, 1920
Max Planck Institute for the History of Science, Berlin

Albert Einstein, standing by his bookshelf
Photograph, Berlin, around 1920
Bildarchiv Preußischer Kulturbesitz, Berlin

Session of the Prussian Academy of Sciences
Photograph, Berlin, around 1931
Archiv der Berlin-Brandenburgischen Akademie der
Wissenschaften, Berlin

The Triumph of the Theory of Relativity

Letter from Albert Einstein to George E. Hale
Zurich, 14 October 1913, reproduction
The Huntington Library, San Marino, CA
Hale Papers

Stereo comparator by Carl Pulfrich
Prototype, Zeiss Company, 1899
Deutsches Museum, Munich

Letter from Erwin F. Freundlich to Albert Einstein
Neubabelsberg, 17 June 1917, facsimile
The Hebrew University of Jerusalem – Einstein Archives
Call Nr. 11–137

Draft of letter from Wilhelm von Siemens to
Albert Einstein
Berlin, 21 January 1918, reproduction
Archiv zur Geschichte der Max-Planck-Gesellschaft,
Berlin
I. Abt., Rep. 1A, Nr. 1656, Bl. 31

Photographs of the total solar eclipse taken by Arthur
S. Eddington on Principe Island, 29 May 1919
Philosophical Transactions, Series A, vol. 220, 1920
Bayerische Staatsbibliothek, Munich

Arthur Stanley Eddington
Photograph
Deutsches Museum, Munich

Telegram from Hendrik A. Lorentz to Albert Einstein
[s'Gravenhage, 22 September 1919], reproduction
Museum Boerhaave, Leiden

Albert Einstein: "Was ist Relativitäts-Theorie?"
Manuscript of the article "My Theory", *The Times*,
28 November 1919, facsimile
The Hebrew University of Jerusalem – Einstein Archives
Call Nr. 1–002

Letter from Arnold Berliner to Albert Einstein
Berlin, 29 November 1919, facsimile
The Hebrew University of Jerusalem – Einstein Archives
Call Nr. 7–004

Letter from Arthur S. Eddington to Albert Einstein
[Cambridge], 1 December 1919, facsimile
The Hebrew University of Jerusalem – Einstein Archives
Call Nr. 9–260

Donor list for the "Einstein Donations"
Berlin, 1920
Archiv der Berlin-Brandenburgischen Akademie der
Wissenschaften, Berlin
Astrophysikalisches Observatorium 147, Bl. 1/2

Erich Mendelsohn: Sketch for the Einstein Tower
"Project with underground laboratory," pencil, 1919,
reproduction
Staatliche Museen zu Berlin-Kunstbibliothek

Erich Mendelsohn: Sketch for the Einstein Tower
Front view, pencil, colored chalk, around 1919,
reproduction
Staatliche Museen zu Berlin-Kunstbibliothek

Model of the Einstein Tower
Erich Mendelsohn, 1920
Deutsches Architektur Museum, Frankfurt am Main

Prism from the spectrograph of the Einstein Tower
Astrophysikalisches Institut, Potsdam

Erwin F. Freundlich observing the solar eclipse
Photograph, Sumatra, 1929
Max Planck Institute for the History of Science, Berlin

Towards a New Cosmology

Letter from Willem de Sitter to Albert Einstein
Leiden, 20 March 1917, facsimile
The Hebrew University of Jerusalem – Einstein Archives
Call Nr. 20–545

Albert Einstein: "Kosmologische Betrachtungen zur
allgemeinen Relativitätstheorie" (Cosmological
Considerations in the General Theory of Relativity)
*Sitzungsberichte der Preußischen Akademie der
Wissenschaften*, 1917, reproduction
Staatsbibliothek, Berlin

Albert Einstein: Part 2 of "Geometrie und Erfahrung"
(Geometry and Experience)
Manuscript, Berlin, 1921, facsimile
The Hebrew University of Jerusalem – Einstein Archives
Call Nr. 1–012

Albert Einstein: "Notiz zu der Arbeit von A. Friedmann
'Über die Krümmung des Raumes'" (Note on the
Article of A. Friedmann on the Curvature of Space)
Manuscript, published in *Zeitschrift für Physik*, 1923,
facsimile
The Hebrew University of Jerusalem – Einstein Archives
Call Nr. 1–026

Albert Einstein: Addendum for a new edition of
*Über die spezielle und allgemeine Relativitätstheorie
(Gemeinverständlich)* (On the Special and the General
Theory of Relativity (A Popular Account))
Manuscript, Princeton, 1946, facsimile
The Hebrew University of Jerusalem – Einstein Archives
Call Nr. 1–186

Model of the Mount Wilson Observatory
Modellbau Keyser, Berlin, 2005
Max Planck Institute for the History of Science, Berlin

Wallpaper: Chandra Deep Field South
Photograph
European Southern Observatory, La Silla, 2003

The Search for the Unity of Nature

Postcard from Albert Einstein to Hermann Weyl
[Berlin], 15 April 1918, reproduction
Bibliothek der Eidgenössischen Technischen
Hochschule, Zurich
Hs 91:541

"Einstein's New Theory; Text of His Treatise"
New York Herald Tribune, 31 January 1929,
reproduction
New York State Library, Albany, (NY)

Albert Einstein: "Über den gegenwärtigen Stand der Feldtheorie" (On the Current State of Field Theory)
Manuscript of the contribution to the festschrift for A. Stodola, Zurich, 1929, reproduction
The Pierpont Morgan Library, New York
Misc. Heineman MAH 4378

Letter from Wolfgang Pauli to Albert Einstein
Zurich, 19 December 1929, facsimile
The Hebrew University of Jerusalem – Einstein Archives
Call Nr. 19–163

Albert Einstein and Walther Mayer: "Semi-Vektoren und Spinoren" (Semivectors and Spinors)
Sitzungsberichte der Preußischen Akademie der Wissenschaften. Physikalisch-mathematische Klasse
1932, reproduction
Max Planck Institute for the History of Science, Berlin

Albert Einstein: "Zur Methode der theoretischen Physik" (On the Method of Theoretical Physics)
Manuscript for Spencer Lecture, Oxford, 10 June 1933, reproduction
The Hebrew University of Jerusalem – Einstein Archives
Call Nr. 1–114

Albert Einstein: "Über rotationssymmetrische stationäre Gravitationsfelder" (About Stationary Gravitational Fields Symmetrical under Rotation)
Manuscript fragment, 1936, reproduction
Syracuse University Library, Syracuse (NY)
NI. Bergmann

Albert Einstein: "Das Inertialsystem muss aus der Grundlage der Physik verschwinden" (The Inertial System Must Disappear from the Foundations of Physics)
Manuscript, 1952, facsimile
The Hebrew University of Jerusalem – Einstein Archives
Call Nr. 2–134

The Unfinished Quantum Revolution

Albert Einstein: "Die Plancksche Theorie der Strahlung und die Theorie der spezifischen Wärme" (Planck's Theory of Radiation and the Theory of Specific Heat)
Annalen der Physik, vol. 22, 1907
Staatsbibliothek, Berlin

Walther Nernst at a lecture
Photograph, Berlin, late 1920s
Archiv zur Geschichte der Max-Planck-Gesellschaft, Berlin

Hydrogen liquefier after Walther Nernst
Around 1920
Museum of the History of Science, Oxford

Atoms according to Democritus
Modellbau Wünsche & Borkenhagen, Berlin, after models of the Deutsches Museum, Munich
Max Planck Institute for the History of Science, Berlin

Atomic model after Joseph J. Thomson, 1903/1904
Das Atelier Niesler, Berlin
Max Planck Institute for the History of Science, Berlin

Atomic model after Niels Bohr, 1913
Das Atelier Niesler, Berlin
Max Planck Institute for the History of Science, Berlin

Installation representing the hydrogen atom
Max Planck Institute for the History of Science, Berlin, 2005

The Franck-Hertz experiment
Demonstration, Phywe Company, Göttingen, 2005
Max Planck Institute for the History of Science, Berlin

James Franck
Photograph, around 1925
Bildarchiv Preußischer Kulturbesitz, Berlin
Gustav Hertz in the institute's laboratory
Photograph, around 1913
Private collection

Postcard from Albert Einstein to Arnold Sommerfeld
[Berlin, 8 February 1916], reproduction
Deutsches Museum, Munich
NL 89, 1977-28/A, 78(5)

Postcard from Albert Einstein to Paul Ehrenfest
[Berlin, 5 April 1920], reproduction
The Hebrew University of Jerusalem – Einstein Archives
Call Nr. 9–486

Wave and Particle as a Persistent Contradiction

Letter from Max Laue to Albert Einstein
Berlin, 2 June 1906, facsimile
The Hebrew University of Jerusalem – Einstein Archives
Call Nr. 16–002

Letter from Hendrik A. Lorentz to Albert Einstein,
Leiden, 6 May 1909, reproduction
Brandeis University Libraries, Waltham, MA
Einstein Collection, no. 1939

Letter from Albert Einstein to Max von Laue
[Prague], 10 June 1912, reproduction
Deutsches Museum, Munich
Handschriften, 1951–4

Proposal for Albert Einstein's membership in the Prussian Academy of Sciences
Manuscript in the hand of Max Planck, Berlin, 12 June 1913, reproduction
Archiv der Berlin-Brandenburgischen Akademie der Wissenschaften, Berlin
II–III, Bd. 36, Bl. 36–37

Albert Einstein: "Strahlungs-Emission und -Absorption nach der Quantentheorie" (Emission and Absorption of Radiation in Quantum Theory)
Verhandlungen der Deutschen Physikalischen Gesellschaft, vol. 18, 1916, reproduction
Max Planck Institute for the History of Science, Berlin

Einstein's Nobel Prize diploma
Stockholm, 1922, facsimile
Archiv zur Geschichte der Max-Planck-Gesellschaft, Berlin

Letter from Hendrik A. Lorentz to Albert Einstein
Haarlem, 13 November 1921, reproduction
Museum Boerhaave, Leiden

Letter from Arnold Sommerfeld to Niels Bohr,
Santa Fe, 21 January 1923, reproduction
Niels Bohr Archive, Copenhagen

The Einstein-Bohr Debate
Multi-media presentation by Christoph Lehner and Florian Sander, 2005
Max Planck Institute for the History of Science, Berlin

Albert Einstein: "Quantentheorie des einatomigen idealen Gases" (Quantum Theory of the Single-Atom Ideal Gas)
Manuscript, 1924, facsimile of a page
The Hebrew University of Jerusalem – Einstein Archives
Call Nr. 1–040

Louis de Broglie: *Recherches sur la Théorie des Quanta* (Investigations on Quantum Theory)
University of Paris, Dissertation, 1924, reproduction
Universitätsbibliothek Eichstätt-Ingolstadt, Eichstätt

Letter from Albert Einstein to Erwin Schrödinger
Berlin, 28 February 1925, reproduction
Österreichische Zentralbibliothek für Physik, Vienna
NI. Schrödinger

Images of electron scattering patterns in a crystal lattice
Labor für Festkörperphysik, Eidgenössische Technische Hochschule, Zurich

Photograph of x-ray scattering
Argonne National Laboratory, Argonne, IL

Electrostatic lens for the EM8 electron microscope
Zeiss-AEG Jena, around 1952
Deutsches Museum, Munich

Illustration of cut through the single-field condensing lens and objective with pole shoe lens and specimen gate
Ernst Ruska: "Über die Auflösungsgrenzen des Durchstrahlungs-Elektronenmikroskops", *Optik*, vol. 22, 1965
Fritz-Haber-Institut der Max-Planck-Gesellschaft, Berlin

Electromagnetic lens (cut)
Ernst Ruska, Berlin, 1966
Fritz-Haber-Institut der Max-Planck-Gesellschaft, Berlin

Pole shoe and specimen gate of DEEKO 100
Berlin, 1966
Fritz-Haber-Institut der Max-Planck-Gesellschaft, Berlin

The Paradoxes of Quantum Mechanics

Letter from Albert Einstein to Max Born
29 April 1924, reproduction
Staatsbibliothek, Berlin
Nl. Born, Nr. 188, Bl. 24

Letter from Albert Einstein to Paul Ehrenfest
Berlin, 18 September 1925, facsimile
The Hebrew University of Jerusalem – Einstein Archives
Call Nr. 10-111

Postcard from Albert Einstein to Erwin Schrödinger
Berlin, 26 April 1926, reproduction
Österreichische Zentralbibliothek für Physik, Vienna
Nl. Schrödinger

Letter from Albert Einstein to Max Born
4 December 1926, reproduction
Staatsbibliothek, Berlin
Nl. Born, Nr. 188, Bl. 25

Letter from Niels Bohr to Albert Einstein
Copenhagen, 13 April 1927, facsimile
The Hebrew University of Jerusalem – Einstein Archives
Call Nr. 8-084

Letter from Werner Heisenberg to Albert Einstein
Copenhagen, 10 June 1927, facsimile
The Hebrew University of Jerusalem – Einstein Archives
Call Nr. 12-174

Albert Einstein: Statement on the 200th anniversary
of Newton's death
Manuscript, published in *Nature*, March 1927, facsimile
The Hebrew University of Jerusalem – Einstein Archives
Call Nr. 1-058

Letter from Albert Einstein to Erwin Schrödinger
Old Lyme, 19 June 1935, reproduction
The Hebrew University of Jerusalem – Einstein Archives
Call. Nr. 22-047

Telegram from Niels Bohr to Albert Einstein with
reply draft by Albert Einstein on the back side
March 1949, facsimile
The Hebrew University of Jerusalem – Einstein Archives
Call Nr. 8-098

Einstein-Podolsky-Rosen Experiment
Installation, 2005
Institut für Experimentalphysik, University of Vienna

Albert Einstein: Commentary to an article by
Max Born
Manuscript, 12 January 1954, reproduction
The Hebrew University of Jerusalem – Einstein Archives
Call. Nr. 8-240.1

World War and Revolution

Poster with the Kaiser's appeal to close ranks in war
August 1914, reproduction
Deutsches Historisches Museum, Berlin
PL 004507

Propaganda postcards from World War I
Gerhard Schneider, Wedemark and
Martin Polzin, Laatzen

Appeal "An die Kulturwelt!" (To the Civilized World!)
4 October 1914, reproduction
Württembergische Landesbibliothek, Stuttgart
Bibliothek für Zeitgeschichte, Archiv. Slg. Dokumente
1914-1918

"Aufruf an die Europäer" (Appeal to the Europeans)
[Mid October 1914], reproduction from
Georg Friedrich Nicolai: *Die Biologie des Krieges*
(The Biologie of War)
Zurich, 1917
Gedenkstätte Deutscher Widerstand

"Aufforderung" (Demand)
Leaflet, [early 1915], reproduction
Deutsches Museum, Munich
Nl. Wien

Minutes of the meeting of the Bund Neues Vaterland
(New Fatherland League)
Berlin, 21 March 1915, reproduction.
Bundesarchiv, Koblenz
NL Wehberg, Bd. 14, Bl. 62-64

Letter from Romain Rolland to Albert Einstein
Geneva, 28 March 1915, facsimile
The Hebrew University of Jerusalem – Einstein Archives
Call Nr. 33-003

Albert Einstein: "Meine Meinung über den Krieg"
(My Opinion about the War)
Draft of the contribution to *Das Land Goethes
1914-1916*
Manuscript, 1915, reproduction
Staatsbibliothek, Berlin
Slg. Darmstädter, F 1e 1908(7), Bl. 9-10

Postcard from Albert Einstein to Paul Ehrenfest
[Berlin], 4 February 1917, reproduction
The Hebrew University of Jerusalem – Einstein Archives
Call Nr. 9-396

Letter from Romain Rolland to Albert Einstein
[Villeneuve], 23 August 1917, reproduction
Laurent Besso, Lausanne

Unknown speaker on the steps of the
Kronprinzenpalais
Berlin, Unter den Linden, 9 November 1918
Photograph
Landesarchiv, Berlin

Albert Einstein's lecture notes for the winter semester
1918/19
Facsimile
The Hebrew University of Jerusalem – Einstein Archives
Call Nr. 3-009

Postcard from Albert Einstein to Pauline Einstein
[Berlin], 11 November [1918], facsimile
The Hebrew University of Jerusalem – Einstein Archives
Call Nr. 29-352

Albert Einstein: On the Need for a Legislative National
Assembly
Draft of a speech held at a meeting of the Bund Neues
Vaterland (New Fatherland League)
[Berlin, 13 November 1918], facsimile
The Hebrew University of Jerusalem – Einstein Archives
Call Nr. 28-001

Postcard from Albert Einstein to Paul Ehrenfest
[Berlin], 6 December 1918, reproduction
The Hebrew University of Jerusalem – Einstein Archives
Call Nr. 9-425

The Anti-Relativists

Ernst Gehrcke: *Die Relativitätstheorie eine
wissenschaftliche Massensuggestion* (The Relativity
Theory: A Scientific Mass Suggestion)
Berlin, 1920
Lecture brochure with notes in the hand of Gehrcke
Max Planck Institute for the History of Science, Berlin

"Is Einstein a Plagiarist?"
Front page of *The Dearborn Independent*,
30 April 1921, reproduction
University of Minnesota Libraries, Minneapolis, MN

Albert Einstein: "Meine Antwort. Über die antirelativ-
itätstheoretische G.m.b.H." (My Response. On the
Anti-Relativity Company)
Berliner Tageblatt, 27 August 1920, reproduction
Staatsbibliothek, Berlin

"Erklärung in Sachen Liebknecht-Luxemburg"
(Declaration on the Matter of Liebknecht-Luxemburg)
Leaflet, Spring 1919, reproduction
Institut für Zeitgeschichte, Munich
ED 184 NL Nicolai, Bd.15

Walther Rathenau
Photograph, 1922
akg-images, Berlin

MP 18/1 submachine gun
T. Bergmann Weapons Factory, 1918
Wehrtechnische Studiensammlung der Bundeswehr,
Koblenz

Demonstration after the Rathenau murder
Berliner Lustgarten, 25 June 1922
Photograph
Ullstein Bild, Berlin

Letter from Albert Einstein to Max von Laue
Berlin, 12 July 1922, reproduction
Archiv zur Geschichte der Max-Planck-Gesllschaft,
Berlin
I. Abt., Rep. 34, Nr. 2, Einstein

Death threats against Einstein in the headlines
Newspaper clippings, Summer 1922, reproduction
Max Planck Institute for the History of Science, Berlin
Nl. Gehrcke

"Austritt Einsteins aus der Kaiser-Wilhelm-Akademie"
(Einstein Withdraws from the Kaiser Wilhelm
Academy)
Newspaper clipping, Neue Berliner 12Uhr Zeitung,
29 September 1922, reproduction
Max Planck Institute for the History of Science, Berlin
Nl. Gehrcke

Appeal to prevent a lecture about the theory of
relativity
Flyer, [Fall 1922], reproduction
Max Planck Institute for the History of Science, Berlin
Nl. Gehrcke

Einstein the Public Figure

"Lights All Askew in the Heavens"
The New York Times, 10. November 1919, reproduction
Bibliothek des John-F.-Kennedy-Instituts,
Freie Universität, Berlin

"Eine neue Größe der Weltgeschichte: Albert Einstein"
(A New Celebrity of World History)
Cover of Berliner Illustrirten Zeitung,
14 December 1919, reproduction
Staatsbibliothek, Berlin

Newspaper clippings about Einstein's journey to
France
March – April 1922, reproduction
Max Planck Institute for the History of Science, Berlin
Nl. Gehrcke

Newspaper clippings about Einstein's lectures in Paris
March–April 1922, reproduction
Max Planck Institute for the History of Science, Berlin
Nl. Gehrcke

Disarmament conference of the League of Nations
Geneva, 18 September 1924
Photograph
Bettmann/CORBIS

Press report about Einstein and the League of Nations
Berliner Tageblatt, 16 May 1922, reproduction
Staatsbibliothek, Berlin

List of members of the German Commission for
Intellectual Cooperation of the League of Nations
Les Commissions Nationales de Coopération
Intellectuelle, Geneva, 1932, reproduction
Geheimes Staatsarchiv, Berlin
Rep. 92 Becker, Nr. 8156

Albert Einstein: "Mein Glaubensbekenntnis"
(My Creed)
Gramophone record, 1932
Friedrich-Herneck-Archiv für Geschichte der
Naturwissenschaften, Dresden
Private collection Uwe V. Lobeck

Albert Einstein: Statement on war
Note on a letter from Frida Perlen,
[late December 1928], reproduction
The Hebrew University of Jerusalem – Einstein Archives
Call Nr. 28–068

Albert Einstein: Declaration on international under-
standing for the British magazine No More War
Manuscript, April 1929, facsimile
The Hebrew University of Jerusalem – Einstein Archives
Call Nr. 28–076

Albert Einstein and Sigmund Freud: Warum Krieg?
(Why War?)
Paris, 1933, reproduction
Stiftung Archiv der Akademie der Künste, Berlin
Archiv Bertolt Brecht

Arbeitsplan der Volkshochschule Groß-Berlin
(Schedule of Adult Evening Classes in Greater Berlin)
1920, reproduction
Staatsbibliothek, Berlin

Albert Einstein holds a public lecture in the
Philharmonie
Berlin, 16 October 1932
Photograph
Archiv der Berlin-Brandenburgischen Akademie der
Wissenschaften, Berlin

Appeal "Für die Freiheit der Kunst"
(For the Freedom of Expression)
Vossische Zeitung, 1 October 1925, reproduction
Staatsbibliothek, Berlin

Albert Einstein: Statement on capital punishment
Berliner Tageblatt, 22 January 1927, reproduction
Staatsbibliotek, Berlin

Letter from Albert Einstein to Alfred Apfel
Declaration against the condemnation of
Carl von Ossietzky
Pasadena, 8 February 1932, reproduction
Bundesarchiv, Berlin
Bestand ZStA Potsdam, R 3003/ORA/RG,
Nr. 7 J 35/25, Bd.1, Bl. 82

Announcement of a meeting for the presentation of
joint lists of candidates of the Communist Party and
the Socialdemocratic Party in the forthcoming
elections
Poster, July 1932
Märkisches Museum, Berlin

Einstein's summer house in Caputh
Photograph, around 1930
Stiftung Archiv der Akademie der Künste, Berlin
Archiv Konrad Wachsmann

Rabindranath Tagore visits Einstein
Caputh, 14 July 1930
Photograph
Deutsches Historisches Museum, Berlin
F86/276

Model of Einstein's yawl "Tümmler" (Dolphin)
Reconstruction, Modellbau Keyser, Berlin, 2005
Max Planck Institute for the History of Science, Berlin

Einstein and Judaism

Letter from Albert Einstein to Paul Ehrenfest
[Berlin], 22 March 1919, facsimile
The Hebrew University of Jerusalem – Einstein Archives
Call Nr. 9-427

Albert Einstein: "Die Zuwanderung aus dem Osten"
(Immigration from the East)
Berliner Tageblatt, 30 December 1919, reproduction
Staatsbibliothek, Berlin

Letter from Albert Einstein to the Central-Verein
Deutscher Staatsbürger jüdischen Glaubens (Central
Association of German Citizens of Jewish Faith)
Profession of Jewishness published in Israelitisches
Wochenblatt für die Schweiz, 24 September 1920,
reproduction
Deutsche Bücherei Leipzig

Letter from Albert Einstein to Fritz Haber
Berlin, 9 March 1921, reproduction
The Hebrew University of Jerusalem – Einstein Archives
Call Nr. 12-332

Albert Einstein: "Wie ich Zionist wurde"
(How I became a Zionist)
Jüdische Rundschau, 21 June 1921, reproduction
www.compactmemory.de

Albert Einstein: "Über ein jüdisches Palästina"
(On a Jewish Palestine)
Manuscript for a speech, Berlin, June 1921
The Hebrew University of Jerusalem – Einstein Archives
Call Nr. 28-010

Draft of letter from Albert Einstein to an unnamed
recipient
Answers to questions on scientific and religious truth
Kyoto, 14 December 1922, facsimile
The Hebrew University of Jerusalem – Einstein Archives
Call Nr. 28-013

Einstein speaks to the Jewish Students Conference
in Germany
Photograph, 27 February 1924
Bildarchiv Preußischer Kulturbesitz, Berlin

Albert Einstein:"The Mission of our University"
The New Palestine, 27 March 1925, reproduction
University Library, Haifa
XBM.N45

Emigration

Letter from Albert Einstein to the Prussian Academy
of Sciences
S. S. Belgenland, 28 March 1933, reproduction
Archiv der Berlin-Brandenburgischen Akademie der
Wissenschaften, Berlin
II-III, Bd. 57, Bl. 6

Press release of the Prussian Academy of Sciences
1 April 1933, reproduction
Archiv der Berlin-Brandenburgischen Akademie der
Wissenschaften, Berlin
II-III, Bd. 57, Bl. 16

Letter from Albert Einstein to the Prussian Academy
of Sciences
Le Coq sur Mer, 5 April 1933, reproduction
Archiv der Berlin-Brandenburgischen Akademie der
Wissenschaften, Berlin
II-III, Bd. 57, Bl. 44

Letter from Elsa Einstein to Antonina Vallentin
Scheveningen, 11 April 1933, reproduction
Archiv zur Geschichte der Max-Planck-Gesellschaft,
Berlin
Va. Abt., Rep. 2, Bl. 105/17

Letters from Elsa Einstein to Lili Petschnikoff
1931–1936
Max Planck Institute for the History of Science, Berlin

Lili and Alexander Petschnikoff
Photograph, New York, 1907
Max Planck Institute for the History of Science, Berlin

Letter from Albert Einstein to Fritz Haber
19 May 1933, reproduction
The Hebrew University of Jerusalem – Einstein Archives
Call Nr. 12-378

Letter from Albert Einstein to Max von Laue
Oxford, 7 June 1933, reproduction
Archiv zur Geschichte der Max-Planck-Gesellschaft,
Berlin
III. Abt., Rep. 50, Nr. 536, Bl. 1

Max von Laue
Photograph
Archiv der Berlin-Brandenburgischen Akademie der
Wissenschaften, Berlin

Letter from Albert Einstein to Heinrich Zangger
Le Coq sur Mer, 1 October 1933, reproduction
Zentralbibliothek, Zurich
Nl. Zangger, Schachtel 1a

Report from the Gestapo to the Prussian Minister of
Finance about seized assets, with seizure list
Berlin, 16 December 1933, reproduction
Geheimes Staatsarchiv, Berlin
I. HA, Rep. 151 Finanzministerium, I A, Nr. 7976,
Bl. 117–125

Letter from the Office of the Führer's Deputy
Rudolf Hess to the Prussian Minister of Education
Bernhard Rust
Munich, 1 March 1934, reproduction
Geheimes Staatsarchiv, Berlin
I. HA, Rep. 76 V C, Sekt. 2, Tit. 23, Lit. F, Nr. 2, Bd. 16,
Bl. 147

Einstein's desk
Historical Society of Princeton, NJ

Albert Einstein upon leaving his residence
Photograph, Berlin, around 1930
ullstein bild, Berlin

Letter from Max von Laue to Albert Einstein
Berlin, 10 January 1939, reproduction
The Hebrew University of Jerusalem – Einstein Archives
Call Nr. 16–114

Telegram from the President of the German Academy
of Sciences Johannes Stroux to Albert Einstein
Berlin, 26 July 1946, facsimile
The Hebrew University of Jerusalem – Einstein Archives
Call Nr. 36-074

Letter from Albert Einstein to Arnold Sommerfeld
Princeton, 14 December 1946, reproduction
Deutsches Museum, Munich
NL 89, 1977–28/A, 78(28)

Letter from Albert Einstein to Otto Hahn
Princeton, 28 January 1949, reproduction
Archiv zur Geschichte der Max-Planck-Gesellschaft,
Berlin
III. Abt., Rep. 14A, Nr. 814, Bl. 5-6

Letter from Arnold Eucken to Erich Regener
Göttingen, 11 February 1949, reproduction
Archiv zur Geschichte der Max-Planck-Gesellschaft,
Berlin
III. Abt., Rep. 37, Nr. 5

Einstein in America

Josef Scharl: Portrait of Albert Einstein
Oil on canvas, 1927
Private collection

Josef Scharl with Albert Einstein
Photograph, Princeton, around 1950

Susanne Fiegel, Velden
Showcase of Lotte Jacobi's Studio with 4 photographs
of Albert Einstein
New York, 1938
Hope Zane, New Hampshire (Photo: Jeffrey Nintzel)

Lotte Jacobi: Self-Portrait
Photograph, around 1930
akg-images, Berlin

Eric Schaal: Albert Einstein in front of his house
Photograph, Princeton, 1939
Weidle-Verlag, Bonn

Letter from Albert Einstein to Thomas Mann
Le Coq sur Mer, 29 April 1933, reproduction
Bibliothek der Eidgenössischen Technischen
Hochschule, Zurich
Thomas-Mann-Archiv

Albert Einstein: "Antwort auf einen Artikel von B. D.
Allinson" (Reply to an Article by B. D. Allinson
[recte: Allison])
Manuscript, 27 November 1934, facsimile
Published as "A Reexamination of Pacifism", *Polity*,
1935
The Hebrew University of Jerusalem – Einstein Archives
Call Nr. 28-296

Letter from Albert Einstein to the American Jewish
Congress
Saranac Lake, N.Y., 27 July 1936
Deutsches Historisches Museum, Berlin
Do2 97/1972; BA E05/231, 05/304

Albert Einstein: Memorandum for Charles Burlingham
Manuscript, 9 April 1938, facsimile
The Hebrew University of Jerusalem – Einstein Archives
Call Nr. 28-419

Letter from Hermann Broch to Albert Einstein
St. Andrews, 4 September 1938, facsimile
The Hebrew University of Jerusalem – Einstein Archives
Call Nr. 34-037

Letter from Albert Einstein to Karl Veit
9 January 1939, reproduction
Deutsche Bibliothek, Frankfurt am Main
Deutsches Exilarchiv 1933–1945, Slg. Einstein

Letter from Albert Einstein to the President of the
USA Franklin D. Rooswelt
Long Island, 2 August 1939, facsimile
Franklin D. Roosevelt Library, New York
a64a01, a64a02

Albert Einstein and Leo Szilard sign the letter
to President Roosevelt
Photograph of reenactment, after 1945
Mandeville Special Collections Library, La Jolla, CA

Letter from Albert Einstein to Lotte Loeb
(Emergency Rescue Committee)
16 December 1940, reproduction
Deutsche Bibliothek, Frankfurt am Main
Deutsches Exilarchiv 1933–1945, EB 73/21

Letter from Albert Einstein to William E. Rappard
Princeton, 11 September 1941, reproduction
Schweizerisches Bundesarchiv, Bern
J I.149, 1977/135/18

Recordings of popular music songs
The Buchanan Brothers: "Atomic Power", 1946
Hawkshaw Hawkins: "When They Found the Atomic
Power", 1946
Sons Of The Pioneers: "Old Man Atom", 1947
Rolf Wehnert, Hannover
Reinhard Matthes, Garbsen

Albert Einstein: "Der Ausweg" (The Way Out)
Manuscript of the contribution to
D. Masters and K. Way (eds.) One World or None,
New York, 1946
Facsimile
The Hebrew University of Jerusalem – Einstein Archives
Call Nr. 28–684

Albert Einstein: Presentation of The Black Book:
The Nazi Crime Against the Jewish People
Manuscript for a broadcast recording, 27 March 1946,
facsimile
The Hebrew University of Jerusalem – Einstein Archives
Call Nr. 28–692

Einstein receives an honorary doctorate from Lincoln
University of Pennsylvania
Photograph, 3 May 1946
Langston Hughes Memorial Library,
Lincoln University, PA

Albert Einstein: Address to the conference of
the National Urban League
Manuscript, 16 September 1946, facsimile
The Hebrew University of Jerusalem – Einstein Archives
Call Nr. 28–707

Albert Einstein: Appeal of the Emergency Committee
of Atomic Scientists
Princeton, 17 September 1948, reproduction
Archiv zur Geschichte der Max-Planck-Gesellschaft,
Berlin
Va. Abt., Rep. 2, Nr. 72

Albert Einstein's testament
18 March 1950, print version, reproduction
Landesarchiv, Berlin
Rep. Wiedergutmachungsämter, Nr. 84 WGA 600/55,
Bl. 20 a

Telegram from Azriel Carlebach to Albert Einstein
Tel Aviv, 18 November 1952, reproduction
The Hebrew University of Jerusalem – Einstein Archives
Call Nr. 41–092

Letter from Albert Einstein to Azriel Carlebach
Princeton, 21 November 1952, reproduction
The Hebrew University of Jerusalem – Einstein Archives
Call Nr. 41–093

Draft of letter from Albert Einstein to
William Frauenglass
16 May 1953, facsimile
The Hebrew University of Jerusalem – Einstein Archives
Call Nr. 34–636

Letter from Albert Einstein to the District Mayor
of Berlin-Neukölln Kurt Exner
Princeton, 15 May 1954, reproduction
Albert-Einstein-Gymnasium, Berlin-Neukölln

Letter from Albert Einstein to Max von Laue
Princeton, 3 February 1955, reproduction
Archiv der Johann-Wolfgang-Goethe-Universität, Frank-
furt am Main
NI. von Laue

Letter from Albert Einstein to the teachers and
students of the Albert Einstein School
Princeton, 2 April 1955, reproduction
Albert-Einstein-Gymnasium, Berlin-Neukölln

Albert Einstein: Declaration on Israel's
Independence Day
Manuscript, April 1955, facsimile
The Hebrew University of Jerusalem – Einstein Archives
Call Nr. 28–1098

Albert Einstein
Photograph, 1954
akg-images/AP

Brief von Albert Einstein an Max Laue
Prag, 10. Juni 1912
Deutsches Museum München
Handschriften, 1951-4

EINSTEIN'S WORLD TODAY

Worldviews of Modern Science

"What is Science?"
Film by Jürgen Renn and Michael Dörfler, 2005
Max Planck Institute for the History of Science and
produktion2/cineplus, Berlin

"Images of the World" Part 2
Film on children's workshops by Katja Bödeker and
Henning Vierck, 2004
Comenius-Garten and Max Planck Institute for the
History of Science, Berlin

Putting Relativity to the Test

Particle accelerator for time dilation experiment
Photograph
Max Planck Institute for Nuclear Physics, Heidelberg

Animation illustrating time dilation and relativistic
Doppler shift
Max Planck Institute for the History of Science,
Berlin/Max Planck Institute for Nuclear Physics,
Heidelberg, 2005

Model of the satellite Gravity Probe B
Stanford University/Lockheed Martin/Zentrum für
angewandte Raumfahrttechnologie und Mikrogravitation ZARM, University of Bremen, 2004

Gyroscope and casing from the Gravity Probe B
satellite experiment
Stanford University/Lockheed Martin/Zentrum für
angewandte Raumfahrttechnologie und Mikrogravitation ZARM, University of Bremen, 2004

Test masses from the STEP experiment
Zentrum für angewandte Raumfahrttechnologie und
Mikrogravitation ZARM, University of Bremen, 2005

The Large-Scale Structure of the World: Windows to the Cosmos

Model of the Effelsberg radio telescope
Max Planck Institute for Radio Astronomy, Bonn

Model of the functional principle of the
AMANDA neutrino telescope
Deutsches Elektronen-Synchrotron DESY, 1997

Probe of the AMANDA neutrino telescope
Deutsches Elektronen-Synchrotron DESY, 1997

Probe of the IceCube neutrino telescope
Deutsches Elektronen-Synchrotron DESY, 2005

X-ray image of the spiral galaxy M51
"Whirlpool Galaxy"
NASA

Radio image of the spiral galaxy M51
"Whirlpool Galaxy"
Very Large Array – NRAO

Visible light image of the spiral galaxy M51
"Whirlpool Galaxy"
Science Photo Library

Infrared image of the spiral galaxy M51
"Whirlpool Galaxy"
NASA

X-ray image of the Sun
Science Photo Library

Ultraviolet images of the Sun
NASA

Radio image of the Sun
Very Large Array – NRAO / AUI

The Large-Scale Structure of the World: Intergalactic Structures

Density ratios in the universe
Show model
Max Planck Institute for the History of Science, Berlin, 2005

Millenium Simulation
Computer simulation of the formation of cosmic large-scale structures
Virgo Consortium for Cosmological Simulations / Max Planck Institute for Astrophysics, Garching
Koblenz

The Large-Scale Structure of the World: Cosmic Background Radiation

Dual horn antenna for cosmic microwave background radiation
Jodrell Bank Observatory, The University of Manchester / Instituto de Astrofísica de Canarias, Tenerife, 1985–1999

All-sky image of the cosmic microwave background taken by the Wilkinson Microwave Anisotropy Probe (WMAP)
NASA / WMAP Science Team

Infrared image of the sky with the Milky Way
Two Micron All-Sky Survey (2MASS)

Gravitation as a Challenge: Gravitational Waves

Show model of neutron star material
Max Planck Institute for Extraterrestrial Physics, Garching

Computer simulations for the generation of gravitational waves
Simulation: Max Planck Institute for Gravitational Physics (Albert Einstein Institute), Potsdam
Visualization: W. Benger, Max Planck Institute for Gravitational Physics (Albert Einstein Institute), Potsdam / Konrad Zuse Institute, Berlin, 1998–2005

Gravitation as a Challenge: Gravitational Wave Detectors

Model of the functional principle of GEO600
Max Planck Institute for Gravitational Physics (Albert Einstein Institute), Hanover and Potsdam, 2004

Model of the mirror mounting in GEO600
Max Planck Institute for Gravitational Physics (Albert Einstein Institute), Hanover and Potsdam, 2005

Model of one of the LISA satellites
Max Planck Institute for Gravitational Physics (Albert Einstein Institute), Hanover and Potsdam, 2005

Gravitation as a Challenge: Gravitational Lensing

Animation illustrating the functional principle of gravitational lensing
Max Planck Institute for the History of Science and Joachim Wambsganss, Zentrum für Astronomie, University of Heidelberg, 2005

Demonstration model of gravitational lens
Max Planck Institute for Extraterrestrial Physics / Max Planck Institute of Quantum Optics, Garching, 2005

Black hole as a gravitational lens
Interactive simulation
Hanns Ruder, Department of Theoretical Astrophysics, University of Tübingen, 2004

Galaxy cluster Abell 2218
Image taken by the Hubble Space Telescope
NASA / STScI, 1999

Radio image of the radio source 4C 05.51 representing an Einstein ring
Very Large Array – NRAO / AUI

Gravitational lens G2237 + 0305 "Einstein Cross"
Image taken by the Hubble Space Telescope
NASA / STScI / ESA, 1990

Gravitationally lensed Quasar PG1115+080
Image taken by the Hubble Space Telescope
NASA / STScI / ESA, 1998

Gravitation as a Challenge: Black Holes

Camera and camera mounting from the European Photon Imaging Camera in the XMM-Newton satellite
Max Planck Institute for Extraterrestrial Physics, Garching, 1999

Infrared speckle camera Sharp-I
Max Planck Institute for Extraterrestrial Physics, Garching, 1991–2002

Spiral galaxy NGC 7742
Image taken by the Hubble Space Telescope
NASA / STScI, Science Photo Library

Crab nebula
Image taken by the Hubble Space Telescope
NASA, 1996

Galaxy M87 with black-hole-powered jet
Image taken by the Hubble Space Telescope
NASA / STScI / AURA, 2000

Bubble of hot gas expelled by a black hole at the center of the galaxy NGC 4438
Image taken by the Hubble Space Telescope
NASA / STScI, 2000

Dust disk encircling the center of the galaxy NGC 7052
Image taken by the Hubble Space Telescope
NASA / STScI, Science Photo Library, 1998

Core region of the active galaxy NGC 4261
Image taken by the Hubble Space Telescope
NASA / STScI, Science Photo Library, 1995

Gravitation as a Challenge: Quantum Gravity

Work station of a theoretical physicist
Max Planck Institute for Gravitational Physics (Albert Einstein Institute), Hanover and Potsdam, 2005

The Small-Scale Structure of the World: Particles

Half shell of the OPAL silicon microvertex detector
CERN, Geneva, 1989–2000

Model illustrating the principle of supersymmetry (SUSY)
Das Atelier Niesler, Berlin, 2005

Model illustrating the principle of Higgs field
Das Atelier Niesler, Berlin, 2005

The Small-Scale Structure of the World: Energy and Structure

Coils of the Stellarator Wendelstein 7-X, under construction 2005
Desk model, scale 1:20
Max Planck Institute of Plasma Physics, Greifswald

Coil of the Stellarator Wendelstein 7-AS
Model, scale 1:1
Max Planck Institute of Plasma Physics, Garching

Photovoltaic module of concentrator cell
Hahn-Meitner Institute, Berlin

Photocathode
Deutsches Elektronen-Synchrotron DESY, 2004

TESLA accelerator cavity
Deutsches Elektronen-Synchrotron DESY, 2000

Lead-scintillator calorimeter
Deutsches Elektronen-Synchrotron DESY, 1999

Strange Connections

Virtual tour of the Spiegelsaal
Interactive animation
Siemens AG, CT PP 6, Munich, 2005

Eye tracking system
Interactive demonstration
SensoMotoric Instruments, Teltow, 2004

Exhibits

Animation illustrating superdiffusion
Max Planck Institute for Dynamics and Self-
Organization, Göttingen, 2004

Animation illustrating quantum chaos
Faculty of Physics, University of Marburg, 2005

Interactive animation illustrating photon statistics
Students of the Gymnasium, Unterhaching/Max
Planck Institute of Quantum Optics, Garching, 2005

Interactive animation illustrating a photon-statistical
experiment
Max Planck Institute of Quantum Optics, Garching,
2005

Interactive animation illustrating Bose-Einstein
condensation
University of Bonn

Device for producing Bose-Einstein condensate
University of Constance, 1997
Deutsches Museum, Munich

Paul trap
Max Planck Institute of Quantum Optics, Garching,
1987

Ring trap
Max Planck Institute of Quantum Optics, Garching,

Applications

Prepared end of a segment of the Rheinland Cable
Siemens & Halske, Berlin, around 1920
SiemensForum, Munich – Siemens AG

Small form-factor pluggable (SFP) with transmitter
and receiver diodes
Siemens AG, IC Networks, Berlin

Laser-cutted grids for transmission tubes
Siemens AG, CT MM, Munich, around 1985

Laser-welded mount for fuel rods in nuclear reactor
Siemens AG, CT MM, Munich, around 1990

Demonstration console for lithotripsy
Clyxon Laser GmbH, Berlin, 2005

Network node SURPASS hiT 7050
Siemens AG, IC Networks, Berlin, 2005

Laser-based video transmission
Siemens AG, IC Networks, Berlin, 2005

Laser free transmission line with interrupter
Siemens AG, IC Networks, Berlin, 2005

Atomic clock HP 5061A
Hewlett-Packard Company, Palo Alto, 1969
Physikalisch-Technische Bundesanstalt, Brunswick

Demonstration of cumulated relativistic time
difference (DART)
Physikalisch-Technische Bundesanstalt, Brunswick,
2005

Navigation console with interactive simulation of
satellite-based navigation
MILANO Medien GmbH, Frankfurt am Main/Max
Planck Institute for the History of Science, Berlin, 2005

VDO Dayton MS 5500 XL navigation system
Siemens VDO Automotive AG, Schwalbach, 2005

Model illustrating the functional principle of a GPS
satellite with solar cells
Das Atelier Niesler, Berlin/Hahn-Meitner Institute,
Berlin, 2005

Laser-engraved crystal with representation of the
Spiegelsaal
Crystalix GmbH, Berlin/Siemens AG, CT PP 6,
Munich/FiberTech GmbH, Berlin, 2005

Treasure Chamber –
Original Documents on Einstein's Work

Annalen der Physik
Volume 1905
Universitätsbibliothek der Humboldt-Universität, Berlin

Letter from Hendrik A. Lorentz to Henri Poincaré
Leiden, 8 March 1906
François Poincaré, Paris

Letter from Alfred Bucherer to Albert Einstein
Bonn, 7 September 1908
The Hebrew University of Jerusalem – Einstein Archives
Call Nr. 6–188

Proposal for Albert Einstein's membership in the
Prussian Academy of Sciences
Manuscript in the hand of Max Planck, Berlin,
12 June 1913
Archiv der Berlin-Brandenburgischen Akademie der
Wissenschaften, Berlin
II-III, Bd. 36, Bl. 36–37

Michele Besso: Notes about the "hole argument"
Manuscript, 1913–1914
Laurent Besso, Lausanne

Postcard from Albert Einstein to David Hilbert
[Berlin, 18 November 1915]
Niedersächsische Staats- und Universitätsbibliothek,
Göttingen
Cod. Ms. Hilbert 92b: 2

Letter from Albert Einstein to Arnold Sommerfeld
Berlin, 28 November 1915
Deutsches Museum, Munich
NL 89, 1977–28/A, 78(2)

Letter from Arthur S. Eddington to Albert Einstein
[Cambridge], 1 December 1919
The Hebrew University of Jerusalem – Einstein Archives
Call Nr. 9–260

Letter from Albert Einstein to Max Born
29 April 1924
Staatsbibliothek, Berlin
Nl. Born, Nr. 188, Bl. 24

Albert Einstein: Statement on the 200th anniversary
of Newton's death
Manuscript, published in *Nature*, March 1927
The Hebrew University of Jerusalem – Einstein Archives
Call Nr. 1–058

Drawing for Albert Einstein and Gustav Bucky's
patent application for a fluid level gauge
Around 1934–1936
The Hebrew University of Jerusalem – Einstein Archives
Call Nr. 35–437

Albert Einstein: "Über rotationssymmetrische
stationäre Gravitationsfelder" (About Stationary
Gravitational Fields Symmetrical under Rotation)
Manuscript fragment, 1936
Syracuse University Library, Syracuse (NY)
Nl. Bergmann

Albert Einstein: Addendum for a new edition of *Über
die spezielle und allgemeine Relativitätstheorie
(Gemeinverständlich)* (On the Special and the General
Theory of Relativity (A Popular Account))
Manuscript, Princeton, 1946
The Hebrew University of Jerusalem - Einstein Archives
Call Nr. 1–186

Envelope, addressed to "Dr. Albert Einstein,
Chief Engineer of the Universe"
Troy (NY), January 1947
The Hebrew University of Jerusalem – Einstein Archives
Call Nr. 31–742

Telegram from Niels Bohr to Albert Einstein with
reply draft by Albert Einstein on the back side
March 1949
The Hebrew University of Jerusalem – Einstein Archives
Call Nr. 8–098

Albert Einstein: "Das Inertialsystem muss aus der
Grundlage der Physik verschwinden" (The Inertial
System Must Disappear from the Foundations of
Physics)
Manuscript, 1952
The Hebrew University of Jerusalem – Einstein Archives
Call Nr. 2–134

Albert Einstein: Commentary to an article by
Max Born
Manuscript, 12 January 1954
The Hebrew University of Jerusalem – Einstein Archives
Call. Nr. 8–240.1

Relativity under Suspicion

Ludwig Zehnder: *Der Aether im Lichte der klassischen Zeit und der Neuzeit* (The Aether According to Classic and Contemporary Theories)
Tübingen, 1933
Max Planck Institute for the History of Science, Berlin
Nl. Gehrcke

Ernst Gehrcke: "Über das Uhrenparadoxon in der Relativitätstheorie" (About the Clock Paradox in the Theory of Relativity)
Manuscript, published in *Die Naturwissenschaften*, vol. 9, 1921
Max Planck Institute for the History of Science, Berlin
Nl. Gehrcke

Letter from Franz Karollus to Ernst Gehrcke
Brünn, 25 November 1921, reproduction
Max Planck Institute for the History of Science, Berlin
Nl. Gehrcke

Karl Vogtherr: *Ist die Schwerkraft relativ?* (Is Gravity Relative?)
Karlsruhe, 1926
Max Planck Institute for the History of Science, Berlin
Nl. Gehrcke

Hanns Hörbiger and Philipp Fauth: *Glazial-Kosmogonie* (Glacial Cosmogony)
Leipzig, 1925, reproduction
Max Planck Institute for the History of Science, Berlin

Leonore Ripke-Kühn: "Kant contra Einstein" in: *Hellweg. Westdeutsche Wochenschrift für Deutsche Kunst*, 17 August 1921
Max Planck Institute for the History of Science, Berlin
Nl. Gehrcke

Gusztáv Pécsi: *Liquidierung der Relativitätstheorie. Berechnung der Sonnengeschwindigkeit* (The Liquidation of the Theory of Relativity: The Calculation of the Speed of the Sun)
Regensburg, 1925
Milena Wazeck, Berlin

Franz Karollus: *Wo irrt und was übersieht Einstein?* (Where is Einstein Mistaken and What Does He Overlook?)
Brünn, 1921
Max Planck Institute for the History of Science, Berlin
Nl. Gehrcke

Rudolf Weinmann: *Anti-Einstein*
Leipzig, 1923
Max Planck Institute for the History of Science, Berlin
Nl. Gehrcke

Hans Israel, Erich Ruckhaber, Rudolf Weinmann (eds.): *100 Autoren gegen Einstein* (100 Authors against Einstein)
Leipzig, 1931
Max Planck Institute for the History of Science, Berlin
Nl. Gehrcke

Karl Vogtherr: *Wohin führt die Relativitätstheorie?* (Where is the Theory of Relativity Leading?)
Leipzig, 1923
Max Planck Institute for the History of Science, Berlin
Nl. Gehrcke

Paul Lamberty: *Die Ursache von allem erkannt. Das Ende der Relativitätstheorie* (The Cause of Everything Discovered: The End of the Theory of Relativity)
Haag, 1925
Max Planck Institute for the History of Science, Berlin
Nl. Gehrcke

Bruno Thomas: *Axiom und Dogma in der Relativitäts-theorie* (Axiom and Dogma in the Teory of Relativity)
Vienna and Leipzig, 1933
Max Planck Institute for the History of Science, Berlin
Nl. Gehrcke

Anti-Einstein pamphlets and newspaper clippings from the Ernst Gehrcke estate
Max Planck Institute for the History of Science, Berlin
Nl. Gehrcke

Einstein under Suspicion

Documents from the FBI files on Albert Einstein
Reproductions
http://foia.fbi.gov/foiaindex/einstein.htm

Science as a Challenge

The Russell-Einstein Manifesto
London, 9 July 1955, reproduction
Pugwash United Kingdom, London
Archives

The Göttingen Declaration
Göttingen, 12 April 1957, reproduction
Archiv zur Geschichte der Max-Planck-Gesellschaft, Berlin

"Science News" read by Jan Hofer
Video sequence
Max Planck Institute for the History of Science, Berlin, 2005

News loop
Max Planck Institute for the History of Science, Berlin, 2005

"Insecurity Council"
Series of interviews by Michael Schüring and Marcel Weingärtner
Max Planck Institute for the History of Science and werwiewas Filmproduktion, Berlin, 2005
Interviewees:
Paul J. Crutzen, Max Planck Institute for Chemistry, Mainz
Dieter Deiseroth, Bundesverwaltungsgericht, Leipzig
Yehuda Elkana, Central European University, Budapest
Kathrin Grüber, Institut Mensch, Ethik und Wissen-schaft, Berlin
Hans-Olaf Henkel, Wissenschaftsgemeinschaft Gottfried-Wilhelm-Leibniz
Georg W. Kreutzberg, Max Planck Institute of Neurobiology, Martinsried
Hans Lehrach, Max Planck Institute for Molecular Genetics, Berlin
Wolf Singer, Max Planck Institute for Brain Research, Frankfurt am Main
Günter Stock, Schering AG, Berlin
Martin Vingron, Max Planck Institute for Molecular Genetics, Berlin
Albrecht Wagner, Deutsches Elektronen-Synkrotron DESY, Hamburg

Einstein, the Legend

"Einstein in Hollywood"
Scenes from various Hollywood movies
Compilation by Martin Walz, 2005
Max Planck Institute for the History of Science, Berlin

Stamp collection with motives on Einstein and his work
Kenji Sugimoto, Osaka

Competition entries for a coin to celebrate the Einstein Year 2005
Winning design: Heinz Hoyer, Berlin, 2004
Second prize: Victor Huster, Baden-Baden, 2004
Third prize: František Chochola, Hamburg, 2004
Fourth prize: Jordi Regel, Berlin, 2004
Federal Ministry of Finances, Berlin

Einstein and Art

"Kunst nach Einstein" (Art after Einstein)
Image sequence
Compilation by Violeta Sánchez and Philipp Muras, 2005
Max Planck Institute for the History of Science, Berlin

"Alle kennen meine Visage" (Everyone Knows my Visage)
Music and video installation
Karin Leuenberger and Georg Graewe, 2005

Exhibits

Albert Einstein: His Portrait in the Arts

Max Liebermann: Portrait of Albert Einstein
Pencil on paper, Berlin, 1925
Private collection

Josef Scharl: Profile of Albert Einstein
Pencil on paper, Princeton, 1951
Max-Planck-Gesellschaft zur Förderung der
Wissenschaften, Munich

Loulou Albert-Lazard: Albert Einstein en face
Lithograph, signed (pencil), signed and dated by
Einstein in the stone:
"A. Einstein, 28 April 22"
Private collection

Erich Büttner: Portrait of Albert Einstein
Etching, signed (pencil), handwritten
printing note, signed by Einstein (pencil):
"A. Einstein", 1920
Private collection

J. J. Muller: Portrait of Albert Einstein
Etching, tinted, undated, signed in the plate: J. J. M.,
(pencil) with Einstein's signature (ink):
"Albert Einstein"
Private collection

Emil Orlik: Albert Einstein
Etching, signed (pencil), with handwritten dedication
by Einstein (pencil): "To Miss Marga Schulze Gävernitz
to remind you of Albert Einstein 1926"
Private collection

Ferdinand Schmutzer: Portrait of Albert Einstein
Etching, undated
Private collection

Josef Scharl: Einstein's Hands with Spectacles
Pen-and-ink drawing, signed: "Princeton,
Nov. 29, 1951"
Private collection

Hermann Struck: Albert Einstein
Etching, signed and numbered 14 / 150, 1920,
signed by Einstein (pencil): "Albert Einstein"
Private collection

Emil Stumpp: Albert Einstein
Lithograph, signed in the stone: Berlin, Academy 23
June 1927, 2/3, signed in the stone "A. Einstein"
Private collection

J. C. Thonen: Albert Einstein
Etching, signed and named (pencil): "JCT 1920"
Private collection

Josef Scharl: Albert Einstein
Watercolour sketch, signed and named (pencil):
"Princeton, Febr. 9, 1950", signed by Albert Einstein
(ink): "A. Einstein"
Private collection

Erich Büttner: Ex Libris for Albert Einstein
Private collection

Kurt Kroner: First Portrait Bust of Albert Einstein
Photograph, 1922, for Einstein's private collection
Marion Kroner

ENCOUNTERS

Einstein's Independence

The nineteen-year-old Albert Einstein
Photograph, 4 November 1898
Leo Baeck Institute, New York
F 5307C

Certificate of Einstein's dismissal from the
Wurttemberg citizenship
Ulm, 28 January 1896, reproduction
Schweizerisches Bundesarchiv, Bern
E 21/23560

Albert Einstein's Swiss passport
Reproduction
Historisches Museum, Bern

Einstein's Addresses in Berlin

Guestbook from Einstein's sommer house in Caputh
1930–1933, reproduction
Leo Baeck Institute, New York
Einstein Collection, Box 8, Folder 7

Model of Einstein's summer house in Caputh

Albert and Elsa Einstein in their Berlin apartment,
Haberlandstraße 5
Photograph, 1927
Deutsches Historisches Museum, Berlin

Model of Einstein's apartment, Haberlandstraße 5,
4th floor

Ground plans of the apartments on the 4th flour of
the building Haberlandstraße 5
July 1907, reproduction
Landesarchiv, Berlin

Einstein – Engineer of the Universe

Envelope, addressed to "Dr. Albert Einstein, Chief
Engineer of the Universe"
Troy (NY), January 1947, reproduction
The Hebrew University of Jerusalem – Einstein Archives
Call Nr. 31–742

Einstein and Elsa

Silhouette made by Albert Einstein of Albert, Elsa and
the two daughters Ilse and Margot
Model after silhouette
Schweizerische Landesbibliothek, Bern
Schweizerisches Literaturarchiv

Einstein and his Sons

Hans Albert and Eduard Einstein
Photograph, 1917
Leo Baeck Institute, New York
F 5373A

Letter from Albert Einstein to Hans Albert Einstein
[Berlin 1918], reproduction
Max Planck Institute for the History of Science, Berlin

House in the Berner Oberland close to the Jungfrau
Photograph, around 1900
Hulton-Deutsch Collection/CORBIS

Institute Director Einstein and Bureaucracy

Letter from Albert Einstein to Mendelssohn & Co.
Berlin, 22 April 1919, reproduction
Archiv zur Geschichte der Max-Planck-Gesellschaft
I. Abt., Rep. 1A, Nr. 1656, Bl. 52

Receipt from Bernhard Mueller Company for a
typewriter
[April 1919], Berlin, reproduction
Archiv zur Geschichte der Max-Planck-Gesellschaft
I. Abt., Rep. 1A, Nr. 1656, Bl. 53

Letter from Wilhelm von Siemens to
Mendelssohn & Co.
Berlin, 25 April 1919, reproduction
Archiv zur Geschichte der Max-Planck-Gesellschaft
I. Abt., Rep 1A, Nr. 1656, Bl. 54

Einstein and Maja

Albert Einstein and his sister Maja as children
Photograph, 1893
Leo Baeck Institute, New York
F 5331B

Albert Einstein and his sister Maja at the World Fair
in New York
Photograph, 28 Mai 1939
Bettmann/CORBIS

Einstein's Travels

Albert Einstein: Diary of the journey to
South America
Manuscript, 5–11 March 1925, reproduction
The Hebrew University of Jerusalem – Einstein Archives
Call Nr. 29–132

Meiji Seihanjo: Albert and Elsa Einstein at a reception
in Japan
Photograph, 1922
Leo Baeck Institute, New York

Einstein as an Observer

Albert Einstein: "Die Ursache der Mäanderbildung der
Flussläufe und des sogenannten Beer'schen Gesetzes"
(The Cause of the Formation of Meanders in the
Courses of Rivers and the So-Called Beer's Law)
Manuscript, 1926, reproduction
The Hebrew University of Jerusalem – Einstein Archives
Call Nr. 1-051

Einstein as Inventor

Drawing for Albert Einstein and Gustav Bucky's
patent application for a fluid level gauge
Around 1934–1936, reproduction
The Hebrew University of Jerusalem – Einstein Archives
Call Nr. 35-437

Einstein and Besso

Albert Einstein and Michele Besso
Photograph, around 1925
Laurent Besso, Lausanne

Model of the old town of Bern with Einstein and
Besso's route to the Swiss Patent Office

Einstein and the Baltic

Letter from Albert Einstein to Heinrich Zangger
[Ahrenshoop, August 1918]
Zentralbibliothek, Zurich
Nachlass H. Zangger 1a

Albert Einstein in a deck chair at the Baltic Sea
Photograph, 1928
ullstein bild, Berlin

Einstein and Chess

Emanuel Lasker playing check
Photograph
Archiv der Emanuel-Lasker-Gesellschaft, Berlin

"New Chess Theory not for Einstein"
The New York Times, 28 March 1936, reproduction

Albert Einstein: Appreciation of Emanuel Lasker on
the occasion of his 60th birthday
Manuscript, 1928, reproduction
The Hebrew University of Jerusalem – Einstein Archives
Call Nr. 28-060

Einstein and Children

Three little girls visiting Einstein
Photograph, Princeton, 1953
The Hebrew University of Jerusalem – Einstein Archives

Letter from Ann G. Kocin to Albert Einstein
[Philadelphia?], 1951
The Hebrew University of Jerusalem – Einstein Archives
Call Nr. 42-653

Einstein and Mileva

Albert Einstein and Mileva Marić
Photograph, 1903
SV-Bilderdienst, DIZ, Munich

Postcard from Albert and Mileva Einstein to Conrad
Habicht
Munich, [between 1905 and 1915], reproduction
Bibliothek der Eidgenössischen Technischen
Hochschule, Zurich
Hs 1457:40

Einstein's Love Letters

Albert and Elsa Einstein in Berlin
Photograph
Schweizerische Landesbibliothek, Bern
Schweizerisches Literaturarchiv

Mileva Marić
Photograph, 1898
Schweizerische Landesbibliothek, Bern
Schweizerisches Literaturarchiv

Johanna Fantova and Albert Einstein on the
sailing boat
Photograph
Princeton University Library

Albert Einstein with Margarita Konenkova
Photograph, signed by Einstein
CORBIS/SYGMA

Einstein's Decree of Divorce

Albert and Elsa Einstein
Photo, April 1921
Bettmann/CORBIS

District Court of Zurich: Decree of divorce between
Mileva and Albert Einstein
Zurich, 14 February 1919, reproduction
Staatsarchiv des Kantons Zürich
BEZ Zch 6314.43

Einstein and Friedrich Adler

Letter from Friedrich Adler to Albert Einstein
Vienna, 9 March 1917, reproduction
The Hebrew University of Jerusalem – Einstein Archives
Call Nr. 6-001

Albert Einstein: Draft for a mercy petition for
Friedrich Adler to the Austrian Emperor
Around 1917, reproduction
Max Planck Institut for the History of Science, Berlin
Nl. Gehrcke

Friedrich Adler
Photograph, 1930
ullstein bild, Berlin

Einstein on Honours and Anniversaries

Diploma of Einstein's honorary doctorate from the
Medical Faculty of the University of Rostock
November 1919, reproduction
Universitätsarchiv, Rostock
MD 150/1919

Letter from Albert Einstein to Max von Laue
Princeton, 3 February 1955, reproduction
Archiv der Johann-Wolfgang-Goethe-Universität,
Frankfurt am Main
Nl. von Laue

Draft of letter from Albert Einstein to André Mercier
Letter, 10 February 1955, reproduction
The Hebrew University of Jerusalem – Einstein Archives
Call Nr. 1-203

Einstein und die Folgen. Heidelberg: Spektrum der Wissenschaft Verlagsgesellschaft, 2005.

Innovative Structures in Basic Research. Ringberg Symposium October 2000. München: Max-Planck-Gesellschaft, 2002.

Aczel, Amir D.: *God's Equation. Einstein, Relativity, and the Expanding Universe.* New York: Four Walls Eight Windows, 1999.

Aczel, Amir D.: *The Riddle of the Compass. The Invention that Changed the World.* New York et al.: Hartcourt, 2001.

Albrecht, Ulrich, Ulrike Beisiegel, Reiner Braun and Werner Buckel (Eds.): *Der Griff nach dem atomaren Feuer.* Frankfurt am Main et al.: Lang, 1996.

Almeida, Theophilo Nolasco: *Einstein vs. Michelson.* Rio de Janeiro, 1930.

Ashtekar, Abhay, Robert S. Cohen, Don Howard, Jürgen Renn, Sahotra Sarkar and Abner Shimony (Eds.): *Revisiting the Foundations of Relativistic Physics - Festschrift in Honour of John Stachel.* Dordrecht: Kluwer, 2003.

Baracca, Angelo, Stefano Ruffo and Arturo Russo: *Scienza e industria 1848-1914.* Bari: Laterza, 1979.

Barkan, Diana Kormos: *Walther Nernst and the Transition to Modern Physical Science.* Cambridge: Cambridge University Press, 1999.

Barthes, Roland: *Mythologies.* Paris: Éd. du Seuil, 1970.

Bartusiak, Marcia: *Einstein's Unfinished Symphony. Listening to the Sounds of Space-Time.* Washington: Henry, 2000.

Baumgarth, Christa: *Geschichte des Futurismus.* Reinbek bei Hamburg: Rowohlt, 1966.

Beller, Mara: *Quantum Dialogue. The Making of a Revolution.* Chicago et al.: University of Chicago Press, 1999.

Beller, Mara, Robert S. Cohen and Jürgen Renn (Eds.): *Einstein in Context.* Cambridge: Cambridge University Press, 1993.

Bergia, Silvio: *Einstein. Das neue Weltbild der Physik.* Heidelberg: Spektrum der Wissenschaft Verlagsgesellschaft, 2002.

Bernal, John Desmond: *Science in History (4 Vols.).* Middlesex et al.: Penguin Books, 1969.

Bernal, John Desmond: *The Social Function of Science.* Cambridge et al.: MIT Press, 1967.

Bernstein, Aaron: *Naturwissenschaftliche Volksbücher (21 Vols.).* Berlin: Duncker, 1897.

Bevilacqua, Fabio and Lucio Fregonese (Eds.): *Nuova Voltiana. Studies on Volta and His Times (5 Vols.).* Milan: Università degli studi di Pavia, 2000-2004.

Beyerchen, Alan D.: *Scientists Under Hitler. Politics and the Physics Community in the Third Reich.* New Haven et al.: Yale University Press, 1977.

Bigg, Charlotte: *Behind the Lines. Spectroscopic Enterprises in Early Twentieth Century Europe.* Cambridge: University of Cambridge, 2001.

Börner, Gerhard: *The Early Universe - Facts and Fiction.* Berlin et al.: Springer, 2004.

Bordoni, Stefano: *Una indagine sullo stato dell'etere - Un ragazzo tedesco interroga la fisica di fine Ottocento.* La Goliardica Pavese: Università degli Studi di Pavia, 2005.

Born, Max: *Einstein's Theory of Relativity.* New York: Dover Publications, 1962.

Bosch, Hans-Stephan and Alexander Bradshaw: Kernfusion als Energiequelle der Zukunft. *Physikalische Blätter,* 57 (2001), p. 55-60.

Braun, Reiner and David Krieger: *Einstein - Peace Now! Vision and Ideas.* Weinheim et al.: Wiley-VCH, 2005.

Bredekamp, Horst: *The Lure of Antiquity and the Cult of the Machine. The Kunstkammer and the Evolution of Nature, Art and Technology.* Princeton: Wiener, 1995.

Bredekamp, Horst, Jochen Brüning and Cornelia Weber: *Theater der Natur und Kunst. Wunderkammern des Wissens.* Berlin: Henschel, 2000.

Brian, Denis: *Einstein. A Life.* New York et al.: Wiley, 1996.

Brian, Denis: *The Unexpected Einstein - The Real Man Behind the Icon.* Weinheim: Wiley-VCH, 2005.

Brush, Stephen G.: A History of Random Processes. I. Brownian Movement from Brown to Perrin. *Archive for the History of Exact Sciences,* 5 (1968), p. 1-36.

Brush, Stephen G.: *The Kind of Motion We Call Heat. A History of the Kinetic Theory of Gases in the 19th Century.* (2 Vols.). Amsterdam: North Holland, 1986.

Brush, Steven G. and C. W. Francis Everitt: Maxwell, Osborne Reynolds and the Radiometer. *Historical Studies in the Physical Sciences,* 1 (1969), p. 105-125.

Bucky, Peter R.: *The Private Albert Einstein.* Kansas City: Andrews and McMeel, 1992.

Bührke, Thomas: *Albert Einstein.* München: Deutscher Taschenbuchverlag, 2004.

Büttner, Jochen, Jürgen Renn and Matthias Schemmel: Exploring the Limits of Classical Physics. Planck, Einstein, and the Structure of a Scientific Revolution. *Studies in History and Philosophy of Modern Physics,* 34 (2003), p. 37-59.

Byrne, Patrick: *The Significance of Einstein's Use of History of Science. Dialectica,* 34 (1980), p. 263-276.

Calaprice, Alice (Ed.): *Dear Professor Einstein. Albert Einstein's Letters to and from Children.* New York: Prometheus Books, 2002.

Carroll, Sean: The Cosmological Constant. *Living Reviews in Relativity,* 4 (2001), p. 1–77. http://relativity.livingreviews.org/Articles/lrr-2001-1/index.html.

Cassidy, David C.: *Einstein and Our World.* Revised ed. New York: Humanity Books, 2004.

Cassidy, David C.: *Uncertainty. The Life and Science of Werner Heisenberg.* New York: Freeman, 1992.

Cassirer, Ernst: *Determinismus und Indeterminismus in der modernen Physik. Historische und systematische Studien zum Kausalproblem.* Gesammelte Werke, Bd. 19. Text und Anmerkungen von Claus Rosenkranz. Hamburg: Meiner, 2004.

Castagnetti, Giuseppe, Peter Damerow, Werner Heinrich, Jürgen Renn and Tilman Sauer: *Wissenschaft zwischen Grundlagenkrise und Politik.* Einstein in Berlin. Berlin: Max-Planck-Institut für Bildungsforschung. Forschungsbereich Entwicklung und Sozialisation. Arbeitsstelle Albert Einstein, 1994.

Castagnetti, Giuseppe and Hubert Goenner: *Einstein and the Kaiser Wilhelm Institute for Physics (1917–1922). Institutional Aims and Scientific Results.* Preprint 261. Berlin: Max Planck Institute for the History of Science, 2004.

Clark, Ronald W.: *Einstein. The Life and Times.* London: Sceptre, 1996.

Corry, Leo et al.: Belated Decision in the Hilbert-Einstein Priority Dispute. *Science,* 278 (1997) 5341, p. 1270–1273.

Corry, Leo: *David Hilbert and the Axiomatization of Physics (1898–1918).* Dordrecht: Kluwer, 2004.

Costa, Manuel Amoroso: *Introduçao à teoria da relatividade.* 2. ed. Rio de Janeiro: Ed. UFRJ, 1995.

Damerow, Peter, Gideon Freudenthal, Peter McLaughlin and Jürgen Renn: *Exploring the Limits of Preclassical Mechanics.* 2. ed. New York: Springer, 2004.

Darrigol, Olivier: *Electrodynamics from Ampère to Einstein.* Oxford: Oxford University Press, 2000.

Dijksterhuis, Eduard J.: *The Mechanization of the World Picture. Pythagoras to Newton.* Princeton, NJ: Princeton Univ. Press, 1986.

Duhem, Pierre Maurice Marie: *Medieval Cosmology. Theories of Infinity, Place, Time, Void and the Plurality of Worlds.* Chicago: University of Chicago Press, 1985.

Eckert, Michael: *Die Atomphysiker. Eine Geschichte der theoretischen Physik am Beispiel der Sommerfeldschule.* Braunschweig: Vieweg, 1993.

Eckert, Michael and Karl Märker (Eds.): *Arnold Sommerfeld. Wissenschaftlicher Briefwechsel.* Berlin et al.: Verlag für Geschichte der Naturwissenschaften und Technik, 2000 and 2004.

Ehlers, Jürgen and Gerhard Börner: *Gravitation.* Heidelberg et al.: Spektrum, 1996.

Einstein, Albert: *Akademie-Vorträge. Die Wiederabdrucke der Akademie-Vorträge Albert Einsteins.* Berlin: Akademie Verlag, 1979.

Einstein, Albert: *Albert Einstein, the Human Side. New Glimpses from His Archives.* Ed. by Helen Dukas and Banesh Hoffmann. Princeton: Princeton University Press, 1979.

Einstein, Albert: *Albert Einsteins Relativitätstheorie. Die grundlegenden Arbeiten zur Relativitätstheorie.* Ed. by Karl von Meyenn. Braunschweig: Vieweg, 1990.

Einstein, Albert: *The Collected Papers of Albert Einstein,* Vol. 1–9. Princeton: Princeton University Press, 1987–2004.

Einstein, Albert: *Einstein on Peace.* Ed. by Otto Nathan and Heinz Norden. New York: Avenel Books, 1981.

Einstein, Albert: *Einstein's Annalen Papers. The Complete Collection 1901–1922.* Ed. by Jürgen Renn. Berlin: Wiley-VCH, 2005.

Einstein, Albert: *Einstein's Miraculous Year. Five Papers that Changed the Face of Physics.* Ed. by John Stachel. Princeton et al.: Princeton University Press, 2005.

Einstein, Albert: *Letters to Solovine.* New York: Philosophical Library, 1987.

Einstein, Albert: *The Meaning of Relativity.* Expanded ed. Princeton: Princeton University Press, 2005.

Einstein, Albert: *Out of my Later Years. The Scientist, Philosopher and Man Portrayed Through his Own Words.* New York et al.: Wings Books, 1993.

Einstein, Albert: *The Quotable Einstein.* Ed. by Alice Calaprice. Princeton: Princeton University Press, 1996.

Einstein, Albert: *Relativity. The Special and the General Theory; a Popular Exposition.* London et al.: Routledge, 2002.

Einstein, Albert: *The World as I see it.* Ed. by Carl Seelig. New York: Citadel Press Book, 1999.

Einstein, Albert: Zur einheitlichen Feldtheorie. *Sitzungsberichte der Preussischen Akademie der Wissenschaften,* (1929), I, p. 2–7.

Einstein, Albert and Michele Besso: *Le manuscrit Einstein-Besso. De la relativité restreinte à la relativité générale.* Paris: Scriptura, 2003.

Einstein, Albert and Max Born: *The Born-Einstein Letters. Friendship, Politics, and Physics in Uncertain Times ; Correspondence Between Albert Einstein and Max and Hedwig Born from 1916 to 1955.* New York: Macmillan, 2005.

Einstein, Albert and Wanders J. de Haas: Experimenteller Nachweis der Ampèreschen Molekularströme. *Verhandlungen der Deutschen Physikalischen Gesellschaft,* 17 (1915), p. 152-170.

Einstein, Albert and Leopold Infeld: *The Evolution of Physics. The Growth of Ideas from Early Concepts to Relativity and Quanta.* Cambridge: Cambridge University Press, 1971.

Einstein, Albert and Mileva Marić: *The Love Letters.* Ed. by Jürgen Renn and Robert Schulmann. Princeton: Princeton University Press, 1992.

Eisenstaedt, Jean: *Einstein et la relativité générale. Les chemins de l'espace-temps.* Paris: CNRS Editions, 2002.

Elkana, Yehuda: *Anthropologie der Erkenntnis. Die Entwicklung des Wissens als episches Theater einer listigen Vernunft.* Frankfurt am Main: Suhrkamp, 1986.

Filk, Thomas and Domenico Giulini: *Am Anfang war die Ewigkeit. Auf der Suche nach dem Ursprung der Zeit.* München: Beck, 2004.

Fischbeck, Hans-Jürgen and Regine Kollek (Eds.): *Fortschritt wohin? Wissenschaft in der Verantwortung - Politik in der Herausforderung.* Münster: Agenda-Verlag, 1994.

Fischer, Ernst Peter: *Aristoteles, Einstein & Co. Eine kleine Geschichte der Wissenschaft in Porträts.* München: Piper, 1996.

Fischer, Ernst Peter and Klaus Wiegandt (Eds.): *Mensch und Kosmos. Unser Bild des Universums.* Frankfurt am Main: Fischer Taschenbuch Verlag, 2004.

Flückinger, Max: *Albert Einstein in Bern. Das Ringen um ein neues Weltbild ; eine dokumentarische Darstellung über den Aufstieg eines Genies.* Bern: Haupt, 1974.

Fölsing, Albrecht: *Albert Einstein. A Biography.* New York: Viking, 1997.

Frank, Tibor: *Ever Ready to Go. The Multiple Exiles of Leo Szilard.* Preprint 275. Berlin: Max Planck Institute for the History of Science, 2004.

Freudenthal, Gideon: *Atom and Individual in the Age of Newton. On the Genesis of the Mechanistic World View.* Dordrecht et al.: Reidel, 1986.

Fritzsch, Harald: *An Equation that Changed the World. Newton, Einstein, and the Theory of Relativity.* Chicago et al.: University of Chicago Press, 1994.

Galison, Peter: *Einstein's Clocks and Poincaré's Maps.* New York: Norton, 2003.

Galison, Peter: *How Experiments End.* Chicago et al. University of Chicago Press, 1987.

Galison, Peter: *Image and Logic. The Material Culture of Microphysics.* Chicago: University of Chicago Press, 1997.

Galluzzi, Paolo: *Renaissance Engineers from Brunelleschi to Leonardo da Vinci.* Florence: Giunti, 1996.

Gearhart, Clayton A.: Planck, the Quantum, and the Historians. *Physics in Perspective,* 4 (2002), p. 170–215.

Giedion, Sigfried: *Space, Time, Architecture. The Growth of a New Tradition.* Cambridge: Harvard University Press, 1982.

Gillispie, Charles Coulston (Ed.): *Dictionary of Scientific Biography (18 Vols.).* New York: Scribner, 1981.

Giulini, Domenico: Das Problem der Trägheit. *Philosophia Naturalis,* 39 (2002), p. 343–374.

Giulini, Domenico: *Special Relativity. A First Encounter, 100 Years since Einstein.* New York: Oxford University Press, 2005.

Goenner, Hubert: *Einführung in die Kosmologie.* Heidelberg et al.: Spektrum Akademischer Verlag, 1994.

Goenner, Hubert: *Einstein in Berlin. 1914 - 1933.* München: Beck, 2005.

Goenner, Hubert: *Einsteins Relativitätstheorien. Raum, Zeit, Masse, Gravitation.* 4. ed. München: Beck, 2005.

Graham, Loren R.: The Reception of Einstein's Ideas. Two Examples from Contrasting Political Cultures. In: Holton, Gerald and Yehuda Elkana (Eds.): *Albert Einstein. Historical and Cultural Perspectives.* Princeton: Princeton University Press, 1982, p. 107–136.

Greene, Brian: *The Fabric of the Cosmos. Space, Time, and the Texture of Reality.* New York: Knopf, 2004.

Greither, Aloys and Armin Zweite (Eds.): *Josef Scharl 1896–1954.* München: Prestel, 1982.

Gruber, Howard and Katja Bödeker (Eds.): *Creativity, Psychology, and the History of Science.* Dordrecht: Springer, 2005.

Grundmann, Siegfried: *Einsteins Akte. Wissenschaft und Politik – Einsteins Berliner Zeit.* 2. ed. Berlin et al.: Springer, 2004.

Grüning, Michael (Ed.): *Ein Haus für Albert Einstein. Erinnerungen, Briefe, Dokumente.* Berlin: Verlag der Nation, 1990.

Guth, Alan H.: *The Inflationary Universe. The Quest for a New Theory of Cosmic Origins.* Reading, Mass. et al.: Addison-Wesley Publ., 1997.

Habermas, Jürgen: *The Future of Human Nature.* Cambridge: Polity, 2003.

Habermas, Jürgen: *Zeitdiagnosen. Zwölf Essays 1980–2001.* Frankfurt am Main: Suhrkamp, 2003.

Hartl, Gerhard: *Planeten, Sterne, Weltinseln. Astronomie im Deutschen Museum.* München: Deutsches Museum, 1993.

Hasinger, Günther: *The X-Ray Background. Echo of Black Hole Formation?* Preprint series AIP 99/01. Potsdam: Astrophysikalisches Institut, 1999.

Hawking, Stephen: *A Brief History of Time.* New York: Bantam Books, 2005.

Heilbron, John L.: *The Dilemmas of an Upright Man. Max Planck as Spokesman for German Science.* Berkeley et al.: University of California Press, 1986.

Heisenberg, Werner: *Der Teil und das Ganze. Gespräche im Umkreis der Atomphysik.* München: Piper, 1969.

Held, Carsten: *Die Bohr-Einstein-Debatte. Quantenmechanik und physikalische Wirklichkeit.* Paderborn et al.: Schöningh, 1998.

Henderson, Linda Dalrymple: *The Fourth Dimension and Non-Euclidean Geometry in Modern Art.* Princeton: Princeton University Press, 1983.

Hentschel, Klaus: *The Einstein Tower. An Intertexture of Dynamic Construction, Relativity Theory, and Astronomy.* Stanford: Stanford University Press, 1997.

Hentschel, Klaus: *Interpretationen und Fehlinterpretationen der speziellen und allgemeinen Relativitätstheorie durch Zeitgenossen Albert Einsteins.* Basel et al.: Birkhäuser, 1990.

Hermann, Armin: Einstein. *Der Weltweise und sein Jahrhundert; eine Biographie.* München: Piper, 2004.

Herrmann, Dieter B.: *Astronomiegeschichte. Ausgewählte Beiträge zur Entwicklung der Himmelskunde.* Berlin et al.: Paetec, Verl. für Bildungsmedien, 2004.

Hertz, Heinrich: *Die Constitution der Materie.* Ed. by Albrecht Fölsing. Berlin et al.: Springer, 1999.

Highfield, Roger and Paul Carter: *The Private Lives of Albert Einstein.* New York: St. Martin's Press, 1994.

Hoffmann, Banesh: *Relativity and its Roots.* Mineola, NY: Dover Publications, 1999.

Hoffmann, Dieter: *Einsteins Berlin - Auf den Spuren eines Genies.* Weinheim: Wiley-VCH, 2005.

Hoffmann, Dieter, Hubert Laitko and Staffan Müller-Wille (Eds.): *Lexikon der bedeutenden Naturwissenschaftler (3 Vols.).* Heidelberg et al.: Spektrum Akademischer Verlag, 2003–2004.

Hoffmann, Dieter and Robert Schulmann: *Albert Einstein (1879–1955).* Teetz: Hentrich & Hentrich, 2005.

Holton, Gerald: *Einstein, History, and Other Passions.* Woodbury, NY: American Institute of Physics, 1995.

Holton, Gerald: *Science and Anti-Science.* Cambridge et al.: Harvard University Press, 1993.

Holton, Gerald: *Thematic Origins of Scientific Thought. Kepler to Einstein.* Cambridge: Harvard University Press, 1988.

Holton, Gerald and Yehuda Elkana (Eds.): *Albert Einstein. Historical and Cultural Perspectives.* Princeton: Princeton University Press, 1982.

Howard, Don and John Stachel (Eds.): *Einstein Studies (11 Vols.).* Basel et al.: Birkhäuser, 2000.

Howard, Don and John Stachel (Eds.): *Einstein. The Formative Years, 1879–1909.* Basel et al.: Birkhäuser, 2000.

Hu, Danian: *China and Albert Einstein. The Reception of the Physicist and His Theory in China, 1917–1979.* Cambridge: Harvard University Press, 2005.

Humboldt, Alexander von: *Kosmos. Entwurf einer physischen Weltbeschreibung.* Ed. by Ottmar Ette and Oliver Lubrich. Frankfurt am Main: Eichborn, 2004.

Infeld, Eryk (Ed.): *Leopold Infeld. His Life and Scientific Work.* Warszawa: Polish Scientific Publishers, 1978.

Infeld, Leopold: *Leben mit Einstein. Kontur einer Erinnerung.* Wien et al.: Europa-Verl., 1969.

Israel, Hans, Erich Ruckhaber and Rudolf Weinmann (Eds.): *100 Autoren gegen Einstein.* Leipzig: Voigtländer, 1931.

Janssen, Michel: *A Comparison Between Lorentz's Ether Theory and Special Relativity in the Light of the Experiments of Trouton and Noble.* Ann Arbor: UMI, 1995.

Janssen, Michel and Christoph Lehner (Eds.): *Cambridge Companion to Einstein.* New York: Cambridge University Press, in press.

Janssen, Michel and Jürgen Renn: *Untying the Knot. How Einstein Found His Way Back to Field Equations Discarded in the Zurich Notebook.* Preprint 264. Berlin: Max Planck Institute for the History of Science, 2004.

Jerome, Fred: *The Einstein File. J. Edgar Hoover's Secret War Against the World's Most Famous Scientist.* New York: St. Martin's Press, 2002.

Jorda, Stefan (Ed.): *Albert Einstein und das Wunderjahr 1905. Sonderausgabe von Physik Journal 4 (2005) 3.* Weinheim: Wiley-VCH, 2005.

Jungk, Robert: *The Big Machine.* London: Deutsch, 1969.

Jungnickel, Christa and Russell MacCormmach: *Intellectual Mastery of Nature. Theoretical Physics from Ohm to Einstein (2 Vols.).* Chicago: University of Chicago Press, 1986.

Kant, Horst: *Werner Heisenberg and the German Uranium Project. Otto Hahn and the Declarations of Mainau and Göttingen.* Preprint 203. Berlin: Max Planck Institute for the History of Science, 2002.

Karlsch, Rainer: *Hitlers Bombe. Die geheime Geschichte der deutschen Kernwaffenversuche.* München: Deutsche Verlags-Anstalt, 2005.

Katzir, Shaul: *A History of Piezoelectricity. The First Two Decades.* Tel Aviv: Tel Aviv University, 2001.

Bibliography

Kirsten, Christa and Hans-Jürgen Treder (Eds.): *Albert Einstein in Berlin 1913–1933 (2 Vols.).* Berlin: Akademie Verlag, 1979.

Klafter, Joseph et al.: Beyond Brownian Motion. *Physics Today,* 49 (1996), p. 33–39.

Klein, Ursula: *Verbindung und Affinität. Die Grundlegung der neuzeitlichen Chemie an der Wende vom 17. zum 18. Jahrhundert.* Basel et al.: Birkhäuser, 1994.

Kleinert, Andreas: Nationalistische und antisemitische Ressentiments von Wissenschaftlern gegen Einstein. In: Nelkowski, Horst et al. (Eds.): *Einstein Symposion Berlin, aus Anlass der 100. Wiederkehr seines Geburtstages 25. bis 30. März 1979.* Berlin et al.: Springer, 1979, p. 501–516.

Kocka, Jürgen (Ed.): *Die Königlich Preußische Akademie der Wissenschaften zu Berlin im Kaiserreich.* Berlin: Akademie Verlag, 1999.

Kragh, Helge: *Cosmology and Controversy. The Historical Development of Two Theories of the Universe.* Princeton: Princeton University Press, 1996.

Kuhn, Thomas S.: *The Copernican Revolution. Planetary Astronomy in the Development of Western Thought.* Cambridge et al.: Harvard University Press, 2002.

Kuhn, Thomas S.: *The Structure of Scientific Revolutions.* Chicago et al.: University of Chicago Press, 2004.

Laporte, Paul M.: Cubism and Relativity. With a Letter from Einstein. *Art Journal,* 25 (1966), p. 246–249.

Lefèvre, Wolfgang, Jürgen Renn and Urs Schoepflin (Eds.): *The Power of Images in Early Modern Science.* Basel et al.: Birkhäuser, 2003.

Levenson, Thomas: *Einstein in Berlin.* New York et al.: Bantam Books, 2003.

Lissitzky, El: K. und Pangeometrie. In: Einstein, Carl and Paul Westheim (Eds.): *Europa-Almanach.* Potsdam 1925, p. 103–113.

Lissitzky-Küppers, Sophie (Ed.): *El Lissitzky. Life, Letters, Texts.* London et al.: Thames and Hudson, 1992.

Lorentz, Dominique: *Affaires atomiques.* Paris: Les Arènes, 2001.

Lorimer, Duncan and Michael Kramer: *Handbook of Pulsar Astronomy.* New York: Cambridge University Press, 2005.

Lux-Steiner et al.: Strom von der Sonne. *Physikalische Blätter,* 57 (2001), p. 47–53.

Lyotard, Jean-François: *Die Logik, die wir brauchen. Nietzsche und die Sophisten.* Ed. by Patrick Baum and Günter Seubold. Bonn: DenkMal-Verl., 2004.

Mach, Ernst: *Knowledge and Error. Sketches on the Psychology of Enquiry.* Translation from the 5th edition, 1926. Dordrecht: Reidel, 1976.

Mach, Ernst: *Die Mechanik in ihrer Entwicklung. Historisch-kritisch dargestellt.* Ed. by Renate Wahsner and Horst-Heino von Borzeskowski. Berlin: Akademie Verlag, 1988.

Maeterlinck, Maurice: *Geheimnisse des Weltalls.* Berlin et al.: Deutsche Verlags-Anstalt, 1930.

Markl, Hubert: *Schöner neuer Mensch?* München et al.: Piper, 2002.

Miller, Arthur I.: *Albert Einstein's Special Theory of Relativity. Emergence (1905) and Early Interpretation (1905–1911).* New York et al.: Springer, 1998.

Miller, Arthur I.: *Einstein, Picasso. Space, Time, and the Beauty that Causes Havoc.* New York: Basic Books, 2002.

Misner, Charles W., Kip S. Thorne and John A. Wheeler: *Gravitation.* New York: Freeman, 1973.

Mittelstraß, Jürgen (Ed.): *Enzyklopädie Philosophie und Wissenschaftstheorie (4 Vols.).* Stuttgart et al.: Metzler, 2004.

Montesinos, José and Carlos Solís (Eds.): *Largo Campo di Filosofare. Eurosymposium Galileo 2001.* La Orotava: Fundación Canaria Orotava de Historia de la Ciencia, 2001.

Moreira, Ildeu C. and Antonio A. Videira (Eds.): *Einstein e o Brasil.* Rio de Janeiro: Ed. UFRJ, 1995.

Müller, Falk: *Gasentladungsforschung im 19. Jahrhundert.* Berlin et al.: Verlag für Geschichte der Naturwissenschaften und der Technik, 2004.

Müller, Ulrich: *Raum, Bewegung und Zeit im Werk von Walter Gropius und Ludwig Mies van der Rohe.* Berlin: Akademie Verlag, 2004.

Neffe, Jürgen: *Einstein. Eine Biographie.* Reinbek bei Hamburg: Rowohlt, 2005.

Neumann, Thomas (Ed.): *Albert Einstein.* Berlin: Elefanten Press, 1989.

Newton, Isaac: *Die mathematischen Prinzipien der Physik.* Ed. by Volkmar Schüller. Berlin: de Gruyter, 1999.

Nickel, Jens Uwe: Mikrolaser als Photonen-Billards. Wie Chaos ans Licht kommt. *Physikalische Blätter,* 54 (1998), p. 927–930.

Nietzsche, Friedrich: *Vom Nutzen und Nachteil der Historie für das Leben.* Ed. by Michael Landmann. Zürich: Diogenes, 1984.

North, John David: *The Measure of the Universe. A History of Modern Cosmology.* New York: Dover Publications, 1990.

Nye, Mary Jo: *Molecular Reality. A Perspective on the Scientific Work of Jean Perrin.* London et al.: Macdonald, 1972.

Olschki, Leonardo: *Geschichte der neusprachlichen wissenschaftlichen Literatur (3 Vols.).* Vaduz: Kraus Reprint, 1965.

Overbye, Dennis: *Einstein in Love. A Scientific Romance.* New York: Viking Penguin, 2000.

Overduin, James Martin and P. S. Wesson: Dark Matter and Background Light. *Physics Reports,* 402 (2004), p. 267–406.

Padova, Thomas de: *Am Anfang war kein Mond. 40 Science-Stories, wie unser Sonnensystem entstand und das Leben auf die Erde kam.* Stuttgart: Klett-Cotta, 2004.

Pais, Abraham: *"Subtle is the Lord …". The Science and the Life of Albert Einstein.* Oxford et al.: Oxford University Press, 1983.

Pauli, Wolfgang: *Theory of Relativity.* New York et al.: Dover, 1981.

Penrose, Roger: *The Emperors New Mind. Concerning Computers, Minds, and the Laws of Physics.* New York: Penguin Books, 1991.

Perlick, Volker: Gravitational Lensing from a Spacetime Perspective. *Living Reviews in Relativity*, 7 (2004), p. 9. http://www.livingreviews.org/lrr-2004-9.

Perlmutter, Saul, Greg Aldering and G. Goldhaber et al.: Measurement of W and L from 42 High-Redshift Supernovae. *Astrophysical Journal*, 517 (1999), p. 565–586.

Piaget, Jean: *Le développement de la notion de temps chez l'enfant.* 3. éd. Paris: Presses Universitaires de France, 1982.

Piaget, Jean: *Introduction à l'épistémologie génétique (2 Vols.).* 2. éd. Paris: Presses Universitaires de France, 1973–1974.

Pössel, Markus: *Das Einstein-Fenster. Eine Reise in die Raumzeit.* Hamburg: Hoffmann und Campe, 2005.

Randow, Gero von (Ed.): *Jetzt kommt die Wissenschaft. Von Wahrheiten, Irrtümern und kuriosen Erfindungen.* Frankfurt am Main: FAZ-Institut für Management-, Markt- und Medieninformationen, 2003.

Renn, Jürgen: *Auf den Schultern von Riesen und Zwergen. Einsteins unvollendete Revolution.* Berlin: Wiley-VCH, 2005.

Renn, Jürgen (Ed.): *Galileo in Context.* Cambridge: Cambridge University Press, 2001.

Renn, Jürgen (Ed.): *Genesis of General Relativity (4 Vols.).* Dordrecht: Springer, in press.

Renn, Jürgen and Tilman Sauer: Eclipses of the Stars. Mandl, Einstein, and the Early History of Gravitational Lensing. In: Ashtekar, Abhay et al. (Eds.): *Revisiting the Foundations of Relativistic Physics.* Dordrecht: Kluwer, 2003, p. 69–92.

Renn, Jürgen and Tilman Sauer: *Einsteins Züricher Notizbuch. Die Entdeckung der Feldgleichungen der Gravitation im Jahre 1912.* Preprint 28. Berlin: Max Planck Institute for the History of Science, 1995.

Renn, Jürgen and John Stachel: *Hilbert's Foundation of Physics. From a Theory of Everything to a Constituent of General Relativity.* Preprint 118. Berlin: Max Planck Institute for the History of Science, 1999.

Renn, Jürgen and Henning Vierck: *Künstler, Wisssenschaftler, Kinder und das Nichts.* Ein Werkstattbericht. Berlin: Comenius Garten, 2004.

Rhodes, Richard: *The Making of the Atomic Bomb.* London et al.: Penguin Books, 1988.

Rosenberger, Ferdinand: *Die Geschichte der Physik in Grundzügen (3 Vols.).* Braunschweig: Vieweg, 1882–1890.

Rosenkranz, Ze'ev: *The Einstein Scrapbook.* Baltimore: John Hopkins University Press, 2002.

Salmon, Merrilee H., John Earman, Clark Glymour, James G. Lennox, Peter Machamer, J. A. MacGuire, John D. Norton, Wesley C. Salmon and Kenneth F. Schaffner (Eds.): *Introduction to the Philosophy of Science.* Indianapolis et al.: Hackett, 1999.

Sayen, Jamie: *Einstein in America. The Scientists' Conscience in the Age of Hitler and Hiroshima.* New York: Crown Publications, 1985.

Scheideler, Britta: The Scientist as Moral Authority. Albert Einstein Between Elitism and Democracy, 1914–1933. *Historical Studies in the Physical and Biological Sciences*, 32 (2002), p. 319–346.

Schemmel, Matthias: An Astronomical Road to General Relativity. The Continuity Between Classical and Relativistic Cosmology in the Work of Karl Schwarzschild. In: Renn, Jürgen (Ed.): *The Genesis of General Relativity.* Dordrecht: Springer, in press.

Schiemann, Gregor: *Wahrheitsgewissheitsverlust. Hermann von Helmholtz' Mechanismus im Aufbruch der Moderne.* Darmstadt: Wissenschaftliche Buchgesellschaft, 1979.

Schilpp, Paul Arthur (Ed.): *Albert Einstein. Philosopher – Scientist.* New York: Tudorg, 1970.

Scholz, Erhard: *Herman Weyls Raum-Zeit-Materie and a General Introduction to his Scientific Work.* Basel et al.: Birkhäuser, 2000.

Scholz, Erhard: *An Outline of Weyl Geometric Models in Cosmology.* Preprint. Wuppertal, 2004. http://arXiv.org/astro-ph/0403446.

Schönbeck, Charlotte: *Albert Einstein und Philipp Lenard. Antipoden im Spannungsfeld von Physik und Zeitgeschichte.* Berlin et al.: Springer, 2000.

Schüring, Michael: Der Vorgänger. Carl Neubergs Verhältnis zu Adolf Butenandt. In: Schieder, Wolfgang and Achim Trunk (Eds.): *Adolf Butenandt und die Kaiser-Wilhelm-Gesellschaft. Wissenschaft, Industrie und Politik im Dritten Reich.* Göttingen: Wallstein, 2004, p. 369–403.

Schüring, Michael: Ein Dilemma der Kontinuität. Das Selbst-verständnis der Max-Planck-Gesellschaft und der Umgang mit Emigranten in den 50er Jahren. In: VomBruch, Rüdiger and Brigitte Kaderas (Eds.): *Wissenschaften und Wissenschaftspolitik. Bestandsaufnahmen zu Formationen, Brüchen und Kontinuitäten im Deutschland des 20. Jahrhunderts.* Stuttgart: Steiner, 2002, p. 453–464.

Schutz, Bernard F.: *Gravity from the Ground Up.* Cambridge: Cambridge University Press, 2003.

Schwartz, Joseph and Michael McGuinness: *Einstein for Beginners.* Barton et al.: Icon Books, 1992.

Seelig, Carl: *Albert Einstein. A Documentary Biography.* London: Staples Press, 1956.

Smolin, Lee: *The Life of the Cosmos.* London: Weidenfeld & Nicolson, 1997.

Sommerfeld, Arnold: *Wissenschaftlicher Briefwechsel (2 Vols.).* Ed. by Michael Eckert and Karl Märker. Berlin: Verlag für Geschichte der Naturwissenschaften und Technik, 2000–2004.

Stachel, John: *Einstein from "B" to "Z".* Boston et al.: Birkhäuser, 2002.

Stern, Fritz: *Einstein's German World.* Princeton: Princeton University Press, 1999.

Sugimoto, Kenji (Ed.): *Albert Einstein. A Photographic Biography.* New York: Schocken Books, 1989.

Tauber, Gerald E.: Einstein and Zionism. In: French, Anthony P. (Ed.): *Einstein, a Centenary Volume.* Cambridge: Harvard University Press, 1979, p. 199–207.

Teichmann, Jürgen: *Wandel des Weltbildes. Astronomie, Physik und Meßtechnik in der Kulturgeschichte.* Stuttgart: Teubner, 1999.

Thorne, Kip S.: *Black Holes and Time Warps. Einstein's Outrageous Legacy.* New York et al.: Norton, 1994.

Tolmasquim, Alfredo: *Einstein. O viajante da relatividade na América do Sul.* Rio de Janeiro: Vieira & Lent, 2003.

Toth, Imre: Gott und Geometrie. Eine viktorianische Kontroverse. In: Henrich, Dieter (Ed.): *Evolutionstheorie und ihre Evolution. Vortragsreihe der Universität Regensburg zum 100. Todestag von Charles Darwin.* Regensburg: Mittelbayerische Druckerei- und Verlagsgesellschaft, 1982, p. 141–204.

Tropp, Eduard A., Viktor Frenkel and Artur D. Chernin: *Alexander A. Friedmann. The Man Who Made the Universe Expand.* Cambridge: Cambridge University Press, 1993.

Vierhaus, Rudolf and Bernhard Vom Brocke (Eds.): *Forschung im Spannungsfeld von Politik und Gesellschaft. Geschichte und Struktur der Kaiser-Wilhelm/Max-Planck-Gesellschaft.* Stuttgart: Deutsche Verlags-Anstalt, 1990.

Vogel, Klaus A.: Cosmography. In: Daston, Lorraine and Katherine Park (Eds.): *The Cambridge History of Early Modern Science.* Cambridge: Cambridge University Press, in press.

Vom Brocke, Bernhard and Hubert Laitko (Eds.): *Die Kaiser-Wilhelm-, Max-Planck-Gesellschaft und ihre Institute. Studien zu ihrer Geschichte.* Berlin: de Gruyter, 1996.

Walker, Mark: *German National Socialism and the Quest for Nuclear Power. 1939–1949.* Cambridge et al.: Cambridge University Press, 1989.

Walker, Mark: *Nazi Science. Myth, Truth, and the German Atomic Bomb.* New York: Plenum Press, 1995.

Walther, Herbert: Photonen unter Kontrolle. *Physik Journal,* 4 (2005), p. 45-51.

Wambsganss, Joachim: Gravitational Lensing in Astronomy. *Living Reviews in Relativity,* 1 (1998). http://livingreviews.org/lrr-1998-12.

Wazeck, Milena: Einstein in the Daily Press. A Glimpse into the Gehrcke Papers. In: Eisenstaedt, Jean and A. J. Kox (Eds.): *Universe of General Relativity.* New York: Birkhäuser, 2005, p. 339-357.

Weart, Spencer R.: *Nuclear Fear. A History of Images.* Cambridge: Harvard University Press, 1988.

Weinberg, Steven: *Dreams of a Final Theory.* New York: Vintage Books, 1994.

Wilderotter, Hans (Ed.): *Ein Turm für Albert Einstein. Potsdam, das Licht und die Erforschung des Himmels.* Potsdam: Haus der Brandenburgisch-Preußischen Geschichte, 2005.

Will, Clifford M.: *Was Einstein Right? – Putting General Relativity to the Test.* New York: Basic Books, 1986.

Wohlwill, Emil: *Galilei und sein Kampf für die Copernicanische Lehre (2 Vols.).* Hamburg: Voss, 1909–1926.

Wolters, Gereon: *Mach I, Mach II, Einstein und die Relativitäts-theorie. Eine Fälschung und ihre Folgen.* Berlin et al.: de Gruyter, 1987.

Yam, Philip: Everyday Einstein. *Scientific American,* 9 (2004), p. 50-55.

Zeilinger, Anton: *Einsteins Schleier. Die neue Welt der Quanten-physik.* München: Beck, 2003.

Zuelzer, Wolf: *The Nicolai Case. A Biography.* Detroit: Wayne State University Press, 1982.

http://www.einsteinausstellung.de/
Exhibition "Albert Einstein: Chief Engineer of the Universe"

http://www.einsteinjahr.de/
Einstein Year 2005

http://www.physics2005.org/
World Year of Physics 2005

http://www.mpiwg-berlin.mpg.de
Max Planck Institute for the History of Science

http://www.aei-potsdam.mpg.de/
Max Planck Institute for Gravitational Physics
(Albert Einstein Institute)

http://www.mpq.mpg.de
Max-Planck-Institute of Quantum Optics

http://www.mpa-garching.mpg.de
Max Planck Institute for Astrophysics

www.mpe.mpg.de/institute.html
Max Planck Institute for Extraterrestrial Physics

http://www.mpia-hd.mpg.de/Public/index_en.html
Max Planck Institute for Astronomy

http://public.web.cern.ch/Public/Welcome.html
CERN

http://www.imss.fi.it/index.html
Institute and Museum of the History of Science
Florence

http://www.deutsches-museum.de/
Deutsches Museum München

http://www.mensch-einstein.de/
Einstein-Portal des Rundfunks Berlin-Brandenburg
RBB

http://www.zdf.de/ZDFde/in-
halt/17/0,1872,2254097,00.html
Einstein-Portal des Zweiten Deutschen Fernsehens

http://www.3sat.de/einstein/
Einstein-Portal von 3sat

http://www.dw-world.de/einsteinjahr
Einstein-Portal der Deutschen Welle
http://www3.bbaw.de/bibliothek/digital/index.html
History of *Königlich Preußische Akademie der
Wissenschaften*, Sources

http://echo.mpiwg-berlin.mpg.de
"European Cultural Heritage Online" (ECHO), Sources

http://archimedes2.mpiwg-berlin.mpg.de/
archimedes_templates
History of Mechanics, Sources

http://www.alberteinstein.info/
"Einstein Archives Online"

http://www.einstein.caltech.edu/
"Collected Papers of Albert Einstein"

http://www.living-einstein.de/
"Living Einstein", Einstein Sources

http://einstein-annalen.mpiwg-berlin.mpg.de/home
"Annalen der Physik": Articles of Albert Einstein

http://www.lrz-muenchen.de/~Sommerfeld/
Arnold Sommerfeld (1868-1951): Scientific
Correspondence

http://www.mpiwg-
berlin.mpg.de/KWG/publications.htm#Ergebnisse
Preprints of the Research program "History of the
Kaiser Wilhelm Society in the National Socialist Era"

http://foia.fbi.gov/foiaindex/einstein.htm
Einstein's FBI Files (Freedom of Information Privacy
Act)

http://www.mpiwg-berlin.mpg.de/PREPRINT.HTM
Max Planck Institute for the History of Science:
Preprints

http://www.livingreviews.org/
"Living Reviews", Open Access for Physical Journals

http://arxiv.org/archive/gr-qc
"General Relativity and Quantum Cosmology e-Prints"

http://www-groups.dcs.st-and.ac.uk/~history/
History of Mathematics Archive

http://www.aip.org/history/
Center for History of Physics

http://ppp.unipv.it/
"Pavia Project Physics"

http://hyperphysics.phy-astr.gsu.edu/hbase/hph.html
General Physical Knowledge

http://www.colorado.edu/physics/2000/index.pl
General Physical Knowledge

http://www.einstein-online.info/
Einstein's Theory of Relativity
http://www.colorado.edu/physics/2000/index.pl
"Physics 2000", Interactive Demonstrations of Physics

http://shatters.net/celestia/
"Celestia", Space Simulations

http://www.desy.de/html/home/
Deutsches Elektronen-Synchroton

http://www.zarm.uni-bremen.de/
ZARM, Center of Applied Space Technology and
Microgravity

http://coolcosmos.ipac.caltech.edu/index.html
Infrared-Astronomy

http://www.nrao.edu/whatisra/index.shtml
Radio Astronomy

http://map.gsfc.nasa.gov/m_uni.html
Cosmology

http://archive.ncsa.uiuc.edu/Cyberia/NumRel/
EinsteinLegacy.html
Einstein's Legacy

http://carina.astro.cf.ac.uk/groups/relativity/
research/part1.html
Gravitational waves

http://cosmology.berkeley.edu/Education/BHfaq.html
Black Holes

http://www.damtp.cam.ac.uk/user/gr/public/
Black Holes, Quantum Gravity

http://einstein.stanford.edu/
Gravity Probe B

http://www.ipp.mpg.de/ippcms/de/pr/fusion21/
Nuclear Fusion

http://www.desy.de/html/forschung/
forschungsueberblick.html#elementar
Elementary Particle Physics

http://www.solarserver.de/wissen/photovoltaik.html
Photovoltaik

http://www.iap.uni-bonn.de/P2K/bec/
Bose Einstein Condensation

http://www.ethikrat.org/
Nationaler Ethikrat

http://www.zim.mpg.de/openaccess-berlin/
berlindeclaration.html
Berlin Declaration on Open Access to Knowledge in
the Sciences and Humanities

http://www.fas.org
Federation of American Scientists (US)

http://www.sgr.org.uk
Scientists for Global Responsibility (UK)

http://www.inesap.org
INESAP (International Network of Engineers and
Scientists Against Proliferation, Darmstadt University
of Technology, Germany), Information Bulletins

Abteilung Theoretische Astrophysik, Eberhard Karls Universität, Tübingen

Albert-Einstein-Gymnasium, Berlin-Neukölln

AlltheSky.de, T. Credit & s. Kohle

Archiv der Berlin-Brandenburgischen Akademie der Wissenschaften, Berlin

Archiv zur Geschichte der Max-Planck-Gesellschaft, Berlin

Archives de l'Université Libre, Bruxelles

Archives Générales du Royaume, Bruxelles

Archives Henri Poincaré, Université Nancy 2

Archivio Storico Comunale di Canneto sull'Oglio

Aristophil, Paris

ASM Voghera s.p.a., Voghera

Astrophysikalisches Institut, Potsdam

Laurent Besso, Lausanne

Bayerische Staatsbibliothek, München

Bayerisches Wirtschaftsarchiv, München

Biblioteca dell'Accademia dei Lincei, Roma

Biblioteka Jagielloñska, Kraków

Bibliothek der Berlin-Brandenburgischen Akademie der Wissenschaften, Berlin

Bibliothek der Eidgenössischen Technischen Hochschule, Zürich

Bibliothek der Technischen Universität, Berlin

Bibliothek des John-F.-Kennedy-Instituts, Freie Universität, Berlin

Bildarchiv Preußischer Kulturbesitz, Berlin

Brandeis University Libraries, Waltham (MA)

Bundesarchiv, Koblenz und Berlin

Bundesministerium der Finanzen, Berlin

Burgerbibliothek, Bern

Camera di Commercio di Pavia

Carl Zeiss AG, Optisches Museum, Oberkochen

CERN European Organization for Nuclear Research, Genève

Chandra-X-Ray Observatory

Crystalix GmbH, Berlin

Currier Museum of Art, Manchester (NH)

Deutsche Bücherei, Leipzig

Deutsches Architektur Museum, Frankfurt am Main

Deutsches Elektronen-Synchroton DESY, Hamburg and Zeuthen

Deutsches Historisches Museum, Berlin

Deutsches Museum von Meisterwerken der Naturwissenschaft und Technik, München

Deutsches Patentamt, Berlin

Deutsches Technikmuseum, Berlin

The Einstein Papers Project, Pasadena (CA)

European Space Agency ESA

European Southern Observatory ESO

Fachbereich Physik, Philipps-Universität, Marburg

Federal Bureau of Investigation FBI

FiberTech GmbH, Berlin

Susanne Fiegel, Velchen

Franklin D. Roosevelt Presidential Library and Museum, New York

Fritz-Haber-Institut der Max-Planck-Gesellschaft, Berlin

Gedenkstätte Deutscher Widerstand, Berlin

Geheimes Staatsarchiv Preußischer Kulturbesitz, Berlin

Hahn-Meitner-Institut, Berlin

Harry Ransom Humanities Research Center, University of Texas at Austin

The Hebrew University, Jewish National & University Library, Albert Einstein Archives, Jerusalem

Historical Society of Princeton (NJ)

The Huntington Library, San Marino (CA)

Institut für Energie- und Automatisierungstechnik, Technische Universität, Berlin

Institut für Experimentalphysik, Universität Wien

Institut für Physik, Carl von Ossietzky Universität, Oldenburg

Institut für Zeitgeschichte, München

Instituto de Astrofísica de Canarias, Tenerife

Istituto e Museo di Storia della Scienza, Firenze

Jodrell Bank Observatory of the University of Manchester

Kaller Historical Documents, Marlboro (NJ)

Klems, Berlin

Konrad-Zuse-Zentrum für Informationstechnik, Berlin

Kunstbibliothek, Berlin

Landesarchiv, Berlin

Leo Baeck Institute, New York

U. Lobeck, Dresden

Lotte Jacobi Collection, University of New Hampshire, Durham (NH)

Märkisches Museum, Berlin

Magyar Állami Eötvös Loránd Geofizikai Intézet, Budapest

Mandeville Special Collections Library, La Jolla (CA)

Mathematisch-Physikalischer Salon, Staatliche Kunstsammlungen, Dresden

R. Matthes, Garbsen

Max-Planck-Gesellschaft zur Förderung der Wissenschaften, München

Max-Planck-Institut für Dynamik und Selbstorganisation, Göttingen

Max-Planck-Institut für Extraterrestrische Physik, Garching

Max-Planck-Institut für Gravitationsphysik (Albert-Einstein-Institut), Potsdam and Hannover

Max-Planck-Institut für Plasmaphysik, Garching and Greifswald

Max-Planck-Institut für Quantenoptik, Garching

Max-Planck-Institut für Radioastronomie, Bonn

Max-Planck-Institut für Wissenschaftsgeschichte, Berlin

Münchner Stadtmuseum, München

Jan Murken, München

Sebastian Murken, Mainz

Museo per la Storia dell'Università di Pavia

Museum Boerhaave, Leiden

Museum of the History of Science, Oxford

National Aeronautics and Space Administration NASA

National Radio Astronomy Observatory NRAO

Niedersächsische Staats- und Universitätsbibliothek, Göttingen

Niels Bohr Archive, Copenhagen

New York State Library, Albany (NY)

The New York Times Agency, New York

Österreichische Zentralbibliothek für Physik, Wien

Physikalisches Institut der Eberhard-Karls-Universität, Tübingen

Physikalisch-Technische Bundesanstalt, Braunschweig and Berlin

The Pierpont Morgan Library, New York

François Poincaré, Paris

Politisches Archiv des Auswärtigen Amtes, Berlin

Martin Polzin, Laatzen

Pugwash United Kingdom, Archives, London

Rechen- und Medienzentrum der Universität Lüneburg

Rijksarchief in Noord-Holland, Haarlem

Jürgen Renn, Kleinmachnow

Sammlung zur Laborgerätetechnik – Rainer Friedrich, Mahlow

Gerhard Schneider, Wedemark

Schweizerische Landesbibliothek, Bern

Schweizerisches Literaturarchiv, Bern

Schweizerisches Bundesarchiv, Bern

SensoMotoric Instruments GmbH (SMI), Teltow

Paul Siebertz, München

Ekkehard Sieker, Bad Hönningen

Siemens AG, München and Berlin

SiemensForum, München

Siemens VDO Automotive AG, Schwalbach

Space Telescope Science Institute STScI

Staatsarchiv des Kantons Aargau, Aarau

Staatsarchiv des Kantons Zürich

Staatsbibliothek, Berlin

Stadtarchiv, München

Stadtarchiv, Ulm

Stadtarchiv, Zürich

Standesamt, Ulm

Stanford University, Stanford (CA)

Sterrewacht, Leiden

Stiftung Archiv der Akademie der Künste, Berlin

Kenji Sugimoto, Higashi-Osaka

Syracuse University Library, Syracuse (NY)

Two Micron All-Sky Survey (2MASS)

Universitätsarchiv der Johann Wolfgang Goethe-Universität, Frankfurt am Main

Universitätsbibliothek der Humboldt-Universität, Berlin

Universitätsbibliothek Eichstätt-Ingolstadt, Eichstätt

Universitätsbibliothek, Kassel

Universitätsbibliothek, Mannheim

University of Haifa Library

University of Chicago Library, Chicago (IL)

University of Minnesota Libraries, Minneapolis (MN)

Milena Wazeck, Berlin

R. Wehnert, Hannover

Wehrtechnische Studiensammlung der Bundeswehr, Koblenz

Württembergische Landesbibliothek, Stuttgart

www.compactmemory.de

Zentrum für angewandte Raumfahrttechnologie und Mikrogravitation (ZARM) der Universität Bremen

Zentralbibliothek, Zürich

The Pierpont Morgan Library, New York: 144

Physikalisch-Technische Bundesanstalt, Berlin: 82 r. (Roman März), 190 r.

Phywe Systeme GmbH & Co. KG, Göttingen: 44 l. and r. b., 109 l.

Pinacoteca Tosio Martinengo, Brescia: 49

Poincaré, Paris: 91 r.

Politisches Archiv des Auswärtigen Amtes, Berlin: 197 r. (Roman März)

Princeton University Library, Princeton, NJ: 156

Pugwash United Kingdom, Archives, London: 202

Sammlung zur Laborgerätetechnik, Rainer Friedrich, Mahlow: 41

Science Photo Library, Agentur Focus, Frankfurt am Main: 69 l., 138, 139 c. and r., 173 c., 200, 204 (James King-Holmes), 205 r. (Philippe Psaila)

Schweizerisches Literaturarchiv, Schweizerische Landesbibliothek, Bern: 90 l. a., 93

SensoMotoric Instruments GmbH (SMI), Teltow: 187 r. b.

Siemens AG, München, Berlin: 189 l. (Siemens Archiv), 190 c. (SiemensForum), 191 l. and r. a.

Staatsarchiv des Kantons Aargau, Aarau: 87 r. b.

Staatsarchiv des Kantons Zürich: 98 l. and r. b.

Staatsbibliothek, Berlin: 103, 116 r., 122, 123 l., 136 l., 148

Stadtarchiv, München: 86 l. b.

Stegerphoto.com: 141 r.

Sugimoto, Kenji: 216

Time Magazine, Time Inc., New York: 212

ullstein bild, Berlin: 39 r., 165 l.

United Press International, Washington DC: 157 l.

Universität Marburg, Fachbereich Physik: 187 l.

Universität Tübingen: 91 l., 179 r. b.

University of Minnesota Libraries, Minneapolis, MN: 120 r.

U.S. Navy: 207 a. (Photographer's Mate 3rd Class Douglass M. Pearlman)

Wehrtechnische Studiensammlung der Bundeswehr, Koblenz: 120 r.

Wikipedia: 69 r.

Workshop "Das Nichts", Berlin-Neukölln, 2003: 50 (Private lender)

Württembergische Landesbibliothek, Stuttgart: 102

ZARM, Bremen: 112 r., 171 l.

Zaun, Hans-Otto: 135 l.

Exhibition Imprint

Albert Einstein – Chief Engineer of the Universe

Einstein's Revolution:
Its Prerequisites and Its Consequences.
The Exhibition of the Einstein Year 2005

Patron
Federal Chancellor Gerhard Schröder

Project Management

Organizers
Max Planck Society for the Advancement of Science,
Munich
Max Planck Institute for the History of Science, Berlin

Scientific Concept and General Management
Jürgen Renn

Curator and Exhibition Maker
Stefan Iglhaut

Controlling
Hardo Braun
Wolf Seufert
Manfred Zimpel

Committee
Hardo Braun, Peter Damerow, Stefan Iglhaut,
Rainer Kaufmann, Andrea Noske, Claudia Paaß,
Jürgen Renn, Wolf Seufert, Bernd Tibes,
Milena Wazeck, Gerd Weiberg, Manfred Zimpel

Exhibition Partners

The Hebrew University of Jerusalem
Hanoch Gutfreund, Charlotte Goldfarb, Barbara Wolff,
Ira Rabin, Mara Beller

Deutsches Museum Munich
Alto Brachner, Gerhard Hartl, Christian Sichau,
Stefan Siemer

University of Pavia
Fabio Bevilacqua, Lucio Fregonese, Lidia Falomo,
Carla Garbarino, Lea Cardinali, Enrico Giannetto,
Stefano Bordone

Scientific Team

*Secretarial Office at the Max Planck Institute for the
History of Science*
Ursula Müller, Shadiye Leather-Barrow, Petra Schröter,
Carola Grossmann

Coordination of the Scientific Team
Milena Wazeck

Worldview and Knowledge Acquisition
Peter Damerow, Matteo Valleriani

Children's Knowledge:
Intuitive Understanding of the World
Katja Bödeker, Henning Vierck, Stephanie Giese

Einstein – His Life's Path
Jörg Zaun, Christoph Lehner, Michael Schüring,
Milena Wazeck, Wolf-Dieter Mechler, Dieter Hoffmann,
Giuseppe Castagnetti

Einstein's World Today
Elena Bougleux, Peter Carl, Markus Pössel,
Kurt Sundermeyer

Science as a Challenge
Michael Schüring, Reiner Braun

Einstein's Opponents
Milena Wazeck

Encounters with Einstein
Wendy Coones, Gesa Glück, Dieter Hoffmann,
Horst Kant, Christoph Lehner, Milena Wazeck

Media Stations
Peter Damerow, Malcolm Hyman, Dirk Wintergrün,
Robert Casties, Michael Behr, Julia Damerow,
Library digitalization group,
Max Planck Institute for the History of Science

Film Research
Sandra Schmidt

Image Editing
Hartmut Amon, Edith Hirte, Tanja Starkowski

Pedagogic Program: Simply Einstein
Katja Bödeker, Wendy Coones

Einstein and Art
Violeta Sánchez, Phillipp Muras

Other Members of the Scientific Team
Sabine Bertram, Jochen Büttner, Lindy Divarci, Carmen
Hammer, Horst Kant, Andreas Loos, Simone Rieger,
Matthias Schemmel,
Urs Schoepflin, Ekkehard Sieker

Press and Public Relations
Bernd Wirsing, Ursula Schmidt, Nicole Schuchardt,
Reiner Braun, Ekkehard Sieker, Leyla Dogruel

Project Administration and Management
Claudia Paaß, Sabine Steffan

Exhibition Office

Iglhaut+Partner, Berlin

Management
Stefan Iglhaut

Secretariat Office
Melanie Cakin

Curators
Jörg Zaun, Wolf-Dieter Mechler

Curatorial Assistance / Exhibition Pedagogy
Wendy Coones

Project Management, Sponsoring and Marketing
Anne Maase

Administration of Loan Exhibits
Christoph Schwarz, Elke Kupschinsky

Text Coordination
Vanessa Offen

Image Editing
Edith Hirte

Artwork / Documentation
John Ogli

Design

Scenography
Serge von Arx

Scenography Assistance
Mona Steinke

*Exhibition Architecture / Architectural Design of the
Einstein Mile*
neo.studio, Berlin
Tobias Neumann, Moritz Schneider, Sebastian Blecher,
Sabrina Cegla, Gesa Glück, Sanja Utech

*Exhibition Image / Exhibition Grafics and Typography /
Grafic Design of the Einstein Mile*
Regelindis Westphal Grafik-Design, Berlin
Berno Buff, Barbara Pécot, Norbert Lauterbach,
Sascha Bittner, Anja Gersmann, Antonia Becht

Lighting Design
Michael Flegel

Art Contributions

"Everybody Knows My Face"
Music-/Videoinstallation by Karin Leuenberger and
Georg Graewe

"annus mirabilis"
Composition by Vladimir Ivachkovets

Sound Installations for Einstein's World Today
Composition by Zeitblom

Media Content

Media Design and Production Management
Michael Dörfler

Media Production Assistant
Benja Sachau

Authors / Directors / Film Productions / Animations / Sound Productions
Wolfram Deutschmann with Kick-Film, Berlin
Michael Dörfler with produktion2/cineplus, Berlin
Michael Feser with ArchiMeDes, Berlin
Carsten Hübner
Marco Kuveiller
PICAROmedia Peter Weinsheimer and
Thomas Hammer GbR, Berlin
Violeta Sánchez and Philipp Muras
Florian Sander
Thorsten Schwarz
Ekkehard Sieker
Henning Vierck and Katja Bödeker
Martin Walz
Marcel Weingärtner with
werwiewas Medienproduktion, Berlin
superschool gbr, Berlin
Deutsche Welle, Berlin

Drawings
Laurent Taudin

Cast
Arndt Schwering-Sohnrey as the young Einstein
and Buddy Elias as older Einstein
also featuring Helge Bechert, Mario Irrek,
Max Urlacher, Christof Gareisen, Rainer Reiners,
Tamara Simunovitsch, Jan Uplegger, Tobias Durband

Speakers
John Berwick, Frank Brückner, Erik Hansen,
Markus Hoffmann, Bettina Lohmeyer,
Michael Rotschopf, Brita Sommer, Alexander Weise

Translation of Sound Production
Bonnie Gordon

Sound Production Editing
Peter Craven, Juri Rescheto, Kirstin Schumann

Editing / Post Production Management
Martin Kegel

Audio Guide
audio Konzept, Birge Tetzner
BTL Veranstaltungstechnik, Berlin

Translations

From German into English
Gloria Custance, Robert Culverhouse,
Susan Richter, Lindy Divarci

Exhibition Production

General Planning of Exhibition Construction
DGI Bauwerk, Berlin
Bernd Tibes, Constanze Tibes, Ingo Bunk,
Stephan Ernst, Christiane Geisler

Exhibition Production / Models / Graphics / Reproductions / Exhibition Furniture
Rainer Kaufmann, Berlin

Media Technology Planning
Ground Zero, Berlin
Thomas Buck, Franziskus Scharpff

Planning of Exhibition Engineering Facilities
pin, Berlin
Gerlinde Fasse-Müller, Eugen Löper, Peter Lübcke

Statics Planning
Kaiser, Berlin
Joachim Kaiser, Ricarda Schramm

Fire Prevention Planning
HHP, Berlin
Benedikt Abel, Margot Ehrlicher, Karsten Foth

Exhibition Furnitings, Fixtures and Fittings

Object Furnishings
Thomas Fissler, Leipzig
Falk Lehmann, Peter Borucki, Daniel Klawitter,
Katrin Harder-Klawitter, David Adam, Olaf Brusdeylins,
Andreas Kempe, Claudia Rannow, Mathias Voigt,
Jens Volz,
Museumstechnik, Berlin
Friedel Engels, Charlotte König, Christoph Piontkowitz,
Thomas Kupferstein, Jörg Löbig

Light Fittings
Reinhardt von Bergen-Wedemeyer

Restorers
Ekkehardt Kneer, Berlin
Hildegard Homburger, Berlin
Papieratelier Schröder, Caney, Siedler, Berlin

Restoration of Objects from Deutsches Museum
Maria Przybylo

Exhibition Construction

Exhibition Construction
Display International, Würselen
Zeissig, Springe
Ideea, Berlin
WaltherExpointerieur, Coswig

Model Construction, Exhibits
Rudolf Borkenhagen, Berlin
Phywe, Göttingen
Peik Wünsche, Berlin
Das Atelier Niesler, Berlin
Modellbau Keyser
Ideea, Berlin
Technik & Design, Munich
J. Michel Birn, Berlin
Phänomenta, Flensburg
Till, Gräfelfing
ZARM, Bremen

Construction of Display Cabinets
Museumstechnik, Berlin

Joinery
WaltherExpointerieur, Coswig

Coswig Dry Construction
K. Rogge Spezialbau, Berlin

Building and Masonry
Wagenknecht, Berlin

Metalwork
Gottschalk, Berlin
Wilhelm Frese, Berlin

Electrical Installations
Strohbücker, Berlin

Special Wall / Surface Coatings
Ausstellungsmanufaktur Hertzer, Berlin

Media Hardware
tst, Berlin

Floor Coverings
Anker, Berlin
Schnau, Berlin

Fire Prevention Systems
Sperling, Berlin

Exhibition Graphics / Large Banners
PPS, Berlin
Repro Ringel, Berlin
Druckservice Schellenberg, Berlin

In-house Engineering Systems
Eichelberger, Berlin
Kälte-, Klima- und Gebäudetechnik , Berlin
Bacon, Berlin
Röttig, Berlin
Ax-Air, Berlin

Kronprinzenpalais In-house Engineers
Hartmut Kolberg
Jürgen Zeitler

Exhibition Imprint

Transport and Insurance

Transport of Exhibits
Hasenkamp Internationale Transporte, Köln

Buildings Insurance
AXA Versicherung AG, München
Joachim Clausen, Meitingen

Transport Insurance
Funk Gruppe, Hamburg

Exhibition Operations

Planning and Preparation of Operations
Rainer Kaufmann, Berlin

Exhibition Management
Julia Wendt, Corinna Conradt

Pedagogic Program / Visitors' Office
FührungsNetz des Museumspädagogischen Dienstes, Berlin
Christiane Schrübbers, Almut Weiler, Janine Brandes, Denis Schäfer

Authors of Pedagogic Program
Silke Vorst, Anke Bebiolka, Klaus Heinz, Sybille Braungardt

Surveillance Systems / Security Personnel / Cash Desk / Cloakroom
AMZ, Berlin

Exhibition Shop
Buchhandlung Walther König, Köln

Catering
Einstein Coffeeshops, Berlin

We would like to express our sincere graditude to all security personnel, speakers, guides, and explainers!

**Schlüterhof Opening,
Deutsches Historisches Museum**

Image Installation
jutojo, Berlin
Julie Gayard, Toby Cornish, Johannes Braun

Premiere of "Einstein-Dialog" and other pieces
Music by Luca Lombardi

Direction of Scenic Dialog Nietzsche – Mach – Einstein
Sommer Ulrickson, Alexander Polzin

Cast
Christoph Gareisen, Christian Giese

Organization, Logistics
WE DO / Agentur Einsteinjahr 2005, Berlin

Technical Facilities
Global Sunshine, Berlin

Scientific Cooperation Partners

Institutes of the Max Planck Society
Max Planck Institute for Astronomy, Heidelberg, Jakob Staude
Max Planck Institute for Astrophysics, Garching, Volker Springel
Max Planck Institute for Dynamics and Self-Organization, Göttingen, Dirk Brockmann
Max Planck Institute for Colloids and Interfaces, Golm, Markus Antonietti
Max Planck Institute for Gravitational Physics (Albert-Einstein-Institut), Potsdam/Hannover, Bernard F. Schutz, Karsten Danzmann, Elke Müller, Markus Pössel, Peter Aufmuth, Werner Benger
Max Planck Institute for Nuclear Physics, Heidelberg, Dirk Schwalm
Max Planck Institute for Extraterrestical Physics, Garching, Günther Hasinger, Elmar Pfeffermann, Reiner Hoffmann
Max Planck Institute of Quantum Optics, Garching, Herbert Walther, Gerhard Rempe, Thomas Becker, Stephan Dürr
Max Planck Institute of Plasma Physics, Garching, Isabella Milch
Max Planck Institute for Radio Astronomy, Bonn, Anton Zensus, Rolf Schwartz, Axel Jessner

Other Organizations, Institutes, and Companies
Albert-Einstein-Gymnasium, Berlin-Neukölln, Klaus Lehnert
Albert-Einstein-Grund- und Realschule, Caputh
Astrophysical Institute, Potsdam, Dierck-Ekkehard Liebscher
Center for Applied Space Technology and Microgravity ZARM, University of Bremen, Claus Lämmerzahl, Hansjörg Dittus
Center of Astronomy, University of Heidelberg, Joachim Wambsganss
Centrum Judaicum, Berlin, Hermann Simon
Chinese Academy of Science, Institute for the History of Natural Sciences, Liu Dun, Zaiqing Fang
The Collected Papers of Albert Einstein, Diana Kormos Buchwald
The Comenius-Garten, Berlin-Neukölln, Henning Vierck
European Organization for Nuclear Research CERN, Genf, Maximilian Metzger, Hans Falk Hoffmann, Rolf Landua, Michael Hauschild
Faculty of Physics, Philipps University, Marburg, Hans-Jürgen Stöckmann and Ulrich Kuhl
German Electron Synchrotron DESY, Hamburg/Zeuthen, Thomas Naumann, Ulrike Behrens
Hahn-Meitner Institute, Berlin, Gabriele Lampert and Axel Neisser
Institute for Experimental Physics, University of Vienna, Anton Zeilinger, Markus Aspelmeyer
Institute for Theoretical Astrophysics, Eberhard Karls University, Tübingen, Hanns Ruder, Marc Borchers
Instituto de Astrofísica de Canarias, Tenerife, John Beckman, Robert Watson
Istituto e Museo di Storia della Scienza, Florence, Paolo Galluzzi, Giorgio Strano
Konrad-Zuse-Institut, Berlin
Mathematisch-Physikalischer Salon, State Art Collections, Dresden, Wolfram Dolz, Peter Plaßmeyer
Museum of the History of Science, Oxford, James Bennett
Nobel Foundation, Stockholm, Svante Lindqvist
OpTec Berlin-Brandenburg, Bernd Weidner, Karl-Heinz Schönborn (Clyxon Laser GmbH), Hans Schieber (Crystalix GmbH), Mohammadali Zoheidi (FiberTech GmbH)
Physikalisch-Technische Bundesanstalt, Braunschweig/Berlin, Fritz Riehle, Andreas Bauch, Frank Melchert
Rechen- und Medienzentrum, University of Lüneburg, Martin Warnke
SensoMotoric Instruments (SMI), Niklas Mascher
Siemens AG Corporate Technology, Dietmar Theis, Michael Schumann, Henning Hanebuth
Siemens AG Information and Communications Networks, Berlin, Michael Tochtermann, Detlef Rinder, Rolf-Dieter Kraft
SiemensForum, Karl-Heinz Stritzke
Siemens VDO Automotive AG, Regensburg/Schwalbach, Werner Seier, Urs Weidermann

Anyone who has ever tried to present a rather abstract scientific subject in a popular manner knows the great difficulties of such an attempt. Either he succeeds in being intelligible by concealing the core of the problem and by offering to the reader only superficial aspects or vague allusions, thus deceiving the reader by arousing in him the deceptive illusion of comprehension; or else he gives an expert account of the problem, but in such a fashion that the untrained reader is unable to follow the exposition and becomes discouraged from reading any further.

If these two categories are omitted from today's popular scientific literature, surprisingly little remains. But the little that is left is very valuable indeed. It is of great importance that the general public be given an opportunity to experience – consciously and intelligently – the efforts and results of scientific research. It is not sufficient that each result be taken up, elaborated, and applied by a few specialists in the field. Restricting the body of knowledge to a small group deadens the philosophical spirit of a people and leads to spiritual poverty.

Albert Einstein